I0112985

The Moral Brain

The Moral Brain

A Multidisciplinary Perspective

edited by Jean Decety and Thalia Wheatley

The MIT Press
Cambridge, Massachusetts
London, England

First MIT Press paperback edition, 2017
© 2015 Massachusetts Institute of Technology

All rights reserved. No part of this book may be reproduced in any form by any electronic or mechanical means (including photocopying, recording, or information storage and retrieval) without permission in writing from the publisher.

This book was set in Stone by the MIT Press.

Library of Congress Cataloging-in-Publication Data

Decety, Jean.
The moral brain : a multidisciplinary perspective / Jean Decety and Thalia Wheatley.
    pages   cm. — (Social neuroscience)
Includes bibliographical references and index.
ISBN 978-0-262-02871-4 (hardcover : alk. paper), 978-0-262-53458-1 (pb.)
1. Cognitive neuroscience. 2. Neuroscience—Social aspects. I. Wheatley, Thalia, 1970– II. Title.
QP360.5.D43 2015
612.8′233—dc23

                                                                    2014029653

Contents

Introduction: The Complexity of Moral Cognition Requires Multiple and Converging Levels of Analyses   vii

I   Evolution of Morality

1   **The Evolution of Morality: A Comparative Approach**   3
Laurent Prétôt and Sarah Brosnan

2   **Adaptationist Approaches to Moral Psychology**   19
Andrew W. Delton and Max M. Krasnow

3   **Partner Choice and the Evolution of a Contractualist Morality**   35
Nicolas Baumard and Mark Sheskin

II   Motivations of Morality

4   **Is the Moral Brain Ever Dispassionate?**   51
Jesse Prinz

5   **Devoted Actors and the Moral Foundations of Intractable Intergroup Conflict**   69
Scott Atran and Jeremy Ginges

6   **Why We Cooperate**   87
Jillian Jordan, Alexander Peysakhovich, and David G. Rand

III   The Development of Morality

7   **The Infantile Origins of Our Moral Brains**   105
J. Kiley Hamlin

8   Mechanisms of Moral Development   123
     Joshua Rottman and Liane Young

9   The Neurocognitive Development of Moral Judgments: The Role of
     Executive Function   143
     Ayelet Lahat

10   Girl Uninterrupted: The Neural Basis of Moral Development among
     Adolescent Females   157
     Abigail A. Baird, and Emma V. Roellke

IV   The Affective and Social Neuroscience of Morality

11   Neural Correlates of Human Morality: An Overview   183
     Ricardo de Oliveira-Souza, Roland Zahn, and Jorge Moll

12   The Cognitive Neuroscience of Moral Judgment and Decision
     Making   197
     Joshua D. Greene

13   Neuromodulators and the (In)stability of Moral Cognition   221
     Molly J. Crockett and Regina A. Rini

V   Psychopathic Immorality

14   Immorality in the Adult Brain   239
     Rheanna J. Remmel and Andrea L. Glenn

15   The Moral Brain: Psychopathology   253
     Caroline Moul, David Hawes, and Mark Dadds

VI   Considerations and Implications for Justice and Law

16   Neuroscience versus Phenomenology and the Implications for
     Justice   267
     Thalia Wheatley

17   The Equivocal Relationship between Morality and Empathy   279
     Jean Decety and Jason M. Cowell

Contributors   303
Author index   305
Subject index 321

Introduction: The Complexity of Moral Cognition Requires Multiple and Converging Levels of Analyses

Over the past decade, there has been an explosion of empirical research on moral cognition from a variety of fields. Collectively, this research has captured human moral sensibility as a sophisticated integration of cognitive, emotional, and motivational mechanisms shaped through evolution, development, and culture. Among the most exciting and novel findings and theories, evolutionary biologists and comparative psychologists have shown that moral cognition has evolved to facilitate cooperation and smooth social interactions and that certain components of morality are present in nonhuman animals. Developmental psychologists have devised ingenious paradigms demonstrating that the elements that underpin morality are in place much earlier than we thought and clearly in place before children turn two. Social neuroscientists have begun to map brain circuits implicated in moral decision making and identify the contribution of neuropeptides to moral sensitivity. This research is beginning to uncover how changes in brain chemistry and connectivity realize changes in moral behavior.

The intent of this book is to provide an overview of the current research on the moral brain and examine this fascinating topic from a range of relevant interdisciplinary perspectives. These perspectives include anthropology and neurophilosophy, evolution, development, social neuroscience, psychopathology, and justice and the law. Given the increasing interest in the neuroscience of moral cognition in recent years in both academia and the general public, we believe it is time for a new volume that brings together this new knowledge in a single, concise source.

The Moral Brain addresses the basic questions regarding morality by beginning with a section on the "Evolution of Morality." In the first chapter, Prétôt and Brosnan explore the precursors of human morality that exist in other species such as inequity detection and behavioral contagion and suggest how future research may best determine the extent of these evolutionarily older moral foundations. Although they agree that precursors in other species exist, Delton and Krasnow argue in chapter 2 for uniquely human adaptations that facilitate social exchange, including abilities to pool resources, detect free riders, and evoke group-level condemnation. In chapter 3, Baumard and Sheskin further posit that evolutionary pressures honed a uniquely human contractualist morality—one that maximizes cooperative relationships via a socially tuned sense of fairness. They argue that this sense of fairness predicts moral decisions that appear irrational in terms of a utilitarian framework but are in fact rational if considered through the lens of likely future interaction.

"Motivations of Morality," the second section, focuses on the drives and emotions that fuel moral thought and behavior. Prinz begins the section with a discussion in chapter 4 of whether the moral brain is ever dispassionate. Through a comprehensive review of recent research, he suggests that Hume was right: Moral conclusions cannot be derived from reason alone but are grounded in the passions. In chapter 5, Atran and Ginges explore how principles held so strongly by a culture as to be considered sacred values can lead to the emergence of "devoted actors" willing to commit acts of extreme sacrifice to uphold them. Jordan, Peysakhovich, and Rand view cooperation as a core motivational principle for human morality in chapter 6, echoing earlier chapters proposing that evolutionary pressures honed behavior for social exchange (chapters 2 and 3). They conclude by raising a number of open questions for future research such as how reputational concerns may incentivize cooperation and when and why people who have not been harmed by selfish behavior will still punish the actor, even at considerable personal cost.

"The Development of Morality," from infancy to adolescence, is the focus of the third section. Hamlin begins the section by summarizing recent research that pushes against the traditional view of late-developing moral sensibility. Pointing to her own and others' research, she establishes that infants are sensitive to fairness and prosociality and that even newborns evince a rudimentary form of empathy, suggesting a biological preparedness for a moral sense. This idea of biological preparedness is developed further in chapter 8 by Rottman and Young, who suggest that innate intuitions form the structure of moral cognition, which is then diversified through sociocultural learning. They stress the benefits of developmental research to help uncover the functional architecture of the moral brain. In chapter 9, Lahat looks within this architecture to examine the processes collectively termed "executive function" that subserve moral reasoning. Emerging in the preschool years and developing throughout adolescence and into adulthood, executive function is associated with understanding the distinctions between moral and social conventional violations. Baird and Roellke (chapter 10) round out the section on development by focusing on adolescence as a sensitive period of neural maturation that serves to integrate visceral emotion and social cognition. They draw particular attention to neurobiological modulators (e.g., sex hormones) as a principal organizer of neural activity in adolescence and call for future research to consider these biological changes in tandem with the individual differences that determine how they manifest in moral cognition.

The fourth section of this volume, "The Affective and Social Neuroscience of Morality," explores recent research on the neural mechanisms of moral cognition. In chapter 11, de Oliveira-Souza, Zahn, and Moll demonstrate the multidimensional nature of moral cognition through a comprehensive overview of lesion and neuroimaging research that highlights distinct cortical and subcortical brain areas associated with various component processes. In chapter 12, Greene conceptualizes the neural underpinnings of morality in terms of more general cognitive processes and discusses his dual-process model in which automatic and controlled processes underpin emotional and utilitarian judgments, respectively. Crockett and Rini (chapter 13) complete this section by reviewing evidence that neurotransmitters and hormones influence cognitive processes on which moral decisions depend. They end with a thought-provoking discussion of the myriad normative and ethical questions that emerge from considering how drugs may be used instrumentally to shape moral thought and behavior.

"Psychopathic Immorality" is addressed in the fifth section. Remmel and Glenn (chapter 14) begin by exploring how morality is defined and measured in psychopathy and antisocial personality disorder and then review the literature on how these disorders affect neural mechanisms underlying cognitive processes that, in turn, explain suboptimal judgment and behavior. In chapter 15, Moul, Hawes, and Dadds complement this review by adding a discussion of dysfunctional cognitive processes and traits associated with oppositional defiant disorder and conduct disorder in children, with particular attention to the striking deficits in experiencing and recognizing fear. They conclude with the hope that greater scientific understanding will shift public conception from thinking of psychopaths as "monsters" to a more informed view that spurs the discovery of interventions and treatments.

The final section translates the hope for a more nuanced and biological understanding into "Considerations and Implications for Justice and Law." In chapter 16, Wheatley explores how findings from neuroscience are often at odds with our intuitions and that, in these cases, judicial decisions must increasingly align with scientific evidence. As previous chapters have also intimated, she concludes that a more biological understanding should promote both humility and more humane treatment but that such an understanding does not necessitate a complete overhaul of the judicial system. Decety and Cowell (chapter 17) complete this edited volume with a provocative call to rethink empathy (and its relation to morality) as a constellation of dynamic processes with distinct evolutionary roots rather than as monolithic and stable components of human morality. This view nicely echoes the motif repeated throughout this volume that multiple and converging levels of analysis will best advance our understanding of the myriad biological, cognitive, developmental, and social processes that comprise human moral cognition.

The lesson from all this new knowledge is clear: human moral behavior cannot be separated from human biology, its development, and evolutionary history. As our understanding of the human brain improves, society at large, and justice and the law in particular, are and will be increasingly challenged. Discoveries in affective and social neuroscience will soon impact our legal system in ways that hopefully lead to a more cost-effective, humane, and flexible system than we have today.

I   Evolution of Morality

1 The Evolution of Morality: A Comparative Approach

Laurent Prétôt and Sarah Brosnan

New information that's accumulating daily is blasting away perceived boundaries between human and animals and is forcing a revision of outdated and narrow-minded stereotypes about what animals can and cannot think, do, and feel. We've been too stingy, too focused on ourselves, but now scientific research is forcing us to broaden our horizons concerning the cognitive and emotional capacities of other animals. One assumption in particular is being challenged by this new research, namely the assumption that humans alone are moral beings.
—Marc Bekoff (2009, p. x)

Although studied for centuries in human-focused disciplines such as philosophy and political science, the study of moral behaviors in nonhuman animals (hereafter, animals) has only recently been rediscovered. Even though it is an emerging field, the study of the origins of morality, and whether or not animals are endowed with this capacity, dates from more than a century ago. In *The Descent of Man* (1871/1998), Darwin hypothesized that "any animal whatever, endowed with well-marked social instincts, the parental and filial affections being here included, would inevitably acquire a moral sense or conscience, as soon as its intellectual powers had become as well, or nearly as well developed, as in man." Well before that, the moral philosopher Adam Smith asserted that "animals are not only the causes of pleasure and pain, but are also capable of feeling those sensations" (Smith, 1759/2009).

Given this early exploration into whether or not animals possess a moral sense, why is morality a domain that has been exclusively studied in humans? No doubt this is in part because of our capacity to use language for expressing moral states and sentiments to others, thus providing direct and measurable evidence of the behavior (and likely also due to a bit of homocentrism on the part of humans). Although it is clear that humans engage in moral behavior in a way that is more complex than is seen in other species, there are interesting questions related to moral behaviors that do not primarily focus on humans. This is important; not only does the study of moral behaviors in a particular species allow a better understanding of that species, but by analyzing the origins of moral behaviors in a broader range of taxa, we can better understand the evolution of morality in humans as well.

Of course, this begs the question of how we can study behaviors related to morality in species that do not possess language skills like those of humans. It is this question that we focus on in the current chapter. More specifically, our goal is to provide a brief introduction to how we can use the comparative approach to learn about the evolution of moral behaviors in humans. We focus on three questions. First, how can we study moral behavior in species that are devoid of language? Second, what is the existing evidence for moral behaviors in other species, and what behaviors can be used as indicators in those animals? Finally, assuming that other species do show aspects of moral behaviors, or their precursors, what does this tell us about humans? In this last section we delineate what the evidence to date tells us about the evolution of morality as well as what science currently is or is not equipped to address. We hope that this brief introduction will provide a starting point for those interested in further exploration of the evolution of morality.

How Do You Study Morality in Other Species?

Although human morality has been debated for centuries, moral psychology has only emerged in the last few decades. The initial approach, the cognitive view of morality, focused on moral reasoning (see Piaget, 1932/1965). However, this approach limits the study of morality to those species that are cognitively advanced, such as humans, which makes empirical investigations in other animal taxa nearly impossible. More integrative views that combine approaches focusing on both evolutionary function and cognitive mechanisms have emphasized the study of moral emotions in order to explain moral behaviors. For instance, Haidt's (2001) model suggests that individuals engage in moral reasoning and make decisions based on spontaneous emotional intuitions rather than through verbal discourse and reflection (Haidt, 2007). Haidt refers to these so-called moral emotions as "those emotions that are linked to the interests or welfare either of society as a whole or at least of persons other than the judge or agent" (Haidt, 2003). This focus on emotion, rather than linguistic reflection, opens the possibility of other, nonverbal, species engaging in moral behavior. Based on this definition, Haidt (2008) further defines moral systems as "interlocking sets of values, practices, institutions, and evolved psychological mechanisms that work together to suppress or regulate selfishness and make social life possible." Again, the focus is on the regulation of selfishness and the promotion of social life, both of which are common to many species beyond humans. This synthetic and cross-disciplinary approach gives a more general and flexible view of morality and provides an operational definition that can be applied in species other than humans.

To consider moral behavior comparatively, we need to focus on behavioral mechanisms that are common to both human and animal taxa, such as the moral emotions. Before further discussing the different mechanisms of morality, however, we need to provide information about how one can compare moral behaviors in humans and other animals. A general problem for studying the evolution of behavior is that behaviors are ephemeral and, with rare exceptions, do not fossilize. Consequently, the direct observation of earlier forms of behavior is nearly impossible. In lieu of this, the comparative approach explores behaviors that were present in earlier ancestors by inference from behaviors that occur in currently living species. Behaviors are compared across a variety of species to determine which species show the behavior and whether this pattern is similar to other patterns in the ecology, social organization, or behavior of these species. In this way we can infer at what point in an evolutionary lineage the behavior evolved, by mapping the presence or absence of a behavior onto a phylogenetic tree and determining whether there is a clear point of origin of the trait. We can also determine which social or ecological factors may have been key in driving the evolution of this behavior by determining whether the trait in question correlates with other social or ecological factors in these species. Traits that are shared through common descent, inferred when all or most related species share the trait in question, are referred to as *homologous*, whereas those that emerged independently in species that do not share a common ancestor due to similar social or ecological pressures are referred to as *convergent* (Bshary, Salwiczek, & Wickler, 2007).

Thus, when we study morality or moral behaviors in other species, what we truly are exploring are the "building blocks," or precursors, to moral behavior (Brosnan, 2014a; 2014b; de Waal, 2006; Flack & de Waal, 2000). These are the behaviors that support the evolution of moral behavior as it is seen in humans. Of course, the presence of one of these precursors does not mean that a species has a full sense of morality; rather, it shows us the way in which moral behavior may have evolved. These behaviors may not even be used in the same ways in species other than humans, but they provided the raw material from which the human sense of morality evolved. If we better understand the precursors on which human moral behavior was built, we may be able to extrapolate the important evolutionary pressures that have led to the behavior we see today.

What Is the Evidence for Moral Behavior in Other Species?

Typically, moral behaviors are divided into several categories. The most basic of these, and presumably the most widespread across the animal kingdom, are behaviors related to social regularity (de Waal, 2006; Flack & de Waal, 2000). These are behaviors that promote group cohesion and the smooth functioning of the social group and are, by definition, present in any group-living species. These behaviors have evolved with a multiplicity of mechanisms underlying them. For instance, some social insects solve the issue of group living through a genetically or developmentally constrained caste system, whereas other species, including primates, may solve the same issue through a flexible system that depends on the individual and the context, and may even include empathy. This variability shows us that this basic problem—how to maintain the social group—is so important that solutions have evolved multiple times that solve it in multiple ways. Of course, if we are interested in human behavior, we may be more interested in the flexible system instantiated by the nonhuman primates, but by seeing the commonalities in outcome we get a better understanding of which issues are sufficiently important that they have been solved repeatedly by evolution.

On the other end of the spectrum are behaviors related to judgment, reasoning, and moral decision making. These complex behaviors are likely limited to humans and possibly a few other species. We do see limited evidence for behaviors related to moral judgment in other species. For instance, chimpanzees engage in consolation behavior, in which an uninvolved third party approaches and affiliates with the victim of a fight following the interaction (de Waal, 1996; de Waal & Aureli, 1996; de Waal &

van Roosmalen, 1979). Even within the primates this behavior seems to be limited to the apes, although there is emerging evidence in a few nonprimate species, for example, in ravens (Fraser & Bugnyar, 2010). Although there may be benefits to the consoler (Koski & Sterck, 2007), if consolation is motivated by a feeling that the victim had been inappropriately targeted, this would qualify as moral judgment. Because the possibility of moral judgments in animals is purely speculative at this stage, we do not discuss them further.

Finally, we may also study the precursors to moral behavior through specific behaviors that result in benefiting other members of the group. These are behaviors that support group living, but, unlike behaviors related to social regularity more generally, these behaviors are much less common across the animal kingdom and, in some cases, appear to be targeted. We note one very important caveat: despite the fact that these behaviors result in benefiting others, this does not mean that they are motivated by altruism or other-regarding preferences. As we have alluded to already, there are two types of explanations for any behavior, the underlying function (e.g., why did a behavior evolve?; what problem does it solve?) and the mechanism by which it does so (e.g., what are the genetic, hormonal, neural, developmental, or cognitive pathways through which the behavior is instantiated?). When we discuss behaviors that benefit others, we are referring to the *functional outcome* of the behavior, not the mechanism. That is, we focus on whether this behavior provided a benefit to another individual, not whether the behavior was *intended* to do so. Below we consider the evidence for three particular behaviors that are other-oriented: reciprocity, prosocial behavior, and inequity. Although these behaviors may not be moral in and of themselves, they represent the sorts of behaviors that may have provided the evolutionary foundation for our own moral system.

Reciprocity

One way in which animals benefit other individuals is through reciprocity. Reciprocity, or reciprocal altruism, occurs when individuals alternately provide benefits to one another. This idea, first suggested by Robert Trivers (1971), provides an explanation for the evolution of cooperation because animals receive benefit in the long term from benefits that they provide to others now; reciprocity provides an overall benefit as long as the short-term cost that A undergoes from helping B now is exceeded by the benefit A will receive from B's return help over the long term. Again, reciprocity explains how benefits directed to others might evolve but does *not* mean that the individuals intend to provide benefits or that they understand the ramifications of their actions. Many examples of reciprocity are clearly rather inflexible behavioral patterns that do not indicate any understanding on the part of the actor.

Despite the long history of studying reciprocity in animals, there is little agreement on its prevalence. Using chimpanzees as a case study, despite the evidence of long-term exchange of goods and services in the wild (Gomes & Boesch, 2009), there is surprisingly little evidence of reciprocity in controlled experiments (Brosnan et al., 2009; Melis, Hare, & Tomasello, 2008). That being said, one of the challenges of studying reciprocity in a species with a high degree of cognitive flexibility is that reciprocity is not likely to occur in all circumstances and may require specific conditions that are not often met in typical experiments. For instance, it may be that individuals are more likely to reciprocate when they have the opportunity to choose their partners and when reciprocity can happen on a longer time scale than is typically found in laboratory studies. One captive study that did find evidence of reciprocity in chimpanzees involved observing the correlation between grooming and food sharing within an entire group of chimpanzees across their day. De Waal found that chimpanzees that groomed one another in the morning were more likely to share food in the afternoon. This effect was specific to the grooming recipient sharing food with the groomer in the afternoon, but not the reverse, and the grooming recipient's sharing was specific to the groomer, implying that this was contingent reciprocity rather than generalized good will (de Waal, 1997). The apparent flexibility in the chimpanzees' reciprocity makes this behavior close to that which we associate with moral behavior in humans.

However, lest we give the impression that flexible reciprocity is limited to larger-brained species or nonhuman primates, we note that some of the more impressive examples of reciprocity are found in other species. For example, although we have previously discussed direct reciprocity, another form of reciprocity is generalized reciprocity, in which individuals help one partner in response to help received from another. This mechanism might contribute not only to the maintenance of good individual relationships within an existing social group but possibly even among individuals outside of this immediate group. This may be particularly important for species in which it can be useful to welcome previously unknown individuals into the social group. The only species that has been demonstrated to engage in generalized reciprocity, other than humans (Bartlett & DeSteno, 2006), is rats (Rutte & Taborsky, 2007), which also engage in direct reciprocity that is influenced by the partner's previous behavior (Rutte & Taborsky, 2008). Thus, although not every species that shows reciprocal behavior will comprehend it as such, the evolutionary pressures that led to reciprocal behavior nonetheless selected individuals to behave in ways that benefited their partners. Such behavior provides a foundation on which human moral behavior could have evolved.

Prosocial Behavior
Although some behaviors that benefit others may have evolved in the context of reciprocity, in other cases individuals may help others in situations that are not explicitly reciprocal. Such prosocial behavior[1] in animals occurs both in the realm of food sharing, which is relatively rare in the primates (Jaeggi & van Schaik, 2011), and social interactions. For instance, nonhuman primates use a range of "political" maneuvers and manipulations in order to intervene in conflicts and maintain order and cohesion in a group (Proctor & Brosnan, 2011). For example, policing behavior, which occurs both in the wild (Boehm, 1994) and in captivity (chimpanzees: de Waal, 1982/1998; 1992; macaques: Flack, de Waal, & Krakauer, 2005; Flack, Girvan, de Waal, & Krakauer, 2006), is often impartial, with the apparent goal of minimizing fighting within the group rather than supporting one individual over another or promoting one's own agenda. Additionally, there is extensive anecdotal evidence of animals, particularly apes, helping one another (de Waal, 2006). However, without controlled experiments it is not possible to rule out alternative explanations for these prosocial outcomes, which has led to a recent interest in investigating prosocial behavior experimentally.

These experiments typically rely on giving animals a choice about whether or not to provide food rewards to members of their social group (reviewed in Cronin, 2012). These studies are difficult to interpret. This is likely in part because, as with reciprocity, experimental results do not match the expectations derived from natural observations, possibly because prosocial behavior, too, may not be as likely to occur in the laboratory setting as in natural circumstances. Additionally, procedures typically vary across studies, making them difficult to compare. One interesting finding has been that, at least in chimpanzees, prosocial behavior is more likely in helping situations than in food-sharing situations (e.g., Warneken & Tomasello, 2006), which may be an indication of the contexts in which prosocial behavior is more important in the wild. Additionally, it has been hypothesized that prosocial behavior is more likely to evolve in cooperative breeders, those species in which both the male and female, and often adult offspring, raise current offspring, because of the interdependency inherent in that social system (Hrdy, 2009; van Schaik & Burkart, 2010). Further explorations of the contexts in which prosocial behavior is most common will help to clarify those situations in which behaviors that assist others, a bedrock of moral behavior, are the most essential.

Inequity

One advantage of being a socially living organism is constantly having information about the outcomes of others. This can provide a major advantage if, for instance, a group member develops a better way of processing a food that you can then learn and benefit from. Such social learning has been widely documented across the animal kingdom (Heyes & Galef, 1996). However, the ability to compare one's outcomes to those of others may also be an advantage if one is trying to judge the relative benefit of multiple partners. It has been hypothesized that individuals who can judge their outcomes as compared to those of others have an advantage in choosing cooperative partners, in that they can leave those with whom they earn relatively less benefit and seek out those with whom they might do better (Brosnan, 2011; Fehr & Schmidt, 1999). Such a negative response to inequity has been documented in some species besides humans using an experimental test in which subjects must complete a task with a human experimenter in order to receive a food reward that may be the same as or different from a social partner's reward for the same task. Intriguingly, reactions to inequity occur quite commonly in species that routinely cooperate with nonkin but do not occur in closely related species that do not cooperate, indicating that one function of recognizing inequity is to judge the value of one's cooperative partners (Brosnan, 2011; Price & Brosnan, 2012). Additionally, the inequity response is not seen in cooperatively breeding primates, possibly because of the high costs of finding a new partner in such an interdependent relationship (Freeman et al., 2013).

Of course, once a negative response to inequity has evolved, it is now possible that individuals may also learn to recognize when they get more than a partner. Although reacting against getting more leads to an immediate loss, it may provide a long-term benefit in terms of an improved reputation (Frank, 2001) or a better relationship with that social partner. Individuals who are able to anticipate that their partners will become frustrated and adjust their behavior should be better able to maintain beneficial partnerships (Brosnan & de Waal, 2012). In fact, there is evidence that chimpanzees recognize when they are advantaged as compared to a partner (Brosnan, Talbot, Ahlgren, Lambeth, & Shapiro, 2010), and capuchin monkeys who share a preferred outcome are almost three times as successful at a cooperative task compared to those partnerships in which one monkey dominates the more preferred reward (Brosnan et al., 2006). More than responding negatively to getting less, this response to overcompensation hints to the roots of moral behavior in other species.

Moral Emotions: Empathy

Empathy is the cognitive capacity to feel and perceive others' emotions (Preston & de Waal, 2002). Of course, we cannot measure emotion or perspective taking in other species except through their instantiated behavior (because they cannot tell us what they think), so when we talk about empathy in other species, what we really mean is behaviors that indicate an understanding of another's emotions, desires, or goals. Unfortunately, one challenge of studying empathetic behaviors is that these behaviors are very difficult to elicit in the laboratory, especially without utilizing situations that the animals find aversive. Thus, the majority of evidence comes from anecdotal reports of subjects' behavior (e.g., de Waal, 2006). For instance, apes, but not monkeys, show targeted helping, or the ability to understand what another individual wants and then to provide him or her support to achieve the goal. There are more anecdotes in apes than any other taxa, probably due to a combination of research effort and the assumption that human-like behaviors are most common in humans' closest ancestors. Nonetheless, other species also show evidence of empathetic behavior, such as in situations in which a group member is ill, injured, or dead (McComb, Baker, & Moss, 2006; Simmonds, 2006). One challenge to the anecdotal reports is that there are always alternative explanations that cannot be ruled out without experimental controls. Consequently, researchers have focused on behaviors that may be related to empathy that can be studied experimentally.

Consolation, argued to be an indicator of cognitive empathy, occurs when a victim of a fight is approached by a bystander who provides support (e.g., grooming or other affiliative behaviors). This provides stress reduction to the victim as well as the consoler (Koski & Sterck, 2007). This behavior occurs in apes (Romero, Castellanos, & de Waal, 2010) but has not been seen in monkeys (de Waal & Aureli, 1996), leading to the hypothesis that some empathetic behaviors require the more advanced cognition seen in the apes. Additionally, contagious yawning, a behavior that has been linked to empathy in humans, is also seen in several other species. Notably, chimpanzees shown video of group members and strangers yawning are more likely to yawn when viewing their group members, possibly indicating greater empathy with nonstrangers (Campbell & de Waal, 2011).

Given the difficulty in finding experimental evidence of empathy in primates, it is notable that the best evidence to date comes from rodents (Langford et al., 2006). Church first reported evidence of "emotional reactions" in rats a half-century ago, noting that individuals quit pressing a lever to obtain food if a partner experienced electric shocks, sometimes going without food for extended periods (Church, 1959). Langford et al. (2006) showed that mice watching a conspecific get shocked made the subjects more sensitive to their own pain (also from a shock), and recent evidence shows that this is dependent on the genetic makeup of the mice, indicating a potential genetic basis for empathetic behavior (Chen, Panksepp, & Lahvis, 2009). Rats, too, respond to a conspecifics' pain, intentionally opening a restrainer to free a partner, even if it requires delaying the opportunity to obtain a preferred food (chocolate) and subsequently results in the sharing of that food (Ben-Ami Bartal, Decety, & Mason, 2011). These studies have all received the criticism that the rodents' behavior may be driven more by their interest in stopping their partners' screams of pain or fear (an aversive stimulus) than an interest in helping their partners per se, a possibility that has not been completely ruled out. However, even if this is the case, it is important to keep in mind the difference between function and mechanism. This behavior *functions* to stop the discomfort of a conspecific, regardless of the actor's intent. Thus, assisting a conspecific provides a foundation for the evolution of empathetic intent (if it is not already such), and once the behavior is established, it can evolve to be applied in novel situations (Brosnan, 2014a, 2014b). Therefore, these empathetic responses in other species are of clear interest in our understanding of how such behaviors have evolved in people.

What Does All of This Mean for Us?

Other species show evidence of behaviors that are related to moral behavior, but how does this relate to human behavior? In many cases the behaviors seen in other species bear only limited resemblance to those seen in humans. However, we can learn a lot about the evolutionary foundations of human behavior from studying the behaviors that were likely in place when humans evolved our sense of morality. Evolution is not limitless opportunity, but is constrained by existing traits (Konner, 2002). By better understanding these precursor behaviors, we better understand what limitations were in place prior to the evolution of human morality and what factors were important for, or influenced, the evolution of morality. Moreover, human behavior is complex, and these species represent simpler systems that can become model systems to explore how various environmental and social contexts influence the expression of behaviors related to morality.

Additionally, we now know that human behavior is not unique. The discovery of helping behaviors in distant animal taxa (e.g., rodents) gives us an idea of the range of animal species in which such behaviors may occur. Our next task is to determine the degree to which these behaviors are similar to those in humans. Do other species show the flexibility seen in humans? If so, in what species or contexts does this flexibility occur? These are just a few of the many questions that need to be explored. Of course, one of the challenges for studying these questions in other species is the lack of a way to query them about their own behaviors. However, modern techniques are slowly breaking down this barrier. Modern neuroscience is providing some of the tools for investigating the "invisible" side of animal mental states. For instance, birds display similar neuroaffective mechanisms as humans when listening to birdsong or music (Earp & Maney, 2012). Studies in applied animal behavior have also shown that pain and other affective states (e.g., fear and stress) are likely to be perceived and experienced by fish in a way similar to other animals such as amphibians, reptiles, birds, and mammals (Chandroo, Duncan, & Moccia, 2004;

Sneddon, 2003). These examples provide a promising future for the study of affective and emotional states in a broad range of animal taxa. Some day soon we may be able to go beyond comparing behaviors to discussing shared neural and other mechanisms across humans and other species.

Note

1. Note that we distinguish prosocial behavior, which helps another at little to no cost, from altruistic behavior, which requires a substantial cost. Given the relative rarity of altruistic behavior in other species, we do not consider it further here.

References

Bartlett, M. Y., & DeSteno, D. (2006). Gratitude and prosocial behavior: Helping when it costs you. *Psychological Science, 17*, 319–325.

Bekoff, M. (2009). *Wild Justice: The Moral Lives of Animals* (1st ed.). Chicago: University of Chicago Press.

Ben-Ami Bartal, I., Decety, J., & Mason, P. (2011). Empathy and pro-social behavior in rats. *Science, 334*, 1427–1430.

Boehm, C. (1994). Pacifying interventions at Arnhem Zoo and Gombe. In: R. W. Wrangham, W. C. McGrew, F. B. M. de Waal, & P. G. Heltne (eds.), *Chimpanzees Cultures* (pp. 211–226). Cambridge, MA: Harvard University Press.

Brosnan, S. F. (2011). A hypothesis of the co-evolution of inequity and cooperation. *Frontiers in Decision Neuroscience, 5*, 43–55.

Brosnan, S. F. (2014a). Why an evolutionary perspective is critical to understanding moral behavior in humans. In: M. Bergmann & P. Kain (eds.), *Challenges to Moral and Religious Belief* (pp. 195–219). Oxford: Oxford University Press.

Brosnan, S. F. (2014b). Evidence for moral behaviors in non-human primates. In: M. Christen, C. P. Van Schaik, J. Fischer, M. Huppenbauer, & C. Tanner (eds.), *Empirically Informed Ethics: Morality Between Facts and Norms* (pp. 85–98). Berlin: Springer.

Brosnan, S. F., & de Waal, F. B. (2012). Fairness in animals: Where to from here? *Social Justice Research, 25*, 1–16.

Brosnan, S. F., Freeman, C., & De Waal, F. B. M. (2006). Partner's behavior, not reward distribution, determines success in an unequal cooperative task in capuchin monkeys. *American Journal of Primatology, 68*, 713–724.

Brosnan, S. F., Silk, J. B., Henrich, J., Mareno, M. C., Lambeth, S. P., & Schapiro, S. J. (2009). Chimpanzees (*Pan troglodytes*) do not develop contingent reciprocity in an experimental task. *Animal Cognition, 12*, 587–597.

Brosnan, S. F., Talbot, C., Ahlgren, M., Lambeth, S. P., & Schapiro, S. J. (2010). Mechanisms underlying responses to inequitable outcomes in chimpanzees, *Pan troglodytes*. *Animal Behaviour, 79*, 1229–1237.

Bshary, R., Salwiczek, L., & Wickler, W. (2007). Social cognition in non-primates. In: R. I. M. Dunbar & L. S. Barrett (eds.), *The Oxford Handbook of Evolutionary Psychology* (pp. 83–101). Oxford: Oxford University Press.

Campbell, M. W., & de Waal, F. B. M. (2011). Ingroup-outgroup bias in contagious yawning by chimpanzees supports link to empathy. *PLoS ONE, 6*, e18283.

Chandroo, K. P., Duncan, I. J. H., & Moccia, R. D. (2004). Can fish suffer? Perspectives on sentience, pain, fear and stress. *Applied Animal Behaviour Science, 86*, 225–250.

Chen, Q., Panksepp, J. B., & Lahvis, G. P. (2009). Empathy is moderated by genetic background in mice. *PLoS ONE, 4*, e4387.

Church, R. M. (1959). Emotional reactions of rats to the pain of others. *Journal of Comparative and Physiological Psychology, 52*, 132–134.

Cronin, K. A. (2012). Prosocial behaviour in animals: The influence of social relationships, communication and rewards. *Animal Behaviour, 84*, 1085–1093.

Darwin, C. (1998). *The Descent of Man and Selection in Relation to Sex.* Amherst, NY: Prometheus Books. (Original work published 1871.)

de Waal, F. B. M. (1992). Coalitions as part of reciprocal relations in the Arnhem chimpanzee colony. In A. H. Harcourt & F. B. M. de Waal (Eds.), *Coalitions and Alliances in Humans and Other Animals* (pp. 233–257). Oxford: Oxford University Press.

de Waal, F. B. M. (1996). *Good Natured: The Origins of Right and Wrong in Humans and Other Animals.* Cambridge, MA: Harvard University Press.

de Waal, F. B. M. (1997). The chimpanzee's service economy: Food for grooming. *Evolution and Human Behavior, 18*, 375–386.

de Waal, F. B. M. (1998). *Chimpanzee Politics: Power and Sex Among Apes.* Baltimore: Johns Hopkins University Press. (Original work published 1982.)

de Waal, F. B. M. (2006). *Primates and Philosophers: How Morality Evolved.* Princeton, NJ: Princeton University Press.

de Waal, F. B. M., & Aureli, F. (1996). Consolation, reconciliation and a possible cognitive difference between macaqes and chimpanzees. In: A. E. Russon, K. A. Bard, & S. T. Parker (eds.), *Reaching into Thought: The Minds of the Great Apes* (pp. 80–110). Cambridge: Cambridge University Press.

de Waal, F. B. M., & van Roosmalen, A. (1979). Reconciliation and consolation among chimpanzees. *Behavioral Ecology and Sociobiology, 5*, 55–66.

Earp, S. E., & Maney, D. L. (2012). Birdsong: Is it music to their ears? *Frontiers in Evolutionary Neuroscience, 4*, 1–10.

Fehr, E., & Schmidt, K. M. (1999). A theory of fairness, competition, and cooperation. *Quarterly Journal of Economics, 114*, 817–868.

Flack, J. C., & de Waal, F. B. M. (2000). "Any animal whatever": Darwinian building blocks of morality in monkeys and apes. *Journal of Consciousness Studies, 7*, 1–29.

Flack, J., de Waal, F. B. M., & Krakauer, D. C. (2005). Social structure, robustness, and policing cost in a cognitively sophisticated species. *American Naturalist, 165*, E126–E139.

Flack, J., Girvan, M., de Waal, F. B. M., & Krakauer, D. C. (2006). Policing stabilizes construction of social niches in primates. *Nature, 439*, 426–429.

Frank, R. H. (2001). Cooperation through emotional commitment. In: R. M. Nesse (ed.), *Evolution and the Capacity for Commitment* (pp. 57–76). New York: Russell Sage Foundation.

Fraser, O. N., & Bugnyar, T. (2010). Do ravens show consolation? Responses to distressed others. *PLoS ONE, 5*, e10605.

Freeman, H., Sullivan, J., Hopper, L., Talbot, C., Holmes, A., Schultz-Darken, N., et al. (2013). Different responses to reward comparisons by three primate species. *PLoS ONE, 8*, e76297.

Gomes, C. M., & Boesch, C. (2009). Wild chimpanzees exchange meat for sex on a long-term basis. *PLoS ONE, 4*, e5116.

Haidt, J. (2001). The emotional dog and its rational tail: A social intuitionist approach to moral judgment. *Psychological Review, 108*, 814–834.

Haidt, J. (2003). The moral emotions. In: R. J. Davidson, K. R. Scherer, & H. H. Goldsmith (eds.), *Handbook of Affective Sciences* (pp. 852–870). Oxford: Oxford University Press.

Haidt, J. (2007). The new synthesis in moral psychology. *Science, 316*, 998–1002.

Haidt, J. (2008). Morality. *Perspectives on Psychological Science, 3*, 65–72.

Heyes, C. M., & Galef, B. C. (1996). *Social Learning in Animals: The Roots of Culture.* London: Academic Press.

Hrdy, S. B. (2009). *Mothers and Others: The evolutionary origins of mutual understanding.* Cambridge, MA: Harvard University Press.

Jaeggi, A. V., & van Schaik, C. P. (2011). The evolution of food sharing in primates. *Behavioral Ecology and Sociobiology, 65*, 2125–2140.

Konner, M. (2002). *The Tangled Wing: Biological Constraints on the Human Spirit.* New York: Times Books.

Koski, S. E., & Sterck, E. H. M. (2007). Triadic postconflict affiliation in captive chimpanzees: Does consolation console? *Animal Behaviour, 73*, 133–142.

Langford, D. J., Crager, S. E., Shehzad, Z., Smith, S. B., Sotocinal, S. G., Levenstadt, J. S., et al. (2006). Social modulation of pain as evidence for empathy in mice. *Science, 312*, 1967–1970.

McComb, K., Baker, L., & Moss, C. J. (2006). African elephants show high levels of interest in the skulls and ivory of their own species. *Biology Letters, 2*, 26–28.

Melis, A. P., Hare, B., & Tomasello, M. (2008). Do chimpanzees reciprocate received favours? *Animal Behaviour, 76*(3), 951–962. doi:10.1016/j.anbehav.2008.05.014.

Piaget, J. (1965). *The Moral Judgment of the Child.* New York: Free Press. (Original work published 1932.)

Preston, S. D., & de Waal, F. B. M. (2002). Empathy: Its ultimate and proximate bases. *Behavioral and Brain Sciences, 25*, 1–72.

Price, S. A., & Brosnan, S. F. (2012). To each according to his need? Variability in the responses to inequity in nonhuman primates. *Social Justice Research, 25*, 140–169.

Proctor, D., & Brosnan, S. F. (2011). Political primates: What other primates can tell us about the evolutionary roots of our own political behavior. In: P. K. Hatemi & R. McDermott (eds.), *Man Is by Nature a Political Animal: Evolution, Biology, and Politics* (pp. 47–71). Chicago: Chicago University Press.

Romero, T., Castellanos, M. A., & de Waal, F. B. M. (2010). Consolation as possible expression of sympathetic concern among chimpanzees. *Proceedings of the National Academy of Sciences USA, 107*, 12110–12115.

Rutte, C., & Taborsky, M. (2007). Generalized reciprocity in rats. *PLoS Biology, 5*, e196.

Rutte, C., & Taborsky, M. (2008). The influence of social experience on cooperative behavior of rats (*Rattus norvegicus*): Direct vs. generalized reciprocity. *Behavioral Ecology and Sociobiology, 62*, 499–505.

van Schaik, C. P., & Burkart, J. M. (2010). Mind the gap: Cooperative breeding and the evolution of our unique features. In: P. M. Kappeler & J. Silk (eds.), *Mind the Gap: Tracing the Origins of Human Universals* (pp. 477–497). Heidelberg: Springer.

Simmonds, M. P. (2006). Into the brains of whales. *Applied Animal Behaviour Science, 100*, 103–116.

Smith, A. (2009). *The Theory of Moral Sentiments.* London: Penguin Books. (Original work published 1759.)

Sneddon, L. U. (2003). The evidence for pain in fish: The use of morphine as an analgesic. *Applied Animal Behaviour Science, 83*, 153–162.

Trivers, R. L. (1971). The evolution of reciprocal altruism. *Quarterly Review of Biology*, *46*, 35–57.

Warneken, F., & Tomasello, M. (2006). Altruistic helping in human infants and young chimpanzees. *Science*, *311*, 1301–1303.

2   Adaptationist Approaches to Moral Psychology

Andrew W. Delton and Max M. Krasnow

Keep your word. Love your neighbor. Help those in need. Only marry the child of your father's sister. Stone those who have had sex out of wedlock. Kill outgroup members.

Human moral communities have developed a variety of moral rules, injunctions, prescriptions, and suggestions. How are we to understand the origins and nature of human morality, both at the level of universal building blocks and the level of cultural variation and elaboration? As this volume illustrates, understanding human moral psychology is a truly interdisciplinary endeavor, drawing important contributions from psychologists, neuroscientists, philosophers, biologists, legal scholars, and many others. Our goal in this chapter is to illustrate the utility of taking an adaptationist approach from evolutionary biology to understand universal aspects of moral psychology. We first describe what it means to take an adaptationist approach. We next give several examples of how an adaptationist approach has informed the study of certain aspects of moral psychology. We then briefly conclude with what we see as the value of this approach to the study of moral psychology broadly.

Adaptation

Charles Darwin's *The Origin of Species* is justly celebrated for revolutionizing biology. Although Darwin himself described the book as "one long argument," the book actually develops several important and logically separable hypotheses (see Mayr, 1982). For example, all life is descended from one or a small number of ancestral forms. Additionally, evolution (i.e., descent with modification) is a process of mainly small gradual changes. Most revolutionary, natural selection is *the* driving force causing organisms to become ever more adapted to their environments.

Why is natural selection the best known and most used of these hypotheses? This is because natural selection was the first, and is still the only, non-question-begging explanation for the origin of complex functional organization in animals and other organisms (Williams, 1966). Prior to Darwin, scientists had regularly invoked special creation to explain the existence of complex functionality. William Paley famously asked, if we were to find a working pocket watch on the heath, would we think that random forces conspired to create it or, more parsimoniously, that it was fashioned by the activities of a mind? Similarly, if one comes across a platypus or Venus flytrap, which is more likely, that their ability to survive and reproduce was produced by chance or by an intelligent designer? The theory of natural selection showed, for the first time, a way outside of this false dichotomy (Dawkins, 1986).

Modern evolutionary biologists distinguish between the process of natural selection and its products, *adaptations*. Wings, chemosensory systems, eyes, parental love: all of these are adaptations—features of organisms that came into existence through natural selection because they caused faster genetic replication. Of course, not all features of organisms are adaptations. Some are merely *by-products*, features pulled along incidentally with adaptations. Although bones are white and human blood is red, these traits are by-products, not design features: calcium phosphate helps make sturdy bones, and iron-rich hemoglobin usefully transfers oxygen; their respective features of whiteness and redness do not contribute to these functions. In addition to adaptations and by-products, organisms are also composed of random noise, genetic and developmental insults to an organism's system. For instance, there are a precise number of hairs on your head. This particular number is not a design feature nor is it a by-product shared by all humans like the whiteness of bones; instead, it is simply random noise.

Importantly, by-products and random noise do not lead to complex, functional systems. To the extent that psychologists, including moral psychologists, are interested in complex aspects of human moral nature, they are interested in psychological adaptations. Thus, the conceptual tools from evolutionary biology are additional sources of information that psychologists can use to generate testable hypotheses about the moral mind. Connecting moral psychology with evolutionary biology and adaptationism is one part of the increasingly interdisciplinary nature of moral psychology. We suspect that using the tools of adaptationism will contribute just as much to moral psychology as have connections with neuroscience, the law, and philosophy.

So what kind of moral psychology might selection have designed? Although not explicitly couched in the language of adaptationism, theories of moral development like Piaget's and Kohlberg's implicitly assume that the adaptations that give rise to moral psychology are not specialized for moral thought. Instead, they reflect more general processes such as reasoning and induction. Other work, however, suggests that evolution has equipped the mind with quintessentially morality-specific concepts such as "good" and "bad" (Greene, Sommerville, Nystrom, Darley, & Cohen, 2001; Hamlin, Wynn, & Bloom, 2007; van Leeuwen, Park, & Penton-Voak, 2012). Moreover, evolution may have equipped us with skeletal domains of moral intuitions such as intuitions about community, harm, and fairness (Haidt & Bjorklund, 2008). Our goal in this chapter is to push this argument farther. Natural selection may have designed very specific moral concepts and abilities into the human cognitive architecture, ones more fine-grained than simple right and wrong. Here we present three examples of such fine-grained concepts.

Moral Concepts for Group Cooperation

What constitutes morality and moral psychology? This is a big question and one we have no wish to discuss exhaustively here. Some scholars treat morality as encompassing many aspects of human behavior—altruism, helping, generosity, virtue, and so forth. On this view, helping our kin and our friends is part of morality (e.g., Ridley, 1996). From an evolutionary perspective, psychological mechanisms for aiding kin and friends were likely created by selection pressures such as kin selection and selection for reciprocity. But what makes these selection pressures and the psychologies they create moral? Many other animal minds seem to have been shaped by them, and it is not clear that these animals have anything approaching human moral systems. Theorists taking a restrictive view propose instead that the core of morality are judgments of right and wrong, the moral praise or condemnation that follows, and attendant emotions such as anger, disgust, guilt, and shame (e.g., DeScioli & Kurzban, 2009, 2013).

We are agnostic on this issue. Indeed, we are not sure whether at the margins moral psychology can be cleanly separated from social psychology.

Nonetheless, to make our argument as relevant to the widest audience possible, here we focus mostly on examples that fall under a more narrow reading of the moral domain.

Our first example is drawn from human group cooperation. Humans have evolved a unique kind of cooperative foraging system, often called *resource pooling* or *communal sharing* (Fiske, 1992; Gurven, 2004). Foraging is risky business. On any given day a forager may be a victim of bad luck and return empty handed. Worse, injury and illness can keep a forager disabled for significant lengths of time (Sugiyama, 2004a). Other foragers in one's group, however, may have more food than they can usefully consume. By sharing, foragers can buffer against shortfalls and hardships. Over the long term everyone benefits: yes, you may be giving up some food now to help a person in need, but later, when you are eventually in need, your generosity will be repaid. Such sharing can have huge benefits: in a study of the Shiwiar of the Ecuadorian and Peruvian Amazon, approximately 65% of the sample would be dead had they not been part of a resource-pooling system (Sugiyama, 2004a, 2004b).

Although they provide huge benefits, resource-pooling systems are not trivial to maintain. One problem is that resource-pooling systems (and other types of group cooperation) are vulnerable to free riders, people who take the benefits of a resource-pooling system without contributing to it. Because they take collective benefits but do not pay the costs of creating them, free riders are materially better off than cooperators. This means that, in the absence of any countervailing mechanism, free riders are favored by natural selection over cooperators, and a design for free riding would come to predominate in a population. Of course, humans do engage in resource pooling so somehow the free-rider problem must be solved.

Evolutionary modeling work shows that the free-rider problem can be diminished or eliminated if free riders are either excluded from cooperation or are punished to induce their cooperation (Boyd & Richerson, 1992; Sasaki & Uchida, 2013). Thus, free riders fall under the narrow version of moral psychology—they are likely to be judged as doing wrong and to elicit anger and punitive sentiment. The question we have explored is whether selection has designed a specific mechanism—a free-rider concept—to identify and respond to free riders. In addressing this question we have contrasted the hypothesis that there is a dedicated free-rider concept with a number of alternatives, including economic rationality, statistical learning, and a moral psychology that includes concepts of right and wrong, but nothing so specific as a dedicated free-rider concept.

Why would a dedicated free-rider concept be necessary? Identifying and responding to free riders is a difficult and nuanced problem. Evolutionary analysis reveals that free riders cannot be identified merely by their overt contributions to a communal sharing system. Resource-pooling systems exist to allow people to survive when they are most vulnerable, such as when they are unable to contribute to the common pool (Tooby & Cosmides, 1996). Using a metric of overt contribution to identify free riders would be counterproductive—people in legitimate need of aid would also be classified as free riders. Instead, evolutionary modeling shows that the mind needs a way to "see through" overt behavior and instead make inferences about a person's underlying motivations (Delton & Robertson, 2012).

In a series of experiments we have found that the human mind draws precisely this distinction (Delton, Cosmides, Guemo, Robertson, & Tooby, 2012). These studies used an implicit measure of social categorization. This method uses confusions in memory as a proxy for categorization (Taylor, Fiske, Etcoff, & Ruderman, 1978): thus, if persons A, B, and C are selectively confused with each other, and persons X, Y, and Z are selectively confused with each other, then the two sets are assumed to be represented as distinct categories. When presented with foragers who differ in their motivations—some desiring to exploit a sharing system, others well-intentioned cooperators—experimental subjects sharply distinguished between the two and viewed those desiring to exploit the system as free riders deserving of exclusion or punishment. Importantly, the experimental stimuli were arranged such that free riders and cooperators contributed identical amounts; although free riders purposefully withheld a resource, cooperators attempted to contribute but were prevented by reasons beyond their control. Despite identical contributions, free riders were categorized separately from cooperators based on their motivation and elicited a clearly moralistic response, such as being viewed as deserving exclusion or punishment.

These experiments show that motivations are used as a cue to free riding independent of overt contributions. But perhaps, as predicted by an economic rationality account, overt contributions are the cue of interest, and motivations are only secondarily informative. After all, over the long term, people with exploitive motives are likely to contribute less than people with cooperative motives. Thus, the mind might detect a correlation between motivation and long-term contribution and learn to use motivation as a secondary cue to free riding. This alternative predicts that differences in overt contributions should be used as cue to free riding, perhaps as an even stronger cue than motivation. We tested and experimentally falsified this prediction: when motivation is held constant while overt contributions are manipulated, people did categorize based on this difference, but it did not lead to moralistic responses—no desire to punish or avoid. Instead, people who contribute less were viewed as less competent—an important, but orthogonal, dimension in person perception (Cuddy, Fiske, & Glick, 2008).

Let us consider a final alternative hypothesis. An adaptationist analysis of the logic of resource pooling suggests that the mind has a dedicated free-rider concept for identifying and responding to free riders. Although economic rationality and statistical learning do not fully explain the data, could a more general moral psychology do the job? Perhaps all that is needed is a system that categorizes people as good or bad based on intentional actions as well as other information (Cushman, 2013). Indeed, current evidence suggests that the mind contains reliably developing systems for making general judgments about immorality (Hamlin et al., 2007; van Leeuwen et al., 2012). The question, however, is not whether the mind has a superordinate system for making moral discriminations. The question is instead whether the mind also has finer-grained levels of categorization. These fine-grained levels would exist because different adaptive problems require different solutions. We have tested this issue by comparing free riders to other types of moral violators. In one study, we tested whether free riders were distinguished from what, intuitively, is a very distinct type of moral violation: intentional, unprovoked physical battery. Experimental subjects did indeed categorize free riders separately from batterers and responded more moralistically to batterers (Delton, Cosmides, Guemo, Robertson, & Tooby, 2012).

But still, a free-rider concept may not be needed. The mind may distinguish unprovoked battery from violations involving entitlements and obligations (some theories treat entitlement and obligation as psychologically identical) (Cheng & Holyoak, 1985). Free riders could be construed as violating these general deontological principles but nothing so specific as the rules of a sharing system. To test against this we contrasted free riders with an even more closely matched moral violation: a person who steals a resource communally owned by the group. Thus, both free riders and thieves were in one sense expropriating a group resource for their own benefit. Nonetheless, the logic of resource pooling and the logic of ownership and theft are different adaptive problems. Does the mind categorize them separately? Yes: experimental subjects categorized the two separately and reacted to free riders with greater moralistic punishment.

Although more work needs to be done, the current evidence points to a very specialized piece of conceptual machinery—a specialized free-rider concept. This moral concept was uncovered by taking an adaptationist approach to the human mind, by looking at the kinds of problems humans have regularly solved and generating psychological theories about the kinds of computations required to solve these problems. In other work we have extended this approach to uncover other concepts related to group cooperation, at least some of which are clearly moral. For instance, the mind appears to have a specialized concept for understanding and responding to newcomers to a coalition (Cimino & Delton, 2010; Delton & Cimino, 2010). The mind also appears to contain a concept for identifying and selectively associating with highly valuable cooperation partners (Delton & Robertson, 2012). And the mind appears to have a concept of public goods: not all types of group cooperation produce goods exploitable by free riders; public goods are the subset vulnerable to free riding. Our recent research shows that the mind distinguishes public goods from other cooperatively produced goods and mobilizes relevant moral sentiments to prevent free riding on public goods (Delton, Nemirow, Robertson, Cimino, & Cosmides, 2013).

Moral Concepts for Social Exchange

Our second example is drawn from human dyadic cooperation, often organized as social exchange (Cosmides & Tooby, 2005). When two parties have goods or services that are more valuable to the other than they are to the bearer, then both are better off if they can trade. These items can be traded contemporaneously: if I have an abundance of peanut butter but no jelly, and you have an abundance of jelly but no peanut butter, we can exchange on the spot and both realize gains in trade. Or, exchange can happen over time, such as when I care for you while you are ill now and you care for me when I am ill in the future, allowing both of us to survive our misfortunes. Exchange is a culturally universal (Brown, 1991) major facet of human social life, and the majority of the improvement in the human condition over the last 10,000 years of human history is attributable to the increasing prevalence of such exchange and the specialization and division of labor it enables (Ridley, 2010).

But, realizing gains in trade is a risky business: if you trust me and invest in an exchange (e.g., give me some of your jelly), there is always the chance that I will try to cheat you and not reciprocate (e.g., keeping all my peanut butter for myself). The possibility of cheating suggests that you cannot be uniformly trusting. But neither can you be uniformly *dis*trusting: you can never realize the benefits of exchange without sometimes taking the risk of trusting someone else. Some of this risk can be mitigated by effective partner choice; some partners will present better targets for cooperative investment than others, given their past history or reputation. And some of this risk can be mitigated by effective responses to cheating; either terminate the relationship and thus cut off future loses or deploy sanctions in an attempt to recalibrate the partner's future behavior and thus salvage the future potential for long-term gains in trade. These are adaptive problems, and their adaptive solutions are mechanisms for relationship cultivation, partner selection, and relationship maintenance (see also Cosmides & Tooby, 2005).

It is common wisdom that the outcomes of exchange decisions are moralized: upholding a bargain is morally right, and cheating is morally wrong and often worthy of punishment. But is that as specific as the system gets, merely sketching this moral dimension? Because there are long-enduring benefits, costs, and complexity to social exchange, an adaptationist approach predicts that there should be greater specificity and functional organization in the design of the psychological mechanisms that underpin it. For example, moral responses to exchange violations should be structured to improve long-term profitability for the responder. As such, moral responses to cheating should vary by their relevance to both partner selection and the prospect of improving a dissatisfactory relationship.

We tested these questions in a series of experiments (Krasnow, Cosmides, Pedersen, & Tooby, 2012). We first measured participants' cooperative dispositions. In one study cooperative dispositions were measured by a survey of scenarios in which participants could cheat a business, a friend, a roommate, a co-worker, and so forth, assuming that they could not be caught. A second study measured cooperative dispositions by having participants play a series of one-shot prisoners' dilemma (PD) games, each game ostensibly played with a different partner. (All partners in these games and those described below were actually played by a computer.) In the PD games subjects were randomly paired, and each was given a resource that was more valuable to her or his partner than it was to her- or himself. Participants then each decided whether to keep the resource or give it to his or her partner. In these games the payoff-maximizing decision is to not give. However, if both participants give, they are made better off than if neither gave.

After these measures participants were paired with a partner, and, for some partners, participants were given information about the partner's decisions in the cooperative disposition task. They then played a two-round trust game with this partner. In the game the first mover was given a sum of money that could be evenly split, or he or she could risk trusting the second mover and transfer the money to that person, with the sum multiplying along the way. The second mover could then evenly split this multiplied sum (cooperate by reciprocating the first mover's trust) or could claim the lion's share for him- or herself (cheat). If the second mover cheated, the first mover could spend a small amount of money to impose a large cost on the second mover (punish) or do nothing. Participants first played one round of this game taking the role of first mover and then a second round of the game as the second mover.

The results of these experiments are telling:

Participants exhibited a default openness to exchange. In conditions without any information about a partner's past history of cooperation, the majority trusted their partner.

Participants were sensitive to reputation and used reputation as a cue to how the partner would treat them. There may be individuals who are lousy at exchange with everybody, and they can be avoided by paying attention to how they treat third parties. But every person represents a unique mix of interests, and someone might be a better cooperative partner for me than he or she is for you. If the architecture responsible for trust merely encodes instances of cheating for avoidance and the attribution of moral wrongness, it would miss out on this second class of actor. If instead the architecture used cues hierarchically in the order of their utility in predicting who will most likely cooperate with the self, then being cheated in the past should carry more weight in the regulation of trust than if that same person cheated someone else. And the mind agrees: when given information about their partners' previous decisions to cheat third parties in the survey, participants were much less likely to trust partners with a history of cheating. But when given information about the partner's decisions in two PDs, one with the participant and one with a third party, only the participant-specific information now regulated trust while the third-party information did not. The moral architecture that interprets the violation of exchange norms appears to be designed to discriminate profitable exchange partners and not merely to attribute moral wrongness generally.

Participants deployed punishment to maintain an ongoing relationship. If the moral architecture responsible for attribution of wrongness and punitive sentiment responded to cheating per se, it would not differentially respond to cheaters the person intends to continue cooperating with. But expending punitive effort to reform a relationship you intend to abandon is a waste of resources. And the mind agrees: participants were much more likely to spend money to punish their cheating partner if they intended to cooperate with them in the next round. Participants who abandoned the relationship were much less likely to punish their partner for cheating. Interestingly, convergent evidence for this two-pronged strategy can be found in the domain of criminal justice, where "high-value" offenders are more likely to receive punishment in the form of rehabilitation than are "low-value" offenders (Petersen, Sell, Tooby, & Cosmides, 2012).

In short, the cognitive architecture that produces our moral responses to being wronged and learning that others are wronged does not appear designed for dispassionately evaluating transgressions and thus appears not to result from only a general system of moral right and wrong. Instead, the system appears to contain features that deploy moral approval and disapproval, trust and punitive sanctions, in order to manage the risks of engaging in social exchange in order to effectively profit from gains in trade.

Condemnation and Coordination

Moral communities the world over regularly condemn various classes of behavior. Among the Yanomamo of the Amazon, it is morally condemnable to marry a parallel cousin (i.e., mom's sister's child or dad's brother's child); instead, it is considered appropriate to marry a cross cousin (i.e., mom's brother's child or dad's sister's child) (Chagnon, 1996). Among the !Kung San of the Kalahari desert, not sharing your meat with others is condemnable (Cashdan, 1980), although they feel little compunction wearing clothing that would make pop stars blush. And many modern Americans morally condemn smoking tobacco (Rozin & Singh, 1999), an activity common in many other times and places.

Although moral condemnation is common, it is anything but simple. One complexity of condemnation is that the classes of condemned behavior vary across time and space. A related question, which we discuss here, is how multiple people can successfully coordinate their condemnation. At a very general level moral condemnation serves to change someone's behavior. If I have done something wrong, you might condemn me to get me to refrain from repeating the wrong in the future. But if you condemn me alone, I might simply retaliate against you rather than change myself. Given the potential costs of condemning alone, condemnation is often a social act: multiple people coordinate their condemnation on a specific target (Boehm, 1993; Boyd, Gintis, & Bowles, 2010).

Coordination, however, is difficult to accomplish. In a series of recent papers DeScioli and Kurzban have shown how the difficulties of coordination impact our moralistic judgments (DeScioli & Kurzban, 2009, 2013). They started with an enduring mystery in moral psychology and philosophy: the difference in culpability between sins of omission and sins of commission. Committing a wrong is usually judged more harshly than not preventing a wrong, even if the prevention would have cost nothing. For instance, imagine that Bill is alone in the control room for a train hub. Bill sees a train hurtling down track A and a person standing on track A. Bill could push a button to divert the train to track B, but he does not and the person is killed. How wrong was that? Now imagine that the train is still hurtling down track A, but the person is standing on track B. Bill presses the button and diverts the train to track B and person is killed. How wrong was that? If you are like most people, Bill was more morally wrong in the second scenario where he actively diverted the train.

Why? Logically the distinction appears meaningless. In both cases a death occurred, and Bill could have prevented it. Yet Bill's act of commission is judged more harshly. DeScioli, Bruening, and Kurzban (2011) argue that one of the functions of our moralistic reactions is to help coordinate multiple people in condemnation. Because of this, one factor that should influence the strength of our moral judgments is the ease with which coordination can be achieved. If coordination cannot easily be achieved, it will be too costly to enact condemnation. When condemnation would be costly, there are not benefits in becoming morally outraged (and possibly costs if the retaliation is severe). But if coordination can be achieved, then condemnation will be relatively cheap and moral outrage worthwhile.

What do omission and commission have to do with this? DeScioli and Kurzban argue that sins of omission leave behind little in the way of public evidence that a wrong was committed. Without public evidence that can be used to form common knowledge about the wrong (cf. Pinker, Nowak, & Lee, 2008), sins of omission make coordination difficult. Sins of commission, conversely, do leave behind clear public evidence, making coordination much easier to achieve. DeScioli, Bruening, and Kurzban (2011) designed an ingenious series of experiments to test this hypothesis against the most prominent class of alternative theories. These alternative views propose that direct causal responsibility is the distinction driving the omission–commission effect: sins of commission are directly caused; sins of omission are not. To test against this, DeScioli, Bruening, and Kurzban created scenarios similar to the one described above. They also added a critical test condition, which to simplify and continue with the above examples worked as follows: as Bill watched the train on track A hurtle toward a person on track A, he pressed a button that simply recorded his lack of desire to move the train onto a different track. This was a sin of omission—he did not directly cause the train to kill the person. But it was a *public* sin of omission—there was clear evidence that he allowed the person to be killed. In these studies, when sins of omission leave behind clear public evidence, the omission–commission effect vanishes: sins of public omission elicit almost identical moralistic responses as sins of commission. Moreover, people appear to be sensitive to the fact that they will not elicit condemnation when public coordination is difficult, and thus they behave more self-interestedly in these conditions (DeScioli, Christner, & Kurzban, 2011).

These studies only scratch the surface of the complex phenomenon of moral condemnation. Nonetheless, we believe they illustrate the utility of taking an adaptationist approach. By thinking about the information-processing problems that arise when engaging in moral condemnation, DeScioli and Kurzban were able to generate novel, testable hypotheses.

Parting Thoughts: Why You Should Care about Adaptations for Moral Psychology

In this chapter we have sketched a brief outline of how the adaptationist toolkit has informed the study of human moral psychology in the domains of group cooperation, social exchange, and coordinated condemnation. Our argument, however, is not that adaptationist reasoning is limited to elucidating these few domains of moral thought. The only way for a complex biological structure to endure against the constant destructive forces of entropy and mutation is to reproduce its design faster than its competitors. Thus, to the extent that components of human moral judgment and reasoning are complex in structure, robust in their development, and universal in our species—the components of interest to many moral psychologists—they are very likely to be the product of psychological adaptations. Adaptations are the product of long-term selective regimes: some genetic variants lead to the development of neural systems that, in interaction with environmental regularities, produce fitness-enhancing behavior. These variants thus increase in frequency in the population. The adaptationist approach therefore examines the long-enduring problems that humans have faced to generate hypotheses about the kinds of psychological adaptations that would have evolved to solve these problems (see Delton, Krasnow, Cosmides, & Tooby, 2010, 2011; Krasnow, Delton, Tooby, & Cosmides, 2013). These hypotheses then guide empirical research designed to confirm or falsify them. As illustrated by this volume, insights from neuroscience, philosophy, and the law have begun to inform our understanding of moral psychology. We suggest that the addition of thinking as an adaptationist can improve this understanding.

References

Boehm, C. (1993). Egalitarian behavior and reverse dominance hierarchy. *Current Anthropology, 34*(3), 227–254.

Boyd, R., Gintis, H., & Bowles, S. (2010). Coordinated punishment of defectors sustains cooperation and can proliferate when rare. *Science, 328*, 617–620.

Boyd, R., & Richerson, P. J. (1992). Punishment allows the evolution of cooperation (or anything else) in sizable groups. *Ethology and Sociobiology, 13*(3), 171–195.

Brown, D. E. (1991). *Human universals.* New York: McGraw Hill.

Cashdan, E. (1980). Egalitarianism among hunters and gatherers. *American Anthropologist, 82*, 116–120.

Chagnon, N. (1996). *Yanomamo* (5th ed.). San Diego: Harcourt Brace.

Cheng, P., & Holyoak, K. (1985). Pragmatic reasoning schemas. *Cognitive Psychology, 17*, 391–416.

Cimino, A., & Delton, A. W. (2010). On the perception of newcomers: Toward an evolved psychology of intergenerational coalitions. *Human Nature, 21*(2), 186–202.

Cosmides, L., & Tooby, J. (2005). Neurocognitive adaptations designed for social exchange. In D. M. Buss (Ed.), *The handbook of evolutionary psychology* (pp. 584–627). Hoboken, NJ: Wiley.

Cuddy, A. J. C., Fiske, S. T., & Glick, P. (2008). Warmth and competence as universal dimensions of social perception: The stereotype content model and the BIAS map. *Advances in Experimental Social Psychology, 40*, 61–149.

Cushman, F. (2013). Action, outcome, and value: A dual-system framework for morality. *Personality and Social Psychology Review, 17*(3), 273–292.

Dawkins, R. (1986). *The blind watchmaker.* New York: W. W. Norton.

Delton, A. W., & Cimino, A. (2010). Exploring the evolved concept of newcomer: Experimental tests of a cognitive model. *Evolutionary Psychology, 8*(2), 317–335.

Delton, A. W., Cosmides, L., Guemo, M., Robertson, T. E., & Tooby, J. (2012). The psychosemantics of free riding: Dissecting the architecture of a moral concept. *Journal of Personality and Social Psychology, 102*(6), 1252–1270.

Delton, A. W., Krasnow, M. M., Cosmides, L., & Tooby, J. (2010). Evolution of fairness: Rereading the data. *Science, 329*(5990), 389.

Delton, A. W., Krasnow, M. M., Cosmides, L., & Tooby, J. (2011). The evolution of direct reciprocity under uncertainty can explain human generosity in one-shot encounters. *Proceedings of the National Academy of Sciences USA, 108*(32), 13335–13340.

Delton, A. W., Nemirow, J., Robertson, T. E., Cimino, A., & Cosmides, L. (2013). Merely opting out of a public good is moralized: An error management approach to cooperation. *Journal of Personality and Social Psychology, 105*(4), 621–638.

Delton, A. W., & Robertson, T. E. (2012). The social cognition of social foraging. *Evolution and Human Behavior, 33*, 715–725.

DeScioli, P., Bruening, R., & Kurzban, R. (2011). The omission effect in moral cognition: Toward a functional explanation. *Evolution and Human Behavior, 32*(3), 204–215. doi:10.1016/j.evolhumbehav.2011.01.003.

DeScioli, P., Christner, J., & Kurzban, R. (2011). The omission strategy. *Psychological Science, 22*(4), 442–446.

DeScioli, P., & Kurzban, R. (2009). Mysteries of morality. *Cognition, 112*(2), 281–299.

DeScioli, P., & Kurzban, R. (2013). A solution to the mysteries of morality. *Psychological Bulletin, 139*(2), 477–496.

Fiske, A. P. (1992). The four elementary forms of sociality: A framework for a unified theory of social relations. *Psychological Review, 99*(4), 689–723.

Greene, J. D., Sommerville, R. B., Nystrom, L. E., Darley, J. M., & Cohen, J. D. (2001). An fMRI investigation of emotional engagement in moral judgment. *Science, 293*(5537), 2105–2108.

Gurven, M. (2004). To give and to give not: The behavioral ecology of human food transfers. *Behavioral and Brain Sciences, 27*(4), 543–583.

Haidt, J., & Bjorklund, F. (2008). Social intuitionists answer six questions about moral psychology. In W. Sinnott-Armstrong (Ed.), *Moral psychology: The cognitive science of morality* (pp. 181–254). Cambridge, MA: MIT Press.

Hamlin, J. K., Wynn, K., & Bloom, P. (2007). Social evaluation by preverbal infants. *Nature, 450*, 557–559.

Krasnow, M. M., Cosmides, L., Pedersen, E. J., & Tooby, J. (2012). What are punishment and reputation for? *PLoS ONE, 7*(9), e45662.

Krasnow, M. M., Delton, A. W., Tooby, J., & Cosmides, L. (2013). Meeting now suggests we will meet again: Implications for debates on the evolution of cooperation. *Nature Scientific Reports, 3*, 1747. doi:10.1038/srep01747.

Mayr, E. (1982). *The growth of biological thought.* Cambridge, MA: Harvard University Press.

Petersen, M. B., Sell, A., Tooby, J., & Cosmides, L. (2012). To punish or repair? Evolutionary psychology and lay intuitions about modern criminal justice. *Evolution and Human Behavior, 33*(6), 682–695.

Pinker, S., Nowak, M. A., & Lee, J. J. (2008). The logic of indirect speech. *Proceedings of the National Academy of Sciences USA, 105*(3), 833–838.

Ridley, M. (1996). *The origins of virtue.* London: Viking, Penguin Books.

Ridley, M. (2010). *The rational optimist.* New York: HarperCollins.

Rozin, P., & Singh, L. (1999). The moralization of cigarette smoking in the United States. *Journal of Consumer Psychology, 8*(3), 321–337.

Sasaki, T., & Uchida, S. (2013). The evolution of cooperation by social exclusion. *Proceedings of the Royal Society B: Biological Sciences, 280*, 20122498.

Sugiyama, L. S. (2004a). Illness, injury, and disability among Shiwiar forager-horti-culturists: Implications of health-risk buffering for the evolution of human life history. *American Journal of Physical Anthropology, 123*(4), 371–389.

Sugiyama, L. S. (2004b). Patterns of Shiwiar health insults indicate that provisioning during health crises reduces juvenile mortality. In M. Alvard (Ed.), *Socioeconomic aspects of human behavioral ecology: Research in economic anthropology* (Vol. 23, pp. 379–402). Greenwich, CT: Elsevier.

Taylor, S. E., Fiske, S. T., Etcoff, N. L., & Ruderman, A. J. (1978). Categorical and contextual bases of person memory and stereotyping. *Journal of Personality and Social Psychology, 36*(7), 778–793.

Tooby, J., & Cosmides, L. (1996). Friendship and the banker's paradox: Other pathways to the evolution of adaptations for altruism. In W. G. Runciman, J. Maynard Smith, & R. I. M. Dunbar (Eds.), Evolution of social behaviour patterns in primates and man. *Proceedings of the British Academy, 88*, 119–143.

van Leeuwen, F., Park, J. H., & Penton-Voak, I. S. (2012). Another fundamental social category? Spontaneous categorization of people who uphold or violate moral norms. *Journal of Experimental Social Psychology, 48*, 1385–1388.

Williams, G. C. (1966). *Adaptation and natural selection.* Princeton, NJ: Princeton University Press.

3   Partner Choice and the Evolution of a Contractualist Morality

Nicolas Baumard and Mark Sheskin

Two Views of Human Morality and Its Origins

Utilitarianism and Group Selection

Both in daily life and in empirical investigation, morality is often perceived as the opposite of selfishness. Being moral means helping others, often at a cost to oneself, and the more you help the more moral you are. In this way human morality is described as both generally "consequentialist" (i.e., we judge actions according to their effects) and specifically "utilitarian" (i.e., we most laud those actions that produce the most welfare, in this case referred to as "utility").

Some moral judgments are consistent with utilitarianism. In the trolley dilemma, for instance, people are asked to decide whether it is acceptable to divert a trolley that is going to run over and kill five people onto a side track where only one person will be killed. In response to this dilemma most people agree that it is acceptable to divert the trolley, because it is better to save five lives than one. Similarly, participants in the dictator game, who decide how much experimentally provided money to keep and how much to transfer to another participant, generally transfer around 20 percent of the money. Again, this suggests that people consider the joint welfare of everyone impacted by a decision, although other motivations (e.g., selfishly maximizing one's own welfare) can reduce how morally people behave.

How could natural selection produce behavior in contrast with selfishness? After all, individuals who sacrifice their own fitness to increase others' fitness will, necessarily, have fewer offspring relative to others who just behave selfishly. Historically, one hypothesis has been that unselfish behavior is a result of group selection (for a critical overview, see West, El Mouden, & Gardner, 2010). According to group selection, groups with selfless individuals will outcompete other groups, such that the proportion of selfless individuals might increase in the overall population *even while* it decreases within the group.

However, there are many reasons to think that group selection is not an important factor in the evolution of morality (Abbot et al., 2011; Clutton-Brock, 2009; West et al., 2010) and that moral judgments are best characterized by moral theories other than utilitarianism. In the trolley dilemma, for instance, there are many variations for which people do not choose to save the greater number of people. People are not utilitarian in the "footbridge" variation, in which the only way to save five people is to push a bystander off of a bridge and in the way of the trolley (Greene, Sommerville, Nystrom, Darley, & Cohen, 2001). Furthermore, even the most characteristically utilitarian cases (e.g., the standard switch case) show deviations from utilitarianism: whereas utilitarianism requires that the maximum number of lives must be saved and allows that equal tradeoffs are acceptable, people do not think it is *required* to switch the trolley from five to one workmen, and they do not think it is acceptable to switch when there are equal numbers of workmen on each track (Sheskin & Baumard, unpublished data).

Importantly, antiutilitarian judgments are not limited to the artificial case of the trolley problem. For instance, people oppose organ selling even when they are told that allowing donors to sell their organs and receivers to bid for them would increase the number of lives saved (Tetlock, 2003). Similarly, people oppose a policy that would increase cure rates for one group of patients if it would also reduce cure rates for a second group, even if this second group is much smaller (Baron, 1995; Nord, Richardson, Street, Kuhse, & Singer, 1995). People also object to a distribution system that would sacrifice justice to efficiency (Mitchell, Tetlock, Newman, & Lerner, 2003). Moving from distributive justice to retributive justice, people refuse "overly" harsh punishments that would deter future crimes and thereby produce net benefits (Carlsmith, Darley, & Robinson, 2002; Sunstein, Schkade, & Kahneman, 2000). In the domain of charity most people refuse to risk their lives or to give large amount of money to save others, even if doing so would surely bring about more good than bad from a global perspective (Baron & Miller, 2000; Greene et al., 2001).

Given the many departures from utilitarianism, many psychologists have suggested that our moral cognition is beset with myriad errors or defects (Baron, 1994; Cushman, Young, & Greene, 2010; Sunstein, 2005). Baron

(1994) suggests that nonaltruistic judgments could be the result of "docil-ity" or "overgeneralization." Sunstein (2005) has proposed that nonaltruis-tic rules are "simple heuristics that make us good." They are generally good (e.g., "do no harm"), but sometimes they are mistaken (e.g., "do not kill anyone, even though it may save many people"). According to Cushman et al. (2010), such irrational judgments result from primitive emotional dispo-sitions such as violence aversion, disgust, or empathy. In short this view of moral judgment has to address many departures from utilitarianism, and it often does so by suggestion that our moral psychology is plagued by diverse "moral confabulations" based on "alarm bell emotions."

Contractualism and Partner Choice

In this chapter we advance an alternative view of both moral judgment and its evolutionary origins. Specifically, we propose an alternative solution to the existence of nonutilitarian judgments, in which they are not biases or defects but instead are the signature of a perfectly functional and adaptive system. In other words it has been assumed that the function of the moral system is to maximize welfare, and, with this premise in mind, it looks as if the moral system sometimes fails. But, if we assume a different function, these so called failures might be reconceptualized as perfectly functional. So what could that function be?

Consider the following situations:

• When we help financially, we do not give as much as possible. We give a *quite specific and limited amount*: many people think there is a duty to give some money to charity, but no one feels a duty to donate his or her entire wealth.

• When we share the fruits of a joint endeavor, we do not try to give as much as possible. We share in *a quite specific and limited way*: those who contribute more should receive more.

• When we punish, we do not take as much as possible from the wrong-doer. We take in a *quite specific and limited* way: a year in jail is too much for the theft of an apple and not nearly enough for a murder.

So it appears as though morality is not really about helping as much as possible. Sometimes, being moral means keeping everything for oneself; sometimes, it means giving everything away to others. Morality seems to be about proportioning our interests and others' interests, for instance by proportioning duties and rights, torts and compensations, or contributions and distributions.

How can we conceptualize this logic of proportionality? Many philosophers such as John Locke, Jean-Jacques Rousseau, and John Rawls have proposed a metaphorical contract: humans behave as if they had bargained with others in order to reach an agreement about the distribution of the costs and benefits of cooperation. These "contractualist" philosophers argue that morality is about sharing the benefits of cooperation in a fair way. The contract analogy is both insightful and puzzling. On the one hand it captures the pattern of many moral intuitions: why the distribution of benefits should be proportionate to each cooperator's contribution, why the punishment should be proportionate to the crime, why the rights should be proportionate to the duties, and so on. On the other hand it provides a mere as-if explanation: it is as if people had passed a contract—but of course they hadn't. So where does this seeming agreement come from? In order to answer this question, it is important to go back to the standard evolutionary theory of cooperation.

In the framework of individual selection, A has an interest in cooperating with B if it is the case that the benefits A provides for B will be reciprocated. This "if I scratch your back, you'll scratch my back" is the idea behind reciprocal altruism and the standard reciprocity theory (Trivers, 1971). In this view if I give you one unit, then you give me one unit; if I give you three units, then you give me three units. But imagine that A and B are not equally strong. Imagine that A is much stronger than B and so feels secure reciprocating a benefit of three units with just one unit back. If we assume that B is stuck in her interaction with A, she has no choice but to accept any offer, as unfair or as disproportionate as it may be (André & Baumard, 2011; Schelling, 1960).

However, if we instead imagine that B is *not* stuck with A and has a choice among various collaborators, then she is likely to simply avoid A and instead enter into a mutualism with a fairer individual. This is what biologists call a *biological market* (Noë & Hammerstein, 1994). In this market individuals are in competition to attract the best partners. If they are too selfish, their partner will leave for a more generous collaborator. If, on the other hand, they are too generous, they will end up being exploited. If this "partner choice" model is right, we should see that the only evolutionary stable strategy leads to an impartial distribution of the benefits of cooperation (André & Baumard, 2012).

We summarize the two views of human morality and its origins in figure 3.1. In figure 3.1a, the partner choice model of evolution predicts a fairness-based moral psychology that produces judgments in line with contractualism. In figure 3.1b, the group selection model of evolution predicts a harm/welfare-based moral psychology that produces judgments in line with utilitarianism. It is not our intention to suggest that utilitarian theories of moral psychology are required to endorse group selection, but there is a natural progression from group selection to utilitarianism (i.e., "IF group selection were responsible for our moral psychology, THEN a utilitarian moral psychology would be the plausible result"). Likewise, our argument that moral judgments are contractualist does not require that partner choice is responsible for human moral psychology, but contractualism is a prediction of partner choice.

Three Examples of Morality as Fairness

We are now in a better position to explain why people are "bad utilitarians." In short, they are not utilitarians at all! Rather than trying to maximize group welfare, moral judgments are about allocating welfare in a fair way. Thus, distributive justice does not aim at maximizing overall welfare but at distributing resources in an impartial way (whether or not this is "efficient," that is, producing the most overall welfare); retributive justice does not aim at deterring future crimes but at restoring fairness by diminishing the criminal's welfare or compensating the victim (whether or not this deters crime); helping others does not aim at increasing the welfare of

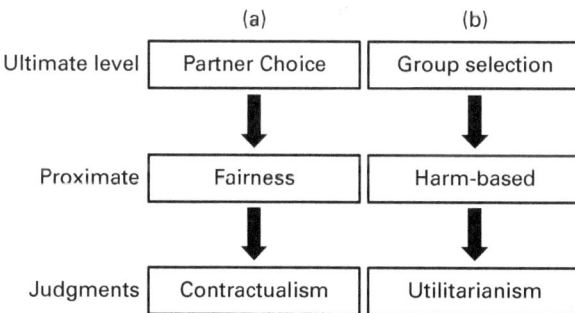

Figure 3.1
Predictions of partner choice and group selection.

the group but at sharing the costs and benefits of mutual aid in a fair way (whether or not helping more would still increase the global welfare). In the sections that follow we discuss these three examples in greater detail.

Distributing Scarce Resources

When distributing scarce resources, such as income in the economy or life-saving resources in healthcare, utilitarianism defends the possibility of sacrificing the welfare of a minority for the greater benefit of a majority (for a review, see Baron, 1994). For a first example Baron and Jurney (1993) presented subjects with six proposed reforms, each involving some public coercion that would force people to behave in ways that maximized joint welfare. In one of the cases most participants thought that a 100 percent gas tax (to reduce global warming) would do more good than harm—with even 48 percent of those opposed to the tax conceding that it was a case of net good—and yet only 39 percent of subjects said they would vote for the tax. Participants justified their resistance by noting that the reforms would harm some individuals (despite helping many others), that a right would be violated, or that the reform would produce an unfair distribution of costs or benefits. As Baron and Jurney (1993) conclude: "Subjects thus admitted to making nonconsequentialist decisions, both through their own judgment of consequences and through the justifications they gave."

In a second set of experiments participants were asked to put themselves in the position of a benevolent dictator of a small island with equal numbers of bean growers and wheat growers. The decision was whether to accept or decline the final offer of the island's only trading partner, as a function of its effect on the incomes of the two groups. Most participants did not accept offers that reduced the income of one group in order to increase the income of the other, even if the reduction was a small fraction of the gain and even if the reduction increased the overall income (for similar results, see Konow, 2001).

Finally, in a third set of experiments a significant proportion of participants refused to reduce cure rates for one group of patients with AIDS in order to increase cure rates in another group, even when the change would increase the overall probability of cure. Likewise, they resisted a vaccine that reduced overall mortality in one group but increased deaths from side effects in another group, even when, again, this decision was best at the global level (Baron, 1995).

Although irrational in a utilitarian framework, this refusal to sacrifice the welfare of some for the benefit of others makes sense if our moral judgments are based on fairness to support mutualism. When being moral is about interacting with others in a mutually advantageous way, then, everything being equal, it is wrong to change a situation in a way that is of benefit only to some of the individuals. Individuals will only agree to situations that are, in fact, mutualisms (i.e., mutually beneficial). Forcing some individuals into a situation that is not to their benefit amounts to stealing from them for the benefit of others.

Punishment and the Need to Restore Justice
Intuitions regarding punishment do not follow utilitarianism and instead are based on restoring fairness. A utilitarian justification for punishing a wrongdoer might be that punishment deters future crime (by the previous offender and/or a new offender) by raising the costs of the crime above the benefits. Rehabilitation or isolation of a criminal can also serve as utilitarian justifications for punishment. However, people are mostly insensitive to each of these factors in their punishment judgments. For example, Carlsmith, Darley, and Robinson (2002) found that people's punishment decisions were strongly influenced by factors related to retribution (e.g., severe punishments for serious offenses) but were not influenced by factors related to deterrence (e.g., severe punishments for offenses that are difficult to detect). These results are particularly striking in that Carlsmith and colleagues (2002) also found that people will endorse deterrence and are able to produce deterrence-based judgments if directed to do so.

More striking are cases in which people pursue retribution even when doing so reduces utility. Baron and Ritov (1993) asked participants to assess penalties and compensation for cases involving no clear negligence, in which there was a rare victim of a medication side effect. For example, one set of cases described a corporation that was being sued because a child died as a result of taking one of its flu vaccines. In one version of the story participants read that a fine would have a positive deterrent effect and make the company produce a safer vaccine. In a different version participants read that a fine would have a perverse effect such that the company would stop making this kind of vaccine altogether (which is a bad outcome, given that the vaccine does more good than harm and that no other firm is capable of making the vaccine). Participants indicated whether they thought a punitive fine was appropriate in either of the cases and whether the fine should differ between the two cases. A majority of participants said that the fine should not differ at all, which suggests that they do not care about the effect of the fine and only care about the magnitude of the harm that was caused. In another test of the same principle, participants assigned penalties to the company even when the penalty was secret, the company was insured, and the company was going out of business, so that (participants were told) the amount of the penalty would have no effect on anyone's future behavior (Baron, 1993; Baron & Ritov, 1993). In all these studies most participants, including a group of judges, "did not seem to notice the incentive issue" (Baron, 1993, p. 124).

Although they clearly deviate from utility maximization, these judgments make sense in a mutualistic framework. If we consider that morality is about demonstrating and enforcing fairness, then a crime creates an unfair relationship between the criminal and the victim. If people care about fairness and have the possibility to intervene, they will thus act to restore the balance of interests either by harming the criminal or by compensating the victim. Data from legal anthropology are in line with this theory. Indeed, many writers have discussed the process of law in stateless societies with such expressions as "restoring the social balance" (Hoebel, 1954). In one of the first ethnographies on law and punishment, *Manual of Nuer Law*, Howell constantly emphasizes that the purpose of the payment is to "restore the equilibrium" between the groups of the killer and killed.

Thus, punishment seems to be motivated by restoring fairness rather than any other purpose (Baumard, 2011). Indeed, Rawls (1955) allows for the disconnect between the many utilitarian justifications for punishment and the retributive basis for people's individual punishment decisions by suggesting that utilitarianism justifies only the institution of punishment (i.e., we have a legal system that punishes people because of the good societal effects of such a system), but that retributivism justifies each individual case of punishment (i.e., we punish an individual person based on the seriousness of his or her crime). By separating the moral foundations of the institution of punishment from the application of punishment in each individual case, this analysis may account for why we recognize and endorse the utility of punishment in general, whereas we base our judgments of individual cases on retribution.

Limited Requirement to Help Others

Perhaps one of the most counterintuitive aspects of utilitarianism is that there is no such thing as a supererogatory action—that is, an action "above and beyond the call of duty." According to one straightforward application of utilitarianism, you are doing something immoral by spending time reading this chapter because it is not the action you could be taking right now to most increase worldwide utility. Instead, you should be doing something like donating (nearly) all of your money to charity and then committing the rest of your life to volunteer work. Contrary to the extreme requirements of utilitarianism, people do not typically think they have a duty to completely sacrifice their own interests to increase the welfare of strangers. This distinction between moral and supererogatory appears early on in moral development (Kahn, 1992) and is present in many moral traditions (Heyd, 1982). Instead of an unlimited duty to help, people perceive well-defined limits to other-directed behavior, and these limits are defined by the logic of fairness. Mutual help is not about being generous or sacrificing for the greatest good but rather about giving others the amount of help we owe them if we want to interact in a mutually advantageous way. For example, there are clear boundaries between failing to do one's duty (i.e., only taking one's fair share, not more), doing one's duty (i.e., taking one's fair share), and going beyond one's duty (i.e., taking less than one's fair share).

The contrast between utilitarianism and fairness can be seen in the classic Peter Singer (1972) article, "Famine, Affluence, and Morality." Singer (1972) observes that we may feel an obligation to save a small child drowning in a shallow pond but no obligation to send money to save millions of lives in Bangladesh. He concludes that this departure from utilitarianism is irrational (see also Greene, 2008; Unger, 1996). This variability, however, makes sense in the theory of fairness. If we help others not for the purpose of increasing the global welfare, but instead because we want to interact with others in a mutually advantageous way, then our duty should take into account the relationship of those involved with a situation. In a systematic analysis of the Singer (1972) case, Unger (1996) argues that the main explanation of the difference for us between the drowning child and the dying Bangladeshis resides in the way we frame the situation. In the case of the famine we consider ourselves in a relationship with millions of Bangladeshis in need of help and millions of Western people who might potentially help; in the case of the drowning child, we are alone with the child.

Consistent with this analysis, experimental studies show that identifying a victim increases the amount of help: when a victim is allocated to us (as charities do in some cases), we feel that we have higher obligations to her: it is I and she and not we and they. Small and Loewenstein (2003) show that even a very weak form of identifiability—determining the victim without providing any personalizing information—increases caring both in the laboratory and in the field. In one striking example Small and Loewenstein (2003) found that people donated more to a housing fund when they were told the recipient had already been chosen, but not who it was, compared to when the recipient had not yet been chosen (see also Kogut & Ritov, 2005a, 2005b). Observations about rescue in war (Varese & Yaish, 2000) likewise show that people feel that they have a greater duty to help when an otherwise identical situation is seen as involving a small group (typically the helper and the person in need) than when it is framed as involving a large group (with many helpers and people who might need help).

This view of mutual help may also help explain why people feel they have more duty toward their friends than toward their colleagues, toward their colleagues than toward their fellow citizens, and so on (Haidt & Baron, 1996). People typically have fewer friends than colleagues and fewer colleagues than fellow citizens, and therefore, they should help their friends more because they constitute a smaller group. More generally, this mutualistic analysis, in contrast to utilitarianism, accounts for the fact that people consider that they have special duties toward their families or their friends and that they are not committed to increasing an abstract greater good (Alexander & Moore, 2012; Kymlicka, 1990). Indeed, the mutualistic theory can demand as much from us as utilitarianism, for example, if we are the only person able to help a family member.

Conclusion

For the sake of the presentation, we have organized the demonstration around three case studies of presumed "biases against utilitarianism": distributive justice, punishment, and supererogation. However, we do not think there is a limited "catalog" (Sunstein, 2005) of biases against utilitarianism. All moral judgments have the same logic: respecting others' interests either by transferring resources to others or by inflicting a cost on those who do not respect others' interests. All moral judgments are the product of a sense of fairness.

Indeed, returning to the trolley dilemma with which we opened the chapter, a contractualist analysis can shed light on why it is acceptable to switch the trolley to a track with one person but not to push a person in front of the trolley. In the former case all parties are on trolley tracks, and it is only a matter of chance that the trolley is heading toward the larger group. Because the trolley might equally have gone down the other set of tracks, there is an important sense in which it is "not distributed yet" and all individuals have an equal right to be saved from it. On the contrary there is no natural way the trolley could have gone on the footbridge, and the trolley is clearly associated with the people on the tracks, and so saving the five amounts to stealing the life of the man on the footbridge. Notice that *identical* mutualistic logic is being applied in both cases. Thus, the switch and footbridge trolley dilemmas do not necessarily highlight separate features of our moral psychology (utilitarian and nonutilitarian) but instead can be accounted for by a single, nonutilitarian, fairness-based principle: when someone has something (e.g., safety from being in the potential path of a trolley), respect it; when people are on a par (e.g., they are all in the potential path of a trolley), then do not favor anyone in particular.

The link between the evolutionary level (the market of cooperative partners) and the proximate psychological level (the sense of fairness) is crucial. Without it, fairness and its logical consequences—the precedence of justice over welfare, the retributive logic of punishment, and the existence of supererogatory actions—look like irrational and unsystematic biases. In an evolutionary framework, by contrast, they are all a unified expression of fairness that serve as an adaptation for our uniquely cooperative social life.

References

Abbot, P., Abe, J., Alcock, J., Alizon, S., Alpedrinha, J. A., Andersson, M., et al. (2011). Inclusive fitness theory and eusociality. *Nature, 471*(7339), E1–E4.

Alexander, L., & Moore, M. (2012). Deontological ethics. In E. N. Zalta (Ed.), *The Stanford Encyclopedia of Philosophy*. Available at <http://plato.stanford.edu/archives/win2012/entries/ethics-deontological/>.

André, J. B., & Baumard, N. (2011). The evolution of fairness in a biological market. *Evolution; International Journal of Organic Evolution, 65*(1), 1447–1456.

André, J. B., & Baumard, N. (2012). Social opportunities and the evolution of fairness. *Journal of Theoretical Biology, 289*, 128–135.

Baron, J. (1993). Heuristics and biases in equity judgments: A utilitarian approach. In B. A. Mellers & J. Baron (Eds.), *Psychological perspectives on justice: Theory and applications* (pp. 109–137). New York: Cambridge University Press.

Baron, J. (1994). Nonconsequentialist decisions. *Behavioral and Brain Sciences, 17*, 1–42.

Baron, J. (1995). Blind justice: Fairness to groups and the do-no-harm principle. *Journal of Behavioral Decision Making, 8*(2), 71–83.

Baron, J., & Jurney, J. (1993). Norms against voting for coerced reform. *Journal of Personality and Social Psychology, 64*(3), 347.

Baron, J., & Miller, J. (2000). Limiting the scope of moral obligations to help: A cross-cultural investigation. *Journal of Cross-Cultural Psychology, 31*(6), 703.

Baron, J., & Ritov, I. (1993). Intuitions about penalties and compensation in the context of tort law. *Making Decisions About Liability and Insurance, 7*(1), 7–33.

Baumard, N. (2011). Punishment is not a group adaptation: Humans punish to restore fairness rather than to support group cooperation. *Mind & Society, 10*(1), 1–26.

Carlsmith, K., Darley, J., & Robinson, P. (2002). Why do we punish? Deterrence and just deserts as motives for punishment. *Journal of Personality and Social Psychology, 83*(2), 284–299.

Clutton-Brock, T. (2009). Structure and function in mammalian societies. *Philosophical Transactions of the Royal Society of London. B, Biological Sciences, 364*(1533), 3229–3242.

Cushman, F., Young, L., & Greene, J. (2010). Our multi-system moral psychology: Towards a consensus view. In J. Doris, G. Harman, S. Nichols, J. Prinz, W. Sinnott-Armstrong, & S. Stich (Eds.), *The Oxford Handbook of Moral Psychology* (pp. 47–71). Oxford: Oxford University Press.

Greene, J., Sommerville, R., Nystrom, L., Darley, J., & Cohen, J. (2001). An fMRI investigation of emotional engagement in moral judgment. *Science, 293*(5537), 2105–2108.

Greene, J. (2008). The secret joke of Kant's soul. In W. Sinnott-Armstrong (Ed.), *Moral Psychology* (Vol. 3, pp. 35–79) Cambridge, MA: MIT Press.

Haidt, J., & Baron, J. (1996). Social roles and the moral judgment of acts and omissions. *European Journal of Social Psychology, 26*(2), 201–218.

Heyd, D. (1982). *Supererogation: Its status in ethical theory*. Cambridge: Cambridge University Press.

Hoebel, A. E. (1954). *The law of primitive man*. Cambridge, MA: Atheneum.

Kahn, P. H., Jr. (1992). Children's obligatory and discretionary moral judgments. *Child Development, 63*(2), 416–430.

Kogut, T., & Ritov, I. (2005a). The "identified victim" effect: An identified group, or just a single individual? *Journal of Behavioral Decision Making, 18*(3), 157–167.

Kogut, T., & Ritov, I. (2005b). The singularity effect of identified victims in separate and joint evaluations. *Organizational Behavior and Human Decision Processes, 97*(2), 106–116.

Konow, J. (2001). Fair and square: The four sides of distributive justice. *Journal of Economic Behavior & Organization, 46*(2), 137–164.

Kymlicka, W. (1990). Two theories of justice. *Inquiry, 33*(1), 99–119.

Mitchell, G., Tetlock, P. E., Newman, D. G., & Lerner, J. S. (2003). Experiments behind the veil: Structural influences on judgments of social justice. *Political Psychology, 24*(3), 519–547.

Noë, R., & Hammerstein, P. (1994). Biological markets: Supply and demand determine the effect of partner choice in cooperation, mutualism and mating. *Behavioral Ecology and Sociobiology, 35*(1), 1–11.

Nord, E., Richardson, J., Street, A., Kuhse, H., & Singer, P. (1995). Who cares about cost? Does economic analysis impose or reflect social values? *Health Policy, 34*(2), 79–94.

Rawls, J. (1955). Two concepts of rules. *Philosophical Review, 64*(1), 3–32.

Schelling, T. C. (1960). *The strategy of conflict.* Cambridge, MA: Harvard University Press.

Singer, P. (1972). Famine, affluence, and morality. *Philosophy & Public Affairs, 1*(3), 229–243.

Small, D. A., & Loewenstein, G. (2003). Helping a victim or helping the victim: Altruism and identifiability. *Journal of Risk and Uncertainty, 26*(1), 5–16.

Sunstein, C. (2005). Moral heuristics. *Behavioral and Brain Sciences, 28*, 531–573.

Sunstein, C., Schkade, D., & Kahneman, D. (2000). Do people want optimal deterrence? *Journal of Legal Studies, 29*(1), 237–253.

Tetlock, P. E. (2003). Thinking the unthinkable: Sacred values and taboo cognitions. *Trends in Cognitive Sciences, 7*, 320–324.

Trivers, R. L. (1971). The evolution of reciprocal altruism. *Quarterly Review of Biology, 46*(1), 35–57.

Unger, P. K. (1996). *Living high and letting die: Our illusion of innocence.* Oxford: Oxford University Press.

Varese, F., & Yaish, M. (2000). The importance of being asked: The rescue of Jews in Nazi Europe. *Rationality and Society, 12*(3), 307–334.

West, S. A., El Mouden, C., & Gardner, A. (2010). Sixteen common misconceptions about the evolution of cooperation in humans. *Evolution and Human Behavior, 32*(4), 231–262.

Williams, B. A. O. (1981). *Moral luck: Philosophical papers, 1973–1980.* Cambridge, New York: Cambridge University Press.

II Motivations of Morality

4   Is the Moral Brain Ever Dispassionate?

Jesse Prinz

Sentiments and Moral Judgments

The neuroscientific investigation of moral judgment is an outgrowth of philosophical inquiry. Many of the key questions under investigation today date back to old theories, beginning especially with a debate that emerged in the eighteenth century—a debate about the role of emotions in moral judgments. At that time the Scottish philosopher David Hume and some other "British moralists" began to argue that moral judgments have a basis in the emotions. Their views coalesced under the title "sentimentalism." Recent work in neuroscience, including fMRI, EEG, and studies of atypical brains, has been brought to bear on this old tradition. Some researchers have sided with sentimentalism, but others have offered grounds of dissent. Here I survey four major lines of resistance and argue that sentimentalism remains the most plausible account of how moral judgments are made.

Sentimentalism

To begin, we should consider the commitments of sentimentalism in comparison to rival theories. This has been a topic of some confusion in recent moral psychology, and it is important to exorcize some misconceptions. We can focus here on Hume's sentimentalism, although there are interestingly different variations among his contemporaries, such as Francis Hutcheson, the Earle of Shaftesbury, and Adam Smith. Hume's sentimentalism is first laid out in the third book of his *Treatise of Human Nature* (1739). There are two dimensions to his position: a view about the limitations of reason and a view about the role of sentiments. Both are relevant to current debates, and both have been somewhat misrepresented.

Hume begins Book Three with a critique of moral rationalism. Rationalists, such as John Balguy, Ralph Cudworth, Samuel Clarke, and William
Wollaston, had claimed that moral obligations could be derived from reason alone. They often drew an analogy between morality and mathemat-
ics. Hume, himself, was not convinced. He offers several arguments that I
do not reconstruct here, but it will be helpful to get their gist. First, Hume
claims that moral judgments are practical in nature: they are experienced
as imperatives to act. Reason, in contrast, is theoretical, not practical; it
describes how things are but not how they should be. We cannot derive an
"ought" from an "is," Hume says, and, when it comes to action, reason is
a slave of the passions. This can be understood in terms of the more recent
idea of belief-desire psychology. To explain why I engage in an action, say
pouring a glass of wine, we need to posit both beliefs (e.g., that there is wine
in the bottle before me) and desire (e.g., the desire to drink). The mere belief
that there is wine before me would not be adequate to explain why I poured
it (suppose I detested wine). So desire, or passion, is needed. Two people can
know the same facts and reason equally well about what those facts entail
but arrive at divergent decisions about what they should do. If moral judgments are calls to action, then it follows that they cannot be derived from
reason alone. Hume supplements this with other arguments. For example,
he points out that no one has come up with a satisfying rational deriva-
tion of a moral rule, and he also identifies the fundamental relations that
are used in reasoning, such as resemblance and contiguity, and argues that
these are inadequate to derive moral rules. He argues further that moral
rationalists unwittingly commit to the absurd view that moral rules govern
simple organisms, such as trees, and he raises the interesting possibility
that, unlike the realm of reason, morality is governed by the norm of appropriateness rather than truth.

I will not assess Hume's arguments against moral rationalism here, but I
want to dispel a misconception. Hume's main claim is that moral conclusions cannot be derived from reason alone. That does not mean reasoning
plays no role in the moral domain or that reasoning is merely a post hoc
mechanism for rationalization (see Haidt, 2001, for this pessimistic view
of reasoning). Hume argues that there are two ways reason can be crucial
to arriving at a moral verdict. These might be termed classificatory and
instruments. He illustrates with an example outside the moral domain: the
decision to reach for a piece of fruit. This decision depends on a desire and,

hence, cannot be explained by reason alone. But it also depends on beliefs, which can be arrived at—or corrected—through reasoning. These beliefs are both classificatory and instrumental. Classificatory beliefs include the assumption that the fruit is edible, ripe, and sweet. Instrumental beliefs encompass the assumption that the fruit can be obtained by reaching in a certain direction, grasping, and pulling. Likewise, if I decide that it would be good to give to Oxfam, I must first classify Oxfam as a charity organization, and I must also form the belief that giving to them will help achieve one of my moral goals (such as the imperative to help people in need). It may take extensive reasoning and research to arrive at the conclusion that it would be good to donate to this particular organization.

Hume also has another argument for the role of reasoning in moral judgments. He observes that we make countless moral judgments, and these are often highly specific (give to Oxfam, consume fair trade products, avoid the temptation to cheat on tax returns, etc.). It is unreasonable to assume that each of these is conditioned independently. Rather, we seem to have some more basic values (help the needy, do not exploit people, be honest and fair), which are used to derive more specific decisions. Reason can play a crucial role in such derivations. So, Hume is not an opponent of reasoning. Rather, he is an advocate of the view that pure reason cannot, on its own, deliver moral judgments.

The second dimension of Hume's account involves the sentiments. If moral judgments cannot be based on reason alone, then what else do they require? Hume's answer is that they require sentiments. Here is a famous passage:

Take any action allowed to be vicious: Wilful murder, for instance. Examine it in all lights, and see if you can find that matter of fact, or real existence, which you call vice. ... You never can find it, till you turn your reflection into your own breast, and find a sentiment of disapprobation, which arises in you, towards this action. Here is a matter of fact; but it is the object of feeling, not of reason. It lies in yourself, not in the object. So that when you pronounce any action or character to be vicious, you mean nothing, but that from the constitution of your nature you have a feeling or sentiment of blame from the contemplation of it. (Hume, 1739: III.i.1)

Hume begins with his first point that reason alone cannot tell us that something is morally good or bad and then makes the bold suggestion that the impression of goodness or badness derives from our sentiments.

The term "sentiment" is a bit antiquated now. We talk instead of emotions. In the seventeenth century the term "emotion" usually referred to a very strong or intense affective state—what we might now call a "passion." "Passion" was used roughly the way we now use "emotion"—the intensity of the terms has switched. "Sentiment" was also used for what we now call emotions, although in the present context sentiments might be more accurately defined as an emotionally based attitude. For example, when we find something amusing or revolting or fascinating, these are all sentiments, based respectively in feelings of mirth, disgust, and interest. In the moral domain Hume talks of the sentiments of "approbation" or praise and "disapprobation" or blame. He does not specify what emotions these sentiments involve, but recent work in psychology offers some suggestions. To view something with disapprobation may involve a family of different emotions. Rozin, Lowry, Imada, and Haidt (1999) used survey methods to show that anger, disgust, and contempt are all associated with moral violations, with each assigned to a different moral domain: crimes against persons, against community, and against nature. This differentiation has been confirmed for anger and disgust in an induction study (Seidel & Prinz, 2013a). There is also evidence linking moral judgment to the self-directed emotions guilt and shame. Tangney, Miller, Flicker, and Barlow (1996) found that both were associated with felt violations of moral standards. There is evidence that guilt and shame are first-person analogues of anger and disgust, respectively (Giner-Sorolla & Espinosa, 2011). On the positive side approbation has been less intensively studied, but much recent work suggests that moral praise may be associated with positive emotions, including admiration and gratitude (e.g., Algoe & Haidt, 2009; Prinz, 2007), elevation (Algoe & Haidt, 2009; Strohminger, Lewis, & Meyer, 2011), uplifting happiness (Seidel & Prinz, 2013a), and pride, which has been linked to prosocial behavior (Bureau, Vallerand, Ntoumanis, & Lafrenière, 2013).

It is important to note that, for Hume, emotions are not *precursors* to moral judgments, which is another easy misconception. It is tempting to think of emotions as "intuitions," which prod us to make one judgment or another, where judgments are something other than emotions (as may be implied in Haidt, 2001). This is not Hume's view. For him, emotions are component parts of our judgments. To judge that something is wrong is to feel an emotion of blame toward it. For example, if you defect during an economic game with me, I may feel angry about that. In this scenario my anger is a sentiment of disapprobation toward your action; it is the judgment that you have done something bad. We sometimes make knee-jerk judgments and then correct them or refuse to endorse them. For example, I might decide that your defection was justified given my own failure to consistently cooperate. I might even be ashamed of my anger, which would be another sentiment of disapprobation—a judgment that my anger was wrong.

In summary, for a sentimentalist like Hume, moral judgments are sentiments, which may or may not be preceded by reasoning but always go beyond reasoning because reason alone cannot dictate how to feel. Philosophers have challenged sentimentalism in many ways over the centuries. For example, many critics point out that one can feel anger or disgust toward something while denying that it is wrong. Empirical work confirms this. For example, people with high socioeconomic status will sometimes judge that odd sexual behavior (e.g., masturbating with a chicken carcass) is disgusting but not wrong (e.g., Haidt et al., 1993). Elsewhere I have responded to this particular objection. My proposal, in brief, is that moral judgments must involve a disposition to both self- and other-directed feelings of blame. If an action that evokes disgust would also evoke shame, then the action is regarded as morally wrong. In addition, we can distinguish our knee-jerk moral responses from the moral responses with which we identify (roughly the ones that we want to have and do not feel ashamed of). This notion of identification can be elaborated by analogy to racism. We can distinguish widespread implicit racism and the antiracist attitudes that many people report on explicit surveys; the antiracist can acknowledge her knee-jerk racism while also saying that it is not an attitude with which she identifies. Returning to the moral case, people with high socioeconomic status may either lack moral attitudes toward unusual sexual behavior because they would not find such behavior shameful for themselves, or else they may fail to identify with such attitudes. In the latter situation we should say that they implicitly regard unusual sexual behavior as wrong, but they consciously reject this response and explicitly believe that unusual sexual behavior is permissible. Note that this explicit attitude is not a matter of cool reason correcting an emotional bias but may itself be passionately felt. Those who believe in sexual tolerance feel strongly that it is wrong to condemn people for sex acts in which no one is harmed.

Emotions and the Moral Brain

Much more can be said about the philosophical debates; many tomes have been written on the topic since Hume's *Treatise*. It is noteworthy that this debate has not resulted in any consensus among philosophers, and that has led some to hope that resolution can come from other methods. In this light I want to focus on evidence from neuroscience. Over the past fifteen years there has been a concerted effort to identify correlates of moral judgment in the brain. Researchers have often interpreted these findings as bearing on the relationship between emotions and moral judgments. Some have defended views that look like heirs to sentimentalism, and others have offered lines of resistance.

Some of the earliest neuroimaging studies of moral judgment focused on the emotions. For example, Moll, Eslinger, and de Oliveira-Souza (2001) found that moral wrongness, as compared to factual wrongness, was associated with frontopolar and orbitofrontal regions (BA 9 and 10) and the temporal pole, among other areas. A follow-up study with a similar design also elicited activation in orbitofrontal area 11 (Moll, de Oliveira-Souza, Bramati, & Grafman, 2002). All three of these frontal areas also showed up in another study, which used photographs of moral transgressions. In another study with morally provocative pictures, Harenski and Hamaan (2006) found activity in Brodmann area 10, the amygdala, and the posterior cingulate. All of these areas are associated with emotion. Indeed, every neuroimaging study involving moral cognition of which I am aware has reported activity in brain structures associated with emotion. Other brain structures have been implicated as well, such as the superior temporal sulcus, which is implicated in social cognition. These may play a role in representing the actions under consideration. In particular, we often restrict moral condemnation to cases where a harm has been perpetrated intentionally, and attribution of intentions may engage brain structures associated with social cognition. This is consistent with sentimentalism. When we represent an action as an intentional harm we are forming what I called a classificatory belief. Imaging results suggest that, in addition to such beliefs, moral judgments involve emotions. Thus, the data conform to the predictions of a sentimentalist theory.

This had led researches to endorse the view that there is a strong relationship between emotions and morality. Indeed, some researchers use language that echoes Hume. For example, in a paper that reviews the first wave of neuroimaging results, Moll, Zahn, de Oliveira-Souza, Krueger, and Grafman (2005: 806) conclude:

Moral cognition depends on elaborated cortical mechanisms for representing and retrieving event knowledge, semantic information and perceptual features. However, morality would be reduced to a meaningless concept if it were stripped from its motivational and emotional aspects.

Critics would be quick to point out, however, that functional imaging is a correlational method, with limited temporal resolution. It is one thing to show that moral judgments, averaging over many cases as trials, tend to involve emotions and quite another to establish what role those emotions play (Huebner, 2013; Prinz, 2014). The sentimentalist is committed to the view that moral judgments are constituted of emotions. If so, it follows that moral judgments cannot be made in the absence of emotions, or prior to emotions. The studies just cited cannot directly confirm these modal and temporal conjectures.

A safe, although disappointingly modest conclusion would be that the aforementioned findings are consistent with sentimentalism. Sentimentalists may gain further confidence knowing that their account predicts such findings. It is less clear that such findings would be predicted by the moral rationalists. Rationalists allow that emotions can arise when we make moral judgments, but they also tend to assume that judgments can occur in the absence of emotions, and one might expect to see this reflected in the scanner, if it were true. Thus, sentimentalists can claim a slight upper hand based on the results mentioned so far. Other results, however, may be harder to square with sentimentalism. I now consider four lines of objection to sentimentalism based on findings from several areas of neuroscience and argue that sentimentalism withstands these objections.

Neuroscientific Objections to Sentimentalism

Dual Processes

As we just saw, some of the earliest neuroimaging studies of moral judgment have been taken to support the claim that moral judgments generally involve the emotions, just as Hume would have predicted. These studies began to appear in 2001. In that same year another pioneering study was published by Joshua Greene and his collaborators. Greene and colleagues (Greene, Sommerville, Nystrom, Darley, & Cohen, 2001) presented participants with moral and nonmoral dilemmas. The moral dilemmas involved situations in which performing some action would result in bringing harm to someone while at the same time preventing harm to a greater number. These were subdivided into "personal" and "impersonal" cases. In personal cases the action in question requires directly and physically aggressing against another human being, and the impersonal cases involve harming someone indirectly as a side effect of some other action. The classic examples are trolley scenarios. In a personal version participants read about someone who must consider pushing a man into the path of a speeding trolley in order to prevent that trolley from hitting five other people. In the impersonal version a person must decide whether to pull a lever that will send a speeding trolley onto a side track, where it will kill one person, instead of the five who would die if no action were taken.

Greene et al. (2001) obtained behavioral results as well as fMRI data, which they used to advance the conjecture that moral decision making is sometimes based on emotions and sometimes based on reasoning, with the implication that the latter decision can be made without emotions. Behaviorally, they found that personal moral dilemmas, such as the pushing case, are regarded as impermissible by most participants and answered quickly. Impersonal moral dilemmas, such as the lever-pulling case, are answered more slowly, and most conclude that the intervention is permissible. Greene et al. call the majority view in the personal cases *deontological*, implying a rigid rule according to which harming people is bad, and they call the majority view in the impersonal case *utilitarian*, meaning that people choose to maximize the number of people saved from harm.

Sentimentalism is called into question by the neuroimaging results. Greene et al. (2001) report that, as compared to impersonal dilemmas, personal dilemmas are associated with greater activations in brain structures associated with emotion, such as the posterior cingulate and the temporal pole. Impersonal dilemmas are associated with greater activity in the dorsolateral prefrontal cortex (dlPFC) and areas associated with working memory and other "cognitive" processes. They interpret this as showing that there are two kinds of processes that lead to moral judgment: an emotional process and a cool, rational process. The emotional process, they conjecture, is an automatic and error-prone carryover from our evolutionary past, in which simple mechanisms evolved to inhibit directly harming conspecifics. Reasoning gives us a way to make moral decisions without relying on emotional instincts.

This dual-process model would threaten sentimentalism by showing that moral judgments can be made without emotions and that indeed such judgments may be more reliable. The data, however, do not support this conclusion. First, when compared to nonmoral dilemmas, impersonal moral dilemmas show an increase in areas associated with emotion. These activations are weaker than the personal cases but significant. Thus, the study confirms that emotions arise for all moral judgments tested.

Second, the dlPFC increase in impersonal cases is completely compatible with sentimentalism. One possibility, is that dlPFC involves cool reasoning, as Greene and colleagues (2001) claim, but in the service of an emotion-based moral rule. Recall that sentimentalism does not eschew reasoning in application of moral rules; it eschews only pure reason. Now consider the rule that says, "try to help people, as many people as possible." This hardly sounds like a dispassionate rule. People feel good about helping. In senti-mentalist terms the rule is constituted by approbation toward maximizing the number of saved lives. To apply this rule one must do some reasoning; in particular one must calculate numerical outcomes. So sentimentalists predict cognitive as well as emotional processing whenever the helping rule is engaged. On a dual-process model, one might expect to see reason without emotion, but instead we see both.

Alternatively, the dlPFC may actually be serving emotional processes during impersonal moral dilemmas. As Greene et al. (2001) imply, the help-ing rule is likely to be drowned out by the rule against harming in personal moral dilemmas. But harming is less salient in impersonal cases because pulling a lever, for example, is not a prototypical example of killing. In those cases the helping norm comes into conflict with the norm against harming. Rather than thinking of this as emotion *versus* reason, it can be regarded as two emotional norms: approbation toward helping and disap-probation toward harming. This emotional conflict slows decision time, and it might explain the dlPFC increase. Although dlPFC is often operative in cool cognitive tasks, it also plays a role in emotional conflicts (Chiew & Braver, 2011). Consistent with this, and contra Greene, transcranial mag-netic stimulation (TMS) to dlPFC actually increases utilitarian decisions, perhaps by disrupting conflict monitoring and allowing the helping norm to kick in as a prepotent emotional response.

Greene et al. (2001) might protest that the foregoing interpretation predicts a neural signature for both negative and positive emotions in the impersonal cases. If helping rules are grounded in approbation, we should see activity in reward centers of the brain, and no such activity is reported in the study. In response, it is worth noting dlPFC itself has been associated with reward (Herrington et al., 2005). A more likely possibility, however, given the aforementioned TMS study, is that reward activity lies elsewhere in the brain and has simply been subtracted out of the Greene et al. analysis. In all their vignettes, both moral and nonmoral, there is a positive outcome under consideration. In the moral scenarios the negative intervention is varied (direct vs. indirect harm), but the positive outcome remains constant, which means that its neural correlates have little chance of appearing in the data analysis. In a more recent study, however, Shenhav and Greene (2010) varied positive outcomes by changing how many lives would be saved. This manipulation revealed classic reward structure activity, such as ventral striatum.

In summary, the initial defense of a dual-process model in Greene et al. (2001) does not establish moral judgments in the absence of emotion. Indeed, the data seem to favor the view that emotions are always operative. Greene (2007) now concedes this, but he suggests that, in impersonal cases, emotions are effects of moral judgments rather than causes. He does not provide evidence for this view. For the sentimentalist, emotions are components of moral judgments in both impersonal and personal cases. This is the simpler hypothesis, and it fits with behavioral work, which suggests that induction of positive emotions increases the judgment that it is good to help people in need (Seidel & Prinz, 2013b).

Rational Overriding

The case for the dual-process theory did not close with the paper of Greene et al. (2001). Greene and collaborators have offered other lines of defense, including behavioral studies (which I do not address here) and further neuroimaging results. Of special interest is a paper by Greene, Nystrom, Engell, Darley, and Cohen published in 2004. There the authors argue that cool rational processes can override emotional processes in some cases.

They focus on difficult personal moral dilemmas. For example, unlike the pushing case, which elicits immediate judgments of impermissibility, they propose we consider a scenario in which a person smothers a crying baby in order to prevent killers from finding a group of people who are hiding, including the baby. This takes a while to answer for most people, and many conclude that it is okay to smother the baby. In the scanner Greene et al. (2004) found two key results about such hard cases: there is dlPFC activation in addition to emotion structures, and the dlPFC activity is especially strong in those who conclude that smothering is permissible. The authors interpret this as showing that cool cognition (which they associate with the dlPFC) can be used to regulate an emotional response (the horror of killing a baby) and to deliver a cool verdict that goes against the dictates of emotion.

This interpretation may not be right, however. As already noted, dlPFC is not necessarily restricted to cool cognition. Moreover, the fact that greater dlPFC activity is associated with utilitarian decisions may simply reflect that fact, agreed on by both Greene and sentimentalists, that such decisions require numerical processing. Moreover, Greene and colleagues (2004) also report that utilitarian decisions are associated with greater activity in two parts of the posterior cingulate (Brodmann areas 31 and 23). Posterior cingulate is regarded as an emotion structure, and BA 23 has been associated with guilt (Basile & Mancini, 2011). There is also a sizable increase in Brodmann area 10, which has been associated with reward computation (e.g., Rogers et al., 1999). The results thus may imply an increase in emotionality for those who choose to kill. Reason may override deontology here, but it seems to do so by recruiting rather than by suppressing emotions.

Emotion Deficits

Another class of objections to sentimentalism derives from populations of individuals who have emotional deficits. Sentimentalism seems to predict that emotional deficits will result in moral deficits, and some authors argue that this prediction fails.

Roskies (2003), for example, argues that patients with lesions in the ventromedial prefrontal cortex (vmPFC) are able to make normal moral judgments despite the fact that their injuries disrupt emotional processing. Subsequent investigations qualified Roskies's interpretation of these cases but bolstered her conclusion. Notably, Koenigs et al. (2007) found that vmPFC patients respond like control groups when given impersonal moral dilemmas, but they are more likely to give utilitarian responses to personal moral dilemmas. This suggests that some moral judgments operate normally, whereas other moral judgments are disrupted, as a dual-process model would predict. The presence of intact judgments without normal emotional responses looks like a direct counterexample to sentimentalism.

The problem, as Koenigs et al. (2007) concede, is that vmPFC patients do not suffer from a lack of emotions as the counterexample requires. Rather, they suffer from problems in emotional modulation. In particular they do not curtail reward-seeking activities when confronted with negative consequences (Bechara, Damasio, Damasio, & Anderson, 1994). This is precisely the structure of a personal moral dilemma: when seeking to save five lives, we are confronted with the horror of pushing a person in front of a trolley. vmPFC patients are comparatively insensitive to that horror, but they are highly motivated to save lives, which is presumably underwritten by intact emotions of approbation. If these patients lacked emotions, their behavior would be random or inert. Moreover, vmPFC patients have a lifetime of emotional learning before their injuries, so they may have a rote memory of moral rules that does not require emotional processing (by analogy a person who loses sight due to a brain injury in adulthood may be able to list colors of familiar objects). When vmPFC injuries occur early in life, comprehension of the moral domain is indeed impaired (Anderson, Bechara, Damasio, Tranel, & Damasio, 1999; Taber-Thomas et al., 2014).

Another psychological disorder of interest is psychopathy. Psychopaths are known to have diminished emotional responses, but there is evidence that they are capable of making moral judgments. The strongest demonstration of this comes from Aharoni, Sinnott-Armstrong, and Kiehl (2012), who report that incarcerated psychopaths respond like other inmates and noncriminals when given a task to distinguish moral and conventional rules. These results are inconclusive, however. First, the examples used might have been too easy because the moral rules were by and large illegal acts and the conventional rules were not. Also, the authors did not ask psychopaths for justifications, which may have revealed aberrant psychological processing. Moreover, a close look at their results reveals that the affective component of psychopathy does in fact predict reduced accuracy on the moral task. Even if they had obtained a true null result, it would do little to negate previous findings, which show that psychopaths perform poorly on moral/conventional discrimination psychopaths (Blair, 1995). In addition, neuroimaging studies suggest that psychopaths process moral information differently from controls, suggesting that they may rely on rote learning of rules rather than on a genuine sense of moral right and wrong (e.g., Glenn, Raine, Schug, Young, & Hauser, 2009; Harenski et al. 2010). To return to the case of color blindness, a congenital achromatope might have verbal knowledge of the colors of familiar objects without any comprehension of what color vocabulary means. Psychopaths give lie to their incomprehension when asked to justify moral judgments, at which point they simply acknowledge that certain actions are forbidden, but they are inarticulate as to why.

Judgments before Emotions

I end with one more attempt to find neuroscientific evidence against sentimentalism. One problem with fMRI, noted above, is that the temporal resolution cannot reveal whether emotions are the causes, components, or effects of moral judgments. Imaging methods with better temporal resolution are needed to settle this question. One method is EEG, which uses electrical potentials on the scalp. A recent study by Yang and colleagues (2013) purports to show that moral judgments arise before emotions kick in.

The design of the study is clever. The authors give their participants a Go/No-Go task in which they need to either select between two answers on a keyboard or refrain from answering. In all cases participants are presented with scenarios that vary on the dimensions immoral–nonmoral or disgusting–not disgusting. In one run through these items participants use the moral probe for the Go/No-Go decision, and if the item is deemed immoral, they select a key for either disgusting or not. In another run through, the disgusting probe is used for the Go/No-Go decision: if and only if a scenario was deemed disgusting, key presses would indicate whether the scenario was immoral or not. Through both runs, motor potentials are recorded. The authors report that the moral question is answered faster than the disgust question, suggesting that participants make moral judgments before they can extrapolate emotional information.

Curiously, the authors do not draw this conclusion about temporal ordering from the fact that moral responses actually show readiness potentials earlier in time than emotional responses. Rather, their inference is based on the fact that moral judgments produce some readiness potentials even in No-Go trials, in the condition where the disgust probe was supposed to render the moral response unnecessary. And conversely, when the moral probe is used to decide between Go and No-Go, there is no motor potential associated with disgust on No-Go, indicating that the moral probe occurred before any disgust response was generated. If disgust had come first, the authors imply, it would have initiated a motor potential before the moral probe suppressed it.

The inference is not unreasonable, but there are other interpretations. In the run where the moral probe was used for the Go/No-Go task, the No-Go trials included items that lack moral significance but vary in disgust. But disgust is a withdrawal emotion, so we should not necessarily expect motor potentials. In the run where the disgust probe decides whether to Go, the No-Go trials include items that are immoral but not disgusting. Moral wrongs that do not involve disgust tend to evoke anger, which is an active emotion, and might trigger motor potentials even when action is supposed to be suppressed.

Crucially too, the temporal data in the study suggest that moral judgment arises more slowly than emotion. First of all consider the runs that show motor potentials in the No-Go trials. These responses, which reflect the assessment that something is wrong but not disgusting, arise only after the potentials in the Go trials, which register that something is disgusting. In addition, the initial readiness potentials for the Go condition for the run where Go means disgusting occur earlier than the initial readiness potentials in the trials where Go means immoral, suggesting again that emotion information may be discerned before moral information. That said, it must also be added that sentimentalism makes no prediction about this particular task. The claim that moral judgments are emotional does not mean that they depend on prior evocation of nonmoral emotions. The association between an immoral action and disgust could be stronger, in some cases, than the association between an action and core biological disgust, especially when verbal materials are used.

The study by Yang et al. (2013) points to the potential value of EEG in the moral domain, but it does not establish the temporal priority of the moral over the emotional. Sentimentalists can breathe a sigh of relief.

Conclusions

We have reviewed here a body of neuroscientific evidence that bears on the relationship between emotions and morality. Neuroimaging studies of morality invariably implicate the emotions, just as sentimentalists would predict, but various lines of evidence have also been taken to suggest the possibility of dispassionate moral judgments. We have considered four efforts to establish this possibility and argued that none of them succeeds. All of them are consistent with the possibility that moral judgments are always grounded in the emotions, and, indeed, that interpretation seems preferable. That said, neuroscience may not, on its own, be able to deliver a final verdict (Huebner, in press), but, when coupled with behavioral results and philosophical argumentation, extant results may best align with a sentimentalist theory (Prinz, 2007, in press). My goal here has been to deflect some recent challenges and embrace the most obvious message from the neuroscience of morals: the moral brain is an emotional brain. This would make Hume smile.

References

Aharoni, E., Sinnott-Armstrong, W., & Kiehl, K. (2012). Can psychopathic offenders discern moral wrongs? *Journal of Abnormal Psychology, 121*, 484–497.

Algoe, B. S., & Haidt, J. (2009). Witnessing excellence in action: The "other-praising" emotions of elevation, gratitude, and admiration. *Journal of Positive Psychology, 4*, 105–127.

Anderson, S. W., Bechara, A., Damasio, H., Tranel, D., & Damasio, A. R. (1999). Impairment of social and moral behavior related to early damage in human prefrontal cortex. *Nature Neuroscience, 2*, 1032–1037.

Basile, B., & Mancini, F. (2011). Eliciting guilty feelings: A preliminary study differentiating deontological and altruistic guilt. *Psychology, 2*, 98–102.

Bechara, A., Damasio, A. R., Damasio, H., & Anderson, S. W. (1994). Insensitivity to future consequences following damage to human prefrontal cortex. *Cognition, 50*, 7–15.

Blair, R. (1995). A cognitive developmental approach to morality: Investigating the psychopath. *Cognition, 57*, 1–29.

Bureau, J., Vallerand, R., Ntoumanis, N., & Lafrenière, M. (2013). On passion and moral behavior in achievement settings: The mediating role of pride. *Motivation and Emotion, 37*, 121–133.

Chiew, K. S., & Braver, T. S. (2011). Neural circuitry of emotional and cognitive conflict revealed through facial expressions. *PLoS ONE, 6*, e17635.

Giner-Sorolla, R., & Espinosa, P. (2011). Social cuing of guilt by anger and of shame by disgust. *Psychological Science, 22*, 49-53.

Glenn, A. L., Raine, A., Schug, R. A., Young, L., & Hauser, M. (2009). Increased DLPC activity during moral decision-making in psychopathy. *Molecular Psychiatry, 14*, 909–911.

Greene, J. D. (2007). The secret joke of Kant's soul. In W. Sinnott-Armstrong (Ed.), *Moral Psychology, Vol. 3: The Neuroscience of Morality: Emotion, Disease, and Development* (pp. 35–79). Cambridge, MA: MIT Press.

Greene, J. D., Sommerville, R. B., Nystrom, L. E., Darley, J. M., & Cohen, J. D. (2001). An fMRI investigation of emotional engagement in moral judgment. *Science, 293,* 2105–2108.

Greene, J. D., Nystrom, L., Engell, A., Darley, J., & Cohen, J. L. (2004). The neural bases of cognitive conflict and control in moral judgment. *Neuron, 44,* 389–400.

Haidt, J. (2001). The emotional dog and its rational tail: A social intuitionist approach to moral judgment. *Psychological Review, 108,* 814–834. ..

Haidt, J., Koller, S., & Dias, M. (1993). Affect, culture, and morality, or is it wrong to eat your dog? *Journal of Personality and Social Psychology, 65,* 613–628.

Harenski, C. L., & Hamaan, S. (2006). Neural correlates of regulating negative emotions related to moral violations. *NeuroImage, 30,* 313–324.

Harenski, C. L., Harenski, K. A., Shane, M. S., & Kiehl, K. A. (2010). Aberrant neural processing of moral violations in criminal psychopaths. *Journal of Abnormal Psychology, 119,* 863–874.

Herrington, J. D., Mohanty, A., Koven, N. S., Fisher, J. E., Stewart, J. L., Banich, M. T., et al. (2005). Emotion-modulated performance and activity in left dorsolateral prefrontal cortex. *Emotion, 5,* 200–207.

Huebner, B. (in press). Do emotions play a constitutive role in moral cognition? *Topoi.* Epub ahead of print available at http://link.springer.com/article/10.1007%2Fs11245-013-9223-6.

Hume, D. (1738). *A Treatise of Human Nature.* L.A. Selby-Biggge & P.H. Nidditch (Eds.). Oxford: Oxford University Press.

Koenigs, M. L., Young, R., Adolphs, D., Tranel, F., Cushman, M., Hauser, M., et al. (2007). Damage to the prefrontal cortex increases utilitarian moral judgements. *Nature, 446,* 908–911.

Moll, J., de Oliveira-Souza, R., Bramati, I. E., & Grafman, J. (2002). Functional networks in emotional moral and nonmoral social judgments. *NeuroImage, 16,* 696–703.

Moll, J., Eslinger, P. J., & de Oliveira-Souza, R. (2001). Frontopolar and anterior temporal cortex activation in a moral judgment task: Preliminary functional MRI results in normal subjects. *Arquivos de Neuro-Psiquiatria, 59,* 657–664.

Moll, J., Zahn, R., de Oliveira-Souza, R., Krueger, F., & Grafman, J. (2005). The neural basis of human moral cognition. *Nature Reviews Neuroscience, 6,* 799–809.

Prinz, J. J. (2007). *The emotional construction of morals*. New York: Oxford University Press.

Prinz, J. J. (in press). Sentimentalism and the moral brain. In S. M. Liao (Ed.), *The significance of neuroscience for morality*. Oxford: Oxford University Press.

Rogers, R. D., Owen, A. M., Middleton, H. C., Williams, E. J., Pickard, J. D., Sahakian, B. J., et al. (1999). Choosing between small, likely rewards and large, unlikely rewards activates inferior and orbital prefrontal cortex. *Journal of Neuroscience, 19,* 9029–9038.

Roskies, A. (2003). Are ethical judgments intrinsically motivational? Lessons from "acquired sociopathy." *Philosophical Psychology, 16,* 51–66.

Rozin, P., Lowry, L., Imada, S., & Haidt, J. (1999). The CAD Triad Hypothesis. *Journal of Personality and Social Psychology, 76,* 574–586.

Seidel, A., & Prinz, J. (2013a). Sound morality: Irritating and icky noises amplify judgments in divergent moral domains. *Cognition, 127,* 1–5.

Seidel, A., & Prinz, J. (2013b). Mad and glad: Musically induced emotions have divergent impact on morals. *Motivation and Emotion, 37,* 629–637.

Shenhav, A., & Greene, J. (2010). Moral judgments recruit domain-general valuation mechanisms to integrate representations of probability and magnitude. *Neuron, 67,* 667–677.

Strohminger, N., Lewis, R. L., & Meyer, D. E. (2011). Divergent effects of different positive emotions on moral judgment. *Cognition, 119,* 295–300.

Taber-Thomas, B. C., Asp, E. W., Koenigs, M., Sutterer, M., Anderson, S. W., & Tranel, D. (2014). Arrested development: Early prefrontal lesions impair the maturation of moral judgment. *Brain, 137,* 1254–1261.

Tangney, J., Miller, R. S., Flicker, L., & Barlow, D. (1996). Are shame, guilt, and embarrassment distinct emotions? *Journal of Personality and Social Psychology, 70,* 1256–1269.

Yang, Q., Yan, L., Luo, J., Li, A., Zhang, Y., Tian, X., et al. (2013). Temporal dynamics of disgust and morality. *PLoS ONE, 8,* e65094.

5   Devoted Actors and the Moral Foundations of Intractable Intergroup Conflict

Scott Atran and Jeremy Ginges

Sacred Values and Devoted Actors

This chapter explores the relationship between: (1) what we and others have termed "sacred values" (Atran & Ginges, 2012; Durkheim, 1912/1955; Eliade, 1959; Ginges, Atran, Medin, & Shikaki, 2007; Rappaport, 1971; Tetlock, 2003) and (2) "fused groups" of imagined kin in which such values may become embedded (Atran, 2010; Swann et al., 2014). Our research hypothesis is that, when fused, value-driven groups perceive existential threats, and they produce "devoted actors" capable of extreme acts of self-sacrifice and violence without regard to likely risks or rewards, costs or consequences.

Devoted Actor Hypothesis   People will become willing to protect sacred or morally important values through costly sacrifice and extreme actions, even being willing to kill and die, particularly when such values are embedded in or fused with group identity, becoming intrinsic to "Who I am" and "Who we are."

Although the term "sacred values" (SVs) intuitively denotes religious belief, in line with recent work we use the term to refer to any preferences regarding objects, beliefs, or practices that people treat as both incompatible or nonfungible with profane or economic goods, as when land becomes "sacred land," and which are part of our conception of "self" and of "who we are." This includes the "secularized sacred" as exemplified, for example, in political notions of "human rights" (Atran & Axelrod, 2008; Smith et al., 2013) or in the transcendent ideological –*isms* that have dominated political life ever since the Enlightenment's secularization of the universal religious mission to redeem and save "humanity" through political revolution (liberalism, socialism, anarchism, communism, fascism, etc.) (Gray, 2007).

Progress in the fields of moral psychology and philosophy has mostly focused on universal "Golden Rule" principles of fairness and reciprocity emotionally supported by empathy and consolation (Baumard, André, & Sperber, 2013; Van Slyke, 2014), rather than on what Darwin referred to as the primary virtue of "morality … patriotism, fidelity, obedience, courage, and sympathy" with which winning groups are better endowed in history's spiraling competition for survival and dominance (Darwin, 1871:163; see Greene, 2009). Nevertheless, a smaller body of work (Baron & Spranca, 1997; Fiske & Tetlock, 1997; Tetlock, 2003) suggests that people resist attempts to compromise sacred values no matter the cost to themselves or others. But even this research has often assumed that uncompromising commitment to SVs, even though possibly heartfelt, is actually impossible in the real world because other pressing material needs may invariably arise that require attention, thus relaxing absolute commitment (Baron & Leshner, 2000). Such values, then, can only be "pseudo-sacred" and ultimately materially negotiable owing to this "reality constraint" (McGraw & Tetlock, 2005).

However, humans sacrifice self-interest, and in extremes they are willing to die and to kill, in the name of abstract and often ineffable values—such as God or national destiny or history (Atran, 2010; Atran & Ginges, 2012). Acts of extreme sacrifice that so frequently punctuate human history suggest that SVs are not so readily reconstrued to allow compromise under the pressure of instrumental pressures. Our fieldwork with suicide terrorists and political and militant leaders and supporters in violent conflict situations provides empirical evidence that ordinary people, when motivated by SVs, become "devoted actors."

Our research indicates that when people act as "devoted actors" they think and behave differently than "rational actors" (however bounded by psychological or ecological constraints). Devoted actors show (1) commitment to a rule-bound logic of moral appropriateness to do what is morally right no matter the likely risks or rewards, rather than following a utilitarian calculus of costs and consequences (Atran, 2006); (2) immunity to material trade-offs coupled with a "backfire effect," where offers of incentives or disincentives to give up SVs heighten refusal to compromise or negotiate (Ginges et al., 2007; Dehghani, Atran, Iliev, Sachdeva, Ginges, & Medin, 2010); (3) resistance to social influence and exit strategies (Atran & Henrich, 2010; Sheikh, Ginges, & Atran, 2013), which leads to unyielding social isolation and opposition as well as to unshakeable social solidarity and which binds genetic strangers to voluntarily sacrifice for one another, even unto death; (4) insensitivity to spatial and temporal discounting, where considerations of distant places and people, and even far past and future events, associated with SVs significantly outweigh concerns with the here and now (Atran, 2010; Sheikh et al., 2013); and (5) brain-imaging patterns consistent with processing SVs as rules rather than as calculations of costs and consequences and with processing perceived violations of SVs as emotionally agitating and resistant to social influence or discounting (Berns et al., 2012, 2013).

Although SVs may operate as necessary moral imperatives to action, they are not sufficient. It is important to understand that group morality does not operate simply from ideological canon or decontextualized principles that drive decisions and actions, but it is almost always embedded and distributed in social groups, most effectively in intimate networks of "imagined kin": brotherhoods, motherlands, fatherlands, homelands, and the like (Atran, 2011). Knowledge of the moral imperatives that drive people to great exertions toward one political goal or another, as well the group dynamics that bind individuals to sacrifice for one another in the name of those values, both appear indispensable to extreme actions where prospects of defeat and death are very high, as with terrorism and revolution.

Devoted Actors Are Deontic Actors

Philosophers of moral virtue suggest that moral values might be deontological (Kant, 1785/2005) or utilitarian (Mill, 1871). Deontic processing is defined by an emphasis on rights and wrongs (Weber, 1864/1958), whereas utilitarian processing is characterized by costs and benefits (von Neumann & Morgenstern, 1944). Models of rational behavior predict many of society's patterns, such as favored strategies for maximizing profit or likelihood for criminal behavior in terms of opportunity costs (Becker, 1968) and important aspects of conflict management (Allison & Zelikow, 1999). But the prospects of crippling economic burdens and huge numbers of deaths do not necessarily sway people from positions on whether going to war or opting for revolution is the right or wrong choice.

For example, in one series of studies, we confronted people in the United States and Nigeria with hypothetical hostage situations and asked them if they would approve of a solution—which was either diplomatic or violent—for freeing the prisoners (Ginges & Atran, 2011). When told that their action would result in all hostages being saved, both groups endorsed the plan presented to them. When asked how many hostages they required to be saved to ensure their support (from 1 to 100), those evaluating the military option said only one hostage needed to be rescued, showing a remarkable insensitivity to scope. In contrast, those evaluating the diplomatic option required a majority of hostages to be rescued.

Commitment to SVs can be key to the success or failure of insurgent or revolutionary movements with far fewer material means than the armies or police arrayed against them (which tend to operate more on the basis of typical "rational" reward structures, such as calculated prospects of increased pay or promotion). Ever since World War II, on average, revolutionary movements have emerged victorious with as little as ten times less firepower and manpower than the state forces arrayed against them (Arreguín-Toft, 2001).

Our research with political leaders and general populations shows that SVs—not political games or economics—underscore seemingly intractable conflicts like those between the Israelis and the Palestinians or Iran and the Western allies that defy the rational give-and-take of business-like negotiation (Atran, Axelrod, & Davis, 2007; Dehghani et al., 2010; Ginges et al., 2007; Ginges, Atran, Sachdeva, & Medin, 2011). Consider the Israeli-Palestinian conflict, where rational cost–benefit analysis says the Palestinians ought to agree to forgo sovereignty over Jerusalem or the claim of refugees to return to homes in Israel in exchange for an autonomous state encompassing their other pre-1967 lands because they would gain more sovereignty and more land than they would renounce. They should support such an agreement even more if the United States and Europe sweetened the deal by giving every Palestinian family substantial, long-term economic assistance. Instead we find that the financial sweetener makes Palestinians more opposed to the deal and more likely to support violence to oppose it, including suicide bombings. Israeli settlers also have rejected a two-state solution that required Israel to give up Judea and Samaria or to "recognize the legitimacy of the right of Palestinian refugees to return" (in an agreement not actually requiring Israel to absorb the refugees). But the Israelis, too, were even more opposed if the deal included additional long-term financial aid or a guarantee of living in peace and prosperity (Ginges et al., 2007).

In another series of studies we find that a relatively small but politically significant portion of the Iranian population believes that acquiring nuclear energy (but not necessarily nuclear weapons) has become a SV in the sense that proposed economic incentives and disincentives backfire by leading to increased and more emotionally entrenched support (Dehghani et al., 2010). Here, it appears that SVs can emerge for issues with relatively little historical background and significance when they become bound up with conflicts over collective identity—the sense of "who we are." For a minority of Iranians (13% in these experiments) the issue had become a sacred subject through association with religious rhetoric and ritual (e.g., Iranian women marching and chanting in favor of "nuclear rights" while waving the Koran). This group, which tends to be close to the regime, now believes a nuclear program is bound up with the national identity and with Islam itself, so that offering material rewards or punishments to abandon the program only increases anger and support for it.

Sacred values do not make people opposed to any sort of compromise. Instead they appear to invoke specific taboos protecting these values against material trade-offs. Offering people materially irrelevant symbolic gestures can work where material incentives do not. For example, Palestinian devoted actors were more willing to consider recognizing the right of Israel to exist if the Israelis offered an official apology for Palestinian suffering in the 1948 war. Similarly, Israeli settlers were less disapproving of compromising sacred land for peace if Hamas and the other major Palestinian groups symbolically recognized Israel (Ginges et al., 2007).

Our survey results were mirrored by our discussions with political leaders (Atran, Axelrod, & Davis, 2007). Mousa Abu Marzook (the deputy chairman of Hamas) said "No" when we proposed a trade-off for peace without granting a right of return. He became angry when we added in the idea of substantial American aid for rebuilding: "No, we do not sell ourselves for any amount." But when we mentioned a potential Israeli apology for 1948, he said: "Yes, an apology is important, as a beginning. It's not enough because our houses and land were taken away from us and something has to be done about that." His response suggested that progress on sacred values might open the way for negotiations on material issues, rather than the reverse. We got a similar reaction from Israeli leader Benjamin Netanyahu. We asked him whether he would seriously consider accepting a two-state solution following the 1967 borders if all major Palestinian factions, including Hamas, were to recognize the right of the Jewish people to an independent state in the region. He answered, "O.K., but the Palestinians would have to show that they sincerely mean it, change their anti-Semitic textbooks." Making these sorts of wholly intangible symbolic but sincere gestures, like recognition of a right to exist or an apology, simply does not compute in any utilitarian calculus. And yet the science suggests that these gestures may be the best way to cut through the knot.

More systematic understanding of what kinds of symbolic gestures involving SVs are likely to be effective in conflict prevention and resolution, including signatures of emotional sincerity, could provide novel possibilities for breakthroughs toward conflict. More recently, in a meeting of senior Iranians, Saudis, Israelis, Americans, and British arranged by members of our team and Lord John Alderdice (Convenor, UK House of Lords) at Oxford on the nuclear issue in early September 2013, we informally monitored expressions of devotion to values, including emotional attachment, and suggested opening negotiations via a symbolic gesture evoking SVs rather than political positions. In response we received a message that Iran's President Rouhani would publicly acknowledge the Holocaust in New York (which U.S. and Israeli officials told us would be a positive development for negotiations).

Devoted Actors as Fused Actors

Thus far our research indicates that sacralizing parochial preferences and prioritizing those sacred values are necessary factors in producing actors willing to sacrifice for a cause. Nevertheless, this is by no means usually sufficient. For example, many millions of people express sympathy with al Qaeda or other forms of violent political expression that support political violence, but relatively few willingly use violence. From a 2001–2007 survey of thirty-five predominantly Muslim nations (with fifty thousand interviews randomly chosen to represent about 90 percent of the Muslim world), a Gallup study projected that 7 percent of the world's 1.3 billion Muslims thought that the 9/11 attacks were "completely justified." If one includes Muslims who considered the attacks "largely justified," their ranks almost double. Adding those who deemed the attacks "somewhat justified" boosts the number to 37 percent, which indicates hundreds of millions of Muslims. (Polls also imply that 20 percent of the American public has a "great deal" of prejudice against

Muslims, two-thirds have "some prejudice" against them, and 6 percent of Americans think that attacks in which civilians may be victims are "completely justified") (Esposito & Mogahed, 2008).

Of these many millions who express support for violence against the out-group, however, there are only thousands willing to actually commit violence. This also appears to be the case in the European Union, where fewer than five thousand suspects have been imprisoned for jihadi activities out of a Muslim population of perhaps 20 million. In the United States fewer than one thousand suspects have been arrested for having anything remotely to do with al Qaeda ideology or support for terrorism after 9/11, with fewer than one hundred cases being considered serious out of an immigrant Muslim population of more than two million.

Numerous case studies show that people usually go on to extreme violence in small, action-oriented groups of friends and family, where the extent of ideological commitment to a cause may vary greatly among individual members of the group (Sageman, 2008). Young jihadis are powerfully bound to each other—they are often campmates, school buddies, soccer pals, and the like—who become die-hard bands of brothers united in what they perceive to be a thrilling and heroic cause. In the book *Talking to the Enemy*, Atran (2010) describes how the "jihadi bug" developed in Hamburg among a group of Middle Eastern youth who became the core of the 9/11 plot; how al Qaeda's viral movement spread among self-styled "Afghan Alumni," Southeast Asian veterans of the Soviet-Afghan War, who had bonded through friendship, marriage, and soccer to blow up tourist spots and hotels in Bali and Jakarta; how an Internet tract, "Iraqi Jihad," culminated in the "organized anarchy" of the Madrid train bombers, whose core group consisted mostly of petty criminals originating from one small Moroccan neighborhood who had little religious education or organized direction; and how ten boys from the same "al-Jihad" soccer team came to be Palestinian martyrs, with parents unaware of what was going on.

There is more to group dynamics, we must realize, than just the weight and mass of people, their behavior, and ideas. There are also the structural relationships among group members that make the group more than the sum of its individual members. It also encompasses the networking among members that distributes thoughts and tasks that no one part may completely control or even understand. Case studies of suicide terrorism and related forms of violent extremism suggest that "people almost never kill and die [just] for the Cause, but for each other: for their group, whose cause makes their imagined family of genetic strangers—their brotherhood, fatherland, motherland, homeland" (Atran, 2010).

In line with these observations a promising new theory holds that when people's collective identities become fused with their personal self-concept, they subsequently display increased willingness to engage in extreme progroup behavior when the group is threatened (Swann, Jetten, Gómez, Whitehouse, & Bastian, 2012). Swann and colleagues have dubbed this powerful form of personal investment in the group "identity fusion," but it is possible that people may fuse not only with groups of people but also with particular issues and values (see figure 5.1).

In an ongoing collaboration with fusion theorists Ángel Gomez and Juan Jiménez, we find highly convergent measures of SVs: resistance to monetary payoffs, alternative benefits to society, and social pressure are

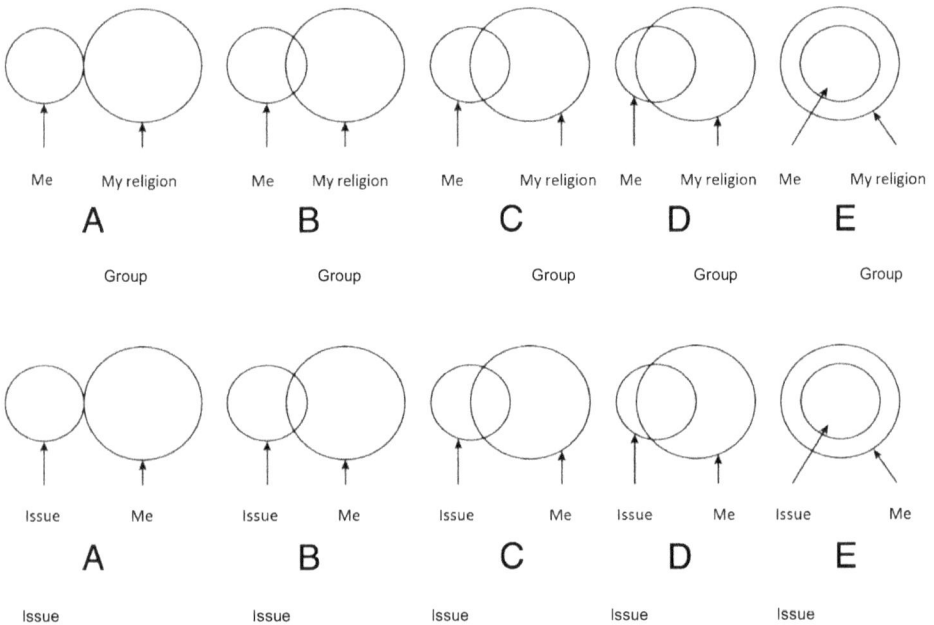

Figure 5.1
(A) Measuring fusion with group. (B) Measuring fusion with issue/value. Responses show a dichotomous distribution: nonfused (A, B, C, D) versus fully fused (E). This is replicated even when the measure is continuous (e.g., sliding a smaller circle into a larger circle on a smart phone).

strongly related to one another. Fusion with values is a complementary but somewhat independent phenomenon. Table 5.1 gives the conditional probabilities calculated from these measures in a recent study of 1600 "pro-choice" and "pro-life" supporters in Spain in early 2014 at a time when rival demonstrations were almost a daily occurrence. These interrelations tend to be maximized among individuals who are fully fused with their group.

Fusion theory is markedly different from various social identity theories in its privileging of group cohesion through social networking and emotional bonding of people and values rather than through processes of categorization and association, thus empowering individuals and their groups with sentiments of exceptional destiny and invincibility. In a recent set of cross-cultural experiments, Swann et al. (2014) found that when "fused people" perceive that group members share core attributes and values, they are more likely to project familial ties common in smaller groups onto the extended group, and this enhances willingness to fight and die for the larger group, echoing field research with militant and terrorist groups (Atran, 2010).

We found that for the relevant group-defining values (pro-life or pro-choice), the greater the fusion with those values, the greater the willingness

Table 5.1

Conditional probabilities of commitments to issues/values (predictors in rows and outcomes in columns)

	Fusion with Value	Resistance to Social Influence	Resistance to Societal Benefit	Resistance to Money Trade-off
Fusion with Value	0% → 100%	60% → 95%	28% → 74%	36% → 80%
Resistance to Social Influence	26% → 75%	0% → 100%	6% → 78%	8% → 84%
Resistance to Societal Benefit	37% → 82%	59% → 99%	0% → 100%	11% → 96%
Resistance to Money Trade-off	39% → 78%	48% → 98%	12% → 88%	0% → 100%

For example, the second row of the fourth column shows that when people were immune to social influence there was an 84 percent chance they would refuse monetary rewards, as compared to just an 8 percent chance when they were not immune to social influence.

to take extreme action and engage in sacrifice (except for risking physical harm to one's own children) (see figure 5.2).

In addition, we found that fusion with values mediates the relation between fusion with family-like groups of "imagined kin" and costly sacrifices. We also found that SVs mediate the relation between fusion with

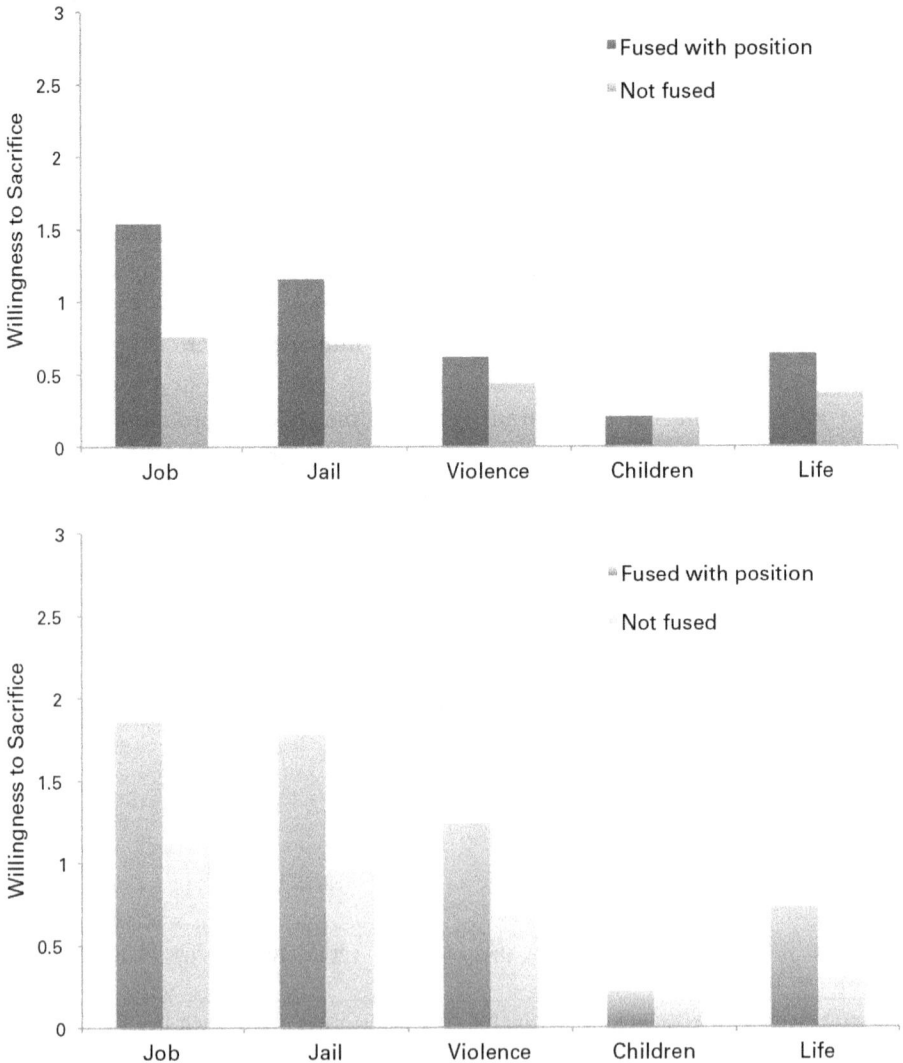

Figure 5.2
(Top) Support for costly sacrifices among Spanish pro-life supporters. (Bottom) Support for costly sacrifices among Spanish pro-choice supporters.

values and costly sacrifices. These findings suggest the Devoted Actor's pathway to costly sacrifices and extreme actions represented in Figure 5.3:

Neural Aspects of the Devoted Actor

In collaboration with Greg Berns and colleagues we recently investigated neural processing of SVs (Berns et al., 2012). Using functional magnetic resonance imaging (fMRI) we presented participants with a set of values, asked them to choose between these values, and then offered them a chance to sell off their choice. In the first "passive phase" of the experiment, participants were presented, under the scanner, with 124 statements involving sixty-two issues that ranged from items dealing with religious issues (e.g., belief in God) and moral issues (e.g., harming an innocent person) to the mundane (e.g., a preference for Macs over PCs). In subsequent phases participants were asked to choose between two pairs of statements (e.g., "You believe in God"/"You do not believe in God") and were then asked if they would be willing to sell off their belief. For example, participants who did not believe in God (or who were "Mac people") were asked to nominate a dollar amount to sign a report disavowing their preference. Participants were given the option of opting out—refusing to nominate a monetary amount, which was taken as one indication of a SV. Out of the scanner people were given the option of auctioning off their belief for any amount of money between $1 and $100. The higher the amount of money, the less chance they had of winning the money. Again, a decision not to participate was taken as an indication of a claim to a SV.

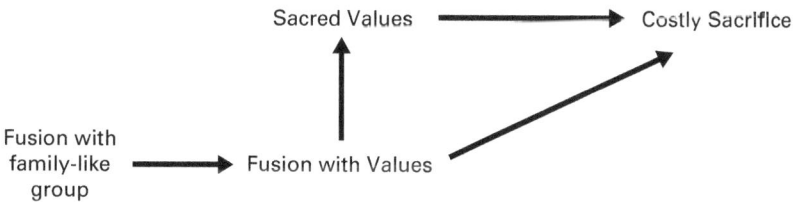

Figure 5.3

Theoretical model of devoted action (based on empirical findings discussed). Other pathways and feedback loops among these main variables are possible, even likely, and we are currently investigating causal relations among SVs, fusion, and devoted action among various groups in conflict.

We were interested in distinguishing between two interpretations of SVs. One interpretation of refusals to sell off SVs is simply that people have not been offered enough money to do so. If a refusal to sell off a value was indicative of greater utility of that value, then passive processing of that value should be associated with greater activation in brain regions associated with processing utility, such as the ventromedial prefrontal cortex (VMPFC), striatum/nucleus accumbens, and parietal cortex. We found instead that SVs were associated with increased activity in the left temporoparietal junction (TPJ) and ventrolateral prefrontal cortex (VLPFC), regions associated with semantic rule retrieval. This suggests that SVs affect behavior through the retrieval and processing of deontic rules and not through a utilitarian evaluation of costs and benefits. Moreover, the statements that resulted in more amygdala activation represent the most repugnant items to the individual, which is consistent with the idea that, when SVs are violated, they induce moral outrage (Tetlock, Kristel, Elson, Green, & Lerner, 2000).

In a follow-up study Berns et al. (2013) introduced a social influence manipulation at the stage of the study when people are asked to choose which statement they identified with (e.g., "I believe in God"). People could see the percentage of fellow participants who agreed with them (i.e., subjects see a "thermometer" consisting of a column of five circles, where each half-filled circle represents 10% social support from the subject's reference population). It turned out that susceptibility to social influence (i.e., willingness to change positions on an issue to reflect majority opinion) on a given issue was negatively correlated with activation of the VLPFC and the amygdala—the same brain regions activated for SVs in Berns et al. (2012).

Our theory of the devoted actor as someone who treats preference as sacred and whose identity changes to become fused with values and relevant groups finds support in a recent cross-cultural neuroimaging study. Here narratives invoking SVs are processed differently from narratives that do not evoke the sacred (Gimbel et al., 2013). The core finding is that sacred narratives are associated with increased activation in the posterior medial cortices (PMC), medial prefrontal cortex (MPFC), and temporoparietal junction (TPJ). PMC, TPJ, and MPFC may be involved in integrating emotion, cognition, and memory into complex models of the self in relation to the community and the world at large. The PMC is also recruited during the experience of complex social emotions such as admiration and compassion, which require a complex conceptualization of the social and moral consequences of actions (Immordino-Yang, McColl, Damasio, & Damasio, 2009). This brain region is unique in its wide-ranging connections to the rest of the brain, putting it in a special position to coordinate the integration of information from various distal brain regions involved in emotion, memory, and perception in order to construct complex abstract meaning structures, such as those that are central to identity and culturally derived values.

Conclusion

A common approach of political scientists, economists, and policymakers assumes that individual and collective decision making are both motivated by a desire to maximize pleasure and minimize pain, preferably in the here and now. This approach has the benefit of elegance; it is attractive to scientists and policy makers alike because it suggests consistent modes of decision making, and thus of dispute resolution, across cultures and contexts. Yet many critical choices in life, such as committing to a cause, a nation, or to God, involve sacrifice of individual rewards for a greater good that may not be immediately attainable or even reasonable or ever likely. Arguably, this feature of human nature facilitated creation of complex cultures and political structures that require sublimation of individual and genetic interests to a greater group and cause under the evolutionary imperative "cooperate to compete" (Atran & Henrich, 2010).

Unlike other creatures, humans form the groups to which they belong in abstract terms. Often they make their greatest exertions and sacrifices not just in order to preserve their own lives or kith and kin but for the sake of an idea—the conception they have formed of themselves, of "who we are" (Hobbes, 1651/1901). Thus, for Darwin (1871), moral virtue was most clearly associated not with intuitions, beliefs, and behaviors about fairness and reciprocity, but with a propensity to what we nowadays call "parochial altruism" (Choi & Bowles, 2007; Ginges & Atran, 2009): especially extreme self-sacrifice in war and other intense forms of human conflict, where likely prospects for individual and even group survival had very low initial probability. Heroism, martyrdom, and other forms of self-sacrifice for the group appear to go beyond the mutualistic, Golden Rule principles of fairness, reciprocity, and related conceptions of cooperation and universal justice (Rawls, 1971).

Our current research hypothesis, in line with that of Darwin, is that SVs confer a decisive advantage for those who hold them; that once values are sacralized and associated with conditions of intergroup conflict, people will adhere to them regardless of social pressures, considerations of time, or the benefits associated with other important values and alternative courses of action and exit. When SVs become embedded in fused groups under conditions of perceived threat, then "devoted actors" emerge from those groups to defend or advance those values through extreme actions that lead to intractable conflicts.

Although actions in accordance with SVs appear to defy the logic of *realpolitik* and "business-like" negotiation, SVs may also provide surprising opportunities for symbolic breakthroughs, including sincere displays of recognition and respect, which may open the way to material compromise. Understanding how and why we get these effects, and learning how to leverage them against enduring or spiraling conflict to promote peaceful outcomes, should be priorities for social science research and for policymaking.

Acknowledgments

This research was supported by grants from the National Science Foundation (BCS-827313, SES-0962080), and the MINERVA programs of the Office of Naval Research and the Air Force Office of Scientific Research. We thank Hammad Sheikh for his data analysis and research collaboration.

References

Allison, G., & Zelikow, P. (1999). *Essence of decision* (2nd ed.). New York: Longmans.

Arreguín-Toft, I. (2001). How the weak win wars. *International Security, 26,* 93–128.

Atran, S. (2006). The moral logic and growth of suicide terrorism. *Washington Quarterly, 29,* 127–147.

Atran, S. (2010). *Talking to the enemy: Violent extremism, sacred values, and what it means to be human.* London: Penguin.

Atran, S. (2011). *US Government Efforts to Counter Violent Extremism,* US Senate Armed Services Committee, 2010–2011 (Testimony, Response to Questions); http://www.jjay.cuny.edu/US_Senate_Hearing_on_Violent_Extremism.pdf.

Atran, S., & Axelrod, R. (2008). Reframing sacred values. *Negotiation Journal, 24,* 221–246.

Atran, S., Axelrod, R., & Davis, R. (2007). Sacred barriers to conflict resolution. *Science, 317,* 1039–1040.

Atran, S., & Ginges, J. (2012). Religious and sacred imperatives in human conflict. *Science, 336,* 855–857.

Atran, S., & Henrich, J. (2010). The evolution of religion. *Biological Theory, 5,* 18–30.

Baron, J., & Leshner, S. (2000). How serious are expressions of protected values? *Journal of Experimental Psychology. Applied, 6,* 183–194.

Baron, J., & Spranca, M. (1997). Protected values. *Organizational Behavior and Decision Processes, 70,* 1–16.

Baumard, N., André, J., & Sperber, D. (2013). A mutualistic theory of morality. *Behavioral and Brain Sciences, 36,* 59–122.

Becker, G. (1968). Crime and punishment: An economic approach. *Journal of Political Economy, 76,* 169-217.

Berns, G., Bell, E., Capra, C. M., Prietula, M. J., Moore, S., Anderson, B., Ginges, J., & Atran, S. (2012). The price of your soul: Neural evidence for the non-utilitarian representation of sacred values. *Philosophical Transactions of the Royal Society of London. B, Biological Sciences, 367,* 754–762.

Berns, G., et al. (2013). Moral outrage and the neurobiological antecedents to political conflict. Presented to AFOSR Annual Trust and Influence Review, Dayton, OH, January 16.

Choi, J.-K., & Bowles, S. (2007). The coevolution of parochial altruism and war. *Science, 318,* 636–640.

Darwin, C. (1871). *The descent of man, and selection in relation to sex.* London: John Murray.

Dehghani, M., Atran, S., Iliev, R., Sachdeva, S., Ginges, J., & Medin, D. (2010). Sacred values and conflict over Iran's nuclear program. *Judgment and Decision Making, 5,* 540–546.

Dehghani, M., Iliev, R., Sachdeva, S., Atran, S., Ginges, J., & Medin, D. (2009). Emerging sacred values. *Judgment and Decision Making, 4,* 990–993.

Durkheim, E. (1955). *The elementary forms of religious life.* New York: The Free Press. (Original work published 1912.)

Eliade, M. (1959). *The sacred and the profane.* New York: Harcourt Brace.

Esposito, J., & Mogahed, D. (2008). *Who speaks for Islam?* New York: Gallup Press.

Fiske, A., & Tetlock, P. (1997). Taboo trade-offs. *Political Psychology, 17*, 255–294.

Gimbel, S., Kaplan, J. T., Immordino-Yang, M. H., Tipper, C. M., Gordon, A. S., Dehghani, M., Sagae, K., Damasio, H., & Damasio, A. (2013). Neural responses to narratives framed with sacred values. Presented to Annual Meeting of Society for Neuroscience.

Ginges, J., & Atran, S. (2009). Why do people participate in violent collective action? *Annals of the New York Academy of Sciences, 1167*, 115–123.

Ginges, J., & Atran, S. (2011). War as a moral imperative. *Proceedings of the Royal Society B: Biological Sciences, 278*, 2930–2938.

Ginges, J., Atran, S., Medin, D., and Shikaki, K. (2007). Sacred bounds on rational resolution of violent political conflict. *Proceedings of the National Academy of Sciences USA, 104*, 7357–7360.

Ginges, J., Atran, S., Sachdeva, S., & Medin, D. (2011). Psychology out of the laboratory: The challenge of violent extremism. *American Psychologist, 66*, 507–519.

Gray, J. (2007). *Al Qaeda and what it means to be modern*. London: Faber & Faber.

Greene, J. (2009). The cognitive neuroscience of moral judgment. In M. Gazzaniga (Ed.), *The cognitive neurosciences* (4th ed., pp. 987–999). Cambridge, MA: MIT Press.

Hobbes, T. (1901). *Leviathan*. New York: E.P. Dutton. (Original work published 1651.)

Immordino-Yang, M., McColl, A., Damasio, H., & Damasio, A. (2009). Neural correlates of admiration and compassion. *Proceedings of the National Academy of Sciences USA, 106*, 8021–8026.

Kant, I. (2005). *Groundwork for the metaphysics of morals*. Toronto: Broadview Press. (Original work published 1785.)

McGraw, A. P., & Tetlock, P. (2005). Taboo tradeoffs, relational framing, and the acceptability of exchanges. *Journal of Consumer Psychology, 15*, 2–15.

Mill, J. S. (1871). *Utilitarianism*. London: Longmans.

Rappaport, R. (1971). The sacred in human evolution. *Annual Review of Ecology and Systematics, 2*, 23–44.

Rawls, J. (1971). *A theory of justice*. Cambridge, MA: Harvard University Press.

Sageman, M. (2008). *Leaderless Jihad*. Philadelphia: University of Pennsylvania Press.

Sheikh, H., Ginges, J., & Atran, S. (2013). Sacred values in intergroup conflict: Resistance to social influence, temporal discounting, and exit strategies. *Annals of the New York Academy of Sciences, 1299*, 11–24.

Smith, C., Vaidyanathan, B., Ammerman, N., Casanova, J., Davidson, H., Ecklund, E., Evans, J., Gorski, P., Konieczny, M., Springs, J., Trinitapoli, J., & Whitnah, M. (2013). Roundtable on the sociology of religion. *Journal of the American Academy of Religion. American Academy of Religion, 81*, 1–36.

Swann, W., Jetten, J., Gómez, Á., Whitehouse, H., & Bastian, B. (2012). When group membership gets personal: A theory of identity fusion. *Psychological Review, 119*, 441–456.

Swann, W., Buhrmester, M., Gómez, Á., Jetten, J., Bastian, B., Vázquez, A., Zhang, A. (2014). What makes a group worth dying for? *Journal of Personality and Social Psychology, 106*, 912–926..

Tetlock, P. (2003). Thinking the unthinkable: Sacred values and taboo cognitions. *Trends in Cognitive Sciences, 7*, 320–324.

Tetlock, P. E., Kristel, O. V., Elson, S. B., Green, M. C., & Lerner, J. S. (2000). The psychology of the unthinkable. *Journal of Personality and Social Psychology, 78*, 853–870.

Van Slyke, J. (2014). Moral psychology, neuroscience, and virtue. In K. Timpe & C. Boyd (Eds.), *Virtues and their vices.* New York: Oxford University Press.

von Neumann, J., & Morgenstern, O. (1944). *Theory of games and economic behavior.* Princeton, NJ: Princeton University Press.

Weber, M. (1958). Religious rejections of the world and their directions. In H. Gerth & C. Wright Mills (Eds.), *From Max Weber.* New York: Oxford University Press. (Original work published 1864.)

6   Why We Cooperate

Jillian Jordan, Alexander Peysakhovich, and David G. Rand

What does it mean to be moral? In general, answers to this question can take a normative ("what should be") or positive ("what is") perspective. Traditionally, philosophers have focused on normative answers to this question, whereas psychologists, evolutionary biologists, and social scientists have focused on the positive aspects of morality: How and why did our sense of morality evolve? What psychological processes contribute to our moral judgments? Which moral rules are universal, and which vary across cultures? In this chapter, we ask these positive questions about a particular set of moral behaviors: *cooperative* behaviors.

We define cooperative (or prosocial) behaviors as cases in which an individual pays a personal cost to provide a benefit to another individual or group of individuals. We broadly define *cost*, considering a wide range of potential costs including time and effort, resources and money, and physical harm. In most of the cases that we consider, the cost of cooperation is smaller than the benefit it creates, making cooperation productive: pairs or groups of individuals who can successfully cooperate with each other are better off than those who cannot. Because of the benefits cooperation creates, many people consider there to be a moral obligation to cooperate.

Cooperation is ubiquitous in the world around us, at a large scale with humanitarian organizations such as the Red Cross to conservation groups such as the World Wildlife Fund and on a small scale with people choosing to cover shifts for sick co-workers or to help friends move from one apartment to another. Although a gene's-eye view of evolution can explain cooperation among relatives, cooperation between unrelated individuals poses a puzzle from both the perspective of natural selection and that of rational self-interest. Why should individuals make sacrifices to help potential competitors succeed? What motivates such prosocial behavior?

In recent years many researchers have sought to shed light on these questions by using economic games to investigate cooperative behavior (for reviews, see Camerer, 2003; Rand & Nowak, 2013). In these studies game theoretical approaches are used to create experimental paradigms: people make choices that affect the money earned by themselves and other players. One such paradigm that has become particularly popular is the public goods game (PGG). In a typical PGG four people are placed into a group (typically interacting anonymously through a computer interface), and each is endowed with $10. Each then chooses how much money to contribute to a "common project," with all contributions being doubled and split evenly among the four group members. Under these rules if everyone contributes all of her or his endowment, everyone's money doubles: cooperation is productive. However, individuals personally lose money on contributing, because for each $1 contributed, the contributor receives back only $0.50. Thus, one's earnings are maximized by contributing nothing and "free riding" off the contributions of others. In this way the PGG captures the tensions between individual and collective interests that are the heart of the cooperation. Closely related to the PGG is the prisoner's dilemma (PD), probably the most widely studied economic game. The PD is a two-person version of the PGG in which players make a binary choice between a socially optimal action (often defined as paying a cost to benefit the other player) or an individually optimal action. The PD, PGG, and other related games have been used in thousands of theoretical and experimental studies to understand when and why people cooperate. In this chapter we review key insights that have emerged from this work.

Mechanisms That Promote Cooperation

In social dilemmas such as the PGG and PD, selfishness always earns more than cooperation (hence the dilemma). However, the one-shot anonymous interactions described by these games omit key elements of our world that can allow cooperators to outcompete noncooperators. We refer to alternative interaction structures that improve the long-run payoff of cooperating as mechanisms that promote cooperation (Nowak, 2006; Rand & Nowak, 2013). These mechanisms provide *ultimate* explanations for why people cooperate: if cooperation actually pays off, evolution and rational self-interest can both favor cooperating. We focus on mechanisms that can explain cooperation among unrelated individuals rather than on theories related to kin selection because cooperation among nonkin is the major challenge facing modern human societies.

Direct Reciprocity

Perhaps the most important mechanism for most human interactions is *direct reciprocity:* individuals interact repeatedly and condition their cooperative behavior on the cooperation of their partners (Axelrod, 1984; Fudenberg & Maskin, 1986). Under direct reciprocity the "shadow of the future" motivates individuals to cooperate today in order to receive the benefits of cooperation tomorrow. Direct reciprocity provides the basis for many long-term relationships in humans, such as friendships and work partnerships.

A simple strategy, "tit-for-tat" (TFT), captures the essence of direct reciprocity (Axelrod, 1984). An individual using the TFT strategy begins a repeated interaction by cooperating and then in subsequent interactions copies her partner's last action. Thus, when playing against a TFT player, you have an incentive to cooperate now so that she will cooperate with you next time (provided the returns to cooperation are large enough relative to the chance that you wind up not meeting your partner again).

In most real cooperative interactions mistakes are possible: for example, forces beyond your control may prevent you from cooperating even when you want to, or you may be confused about which action is actually helpful to your partner. If both partners are using TFT, a single such mistake can derail an otherwise cooperative relationship. As a result the presence of errors favors reciprocal strategies that sometimes cooperate even after their partners have defected (Nowak & Sigmund, 1992; Rand, Ohtsuki, & Nowak, 2009).

Theoretical work on the evolution of cooperation in repeated games is complemented by experimental evidence that repetition can indeed lead to the emergence of cooperation. When subjects playing repeated PDs have a sufficiently large probably of interacting again with the same partner in the next round, they learn to cooperate (Blonski, Ockenfels, & Spagnolo, 2011; Dal Bó & Fréchette, 2011; Dreber, Rand, Fudenberg, & Nowak, 2008; Fudenberg, Rand, & Dreber, 2012; Rand, Fudenberg, & Dreber, 2013). Otherwise, when relationships are too short or the payoff for defection is too tempting, they learn to defect. Furthermore, introducing the possibility of mistakes causes players to adopt strategies that are lenient (i.e., that wait for multiple defections before switching to defection themselves) and forgiving (i.e., that are willing to return to cooperation after periods of defection) (Dreber, Fudenberg, & Rand, 2014; Fudenberg et al., 2012; Rand et al., 2013).

Indirect Reciprocity

Direct reciprocity can make cooperation between pairs of individuals advantageous. But what about cooperation at a larger scale than dyadic interactions? This question is answered in part by *indirect reciprocity*, whereby my actions toward you depend on your previous actions toward others (Nowak & Sigmund, 2005). Under indirect reciprocity people earn good reputations when they cooperate with others and thus can expect increased cooperation from future partners. Social norms within a community specify standards for acceptable behavior, and information about individuals' behavior is spread through gossip. Successful social norms often assign good reputations to those who cooperate with others in good reputation and defect with those in bad reputation. Thus, individuals with good reputations are then rewarded because they are more likely to be the recipients of cooperative behavior.

There are two distinct reasons why one might preferentially cooperate with individuals who are known to be cooperative. One might reason that individuals who are known to be cooperative are more likely to reciprocate cooperative behavior. Thus, individuals may discriminate based on reputation as a way to select desirable interaction partners (Barclay & Willer, 2007).

Alternatively, one might cooperate with other cooperators merely to maintain a good reputation and thus receive more cooperation in the future (Pfeiffer, Tran, Krumme, & Rand, 2012). Rather than using a partner's previous behavior as a signal about her future behavior, the partner's previous behavior stipulates what you must do in order to maintain a good reputation yourself. Although perhaps less intuitive, this latter logic is the overwhelming focus of theoretical work on indirect reciprocity.

Experimental work confirms the theorized importance of reputation in promoting human cooperation: people playing economic games learn to cooperate when it is sufficiently likely that others will know about their previous actions (Feinberg, Willer, Stellar, & Keltner, 2012; Milinski, Semmann, Bakker, & Krambeck, 2001; Pfeiffer et al., 2012; Wedekind & Milinski, 2000). Furthermore, evidence suggests that humans are so highly attuned to their reputations that even subtle images of eyespots can increase cooperation by unconsciously priming the sense of being watched (Haley & Fessler, 2005; Sparks & Barclay, 2013). Reputation systems have also been shown to promote cooperation outside the laboratory: blood donation (Lacetera & Macis, 2010) and giving to charity (Karlan & McConnell, 2012) increase when donors' names are published, and people are three times as likely to sign up for an energy blackout reduction program when sign-ups are observable (Yoeli, Hoffman, Rand, & Nowak, 2013).

Institutions

Scaling up even further, *institutions* provide an important tool for maintaining cooperation in large groups (Bowles, Choi, & Hopfensitz, 2003; North, 1990; Ostrom, 1990). Humans often explicitly design institutions for the purpose of incentivizing good behavior. For example, governments create criminal justice systems, often employing police and courts, to prevent antisocial behaviors such as theft and assault. Such legal institutions have a long history in human societies. Smaller organizations also employ formal codes of conduct and often designate specific individuals to enforce the rules.

Institutions may also refer to social structures that create infrastructure for cooperative exchanges, like markets. Markets provide a regulated environment for strangers to engage in productive trades of goods and services (Greif, 1993; Milgrom & North, 1990). In general, institutions can promote cooperation both by deterring bad behavior and by promoting trust that others will cooperate.

The role of institutions in human cooperation has received much less attention among experimentalists using economic games than direct and indirect reciprocity. But several recent studies have begun to demonstrate the power that institutional incentives have over human cooperative behavior in the lab (Andreoni & Gee, 2012; Baldassari & Grossman, 2011; Ouss & Peysakhovich, 2013). More work in this vein is an important direction for future research.

Proximate Psychology of Cooperation

Because of mechanisms such as these, cooperation often pays off in the long run. Yet the suggestion that cooperation is only about maximizing long-term payoffs does not match well with our daily life experiences. A brief moment of introspection clearly indicates that all cooperative behavior does not result from conscious calculations of expected returns. We are surrounded by examples of people who seem to help because they genuinely care for others. And almost all of us have acted altruistically at one time or another without considering future returns, only motivated by our notions of morality and ethical behavior. How can we explain this form of truly costly cooperation?

We argue that much of the explanation comes from distinguishing between ultimate and proximate explanations for cooperative behavior. *Ultimate* explanations describe *why* people cooperate by outlining how cooperation pays off in the long run. In contrast, *proximate* explanations can explain *how* people cooperate by outlining the motivations, emotions, and cognitions that lead to cooperation in the moment. Thus, when we say that an individual cooperates because of direct reciprocity, we are providing an ultimate explanation. In contrast, when we say that an individual cooperates because she genuinely cares for her friend, we are providing a proximate explanation. Ultimate explanations provide reasons why proximate mechanisms would have come to function as they do: it is long-run advantageous to cooperate with our friends, and so we come to care for our friends (as a psychological mechanism to help motivate us to cooperate). These types of proximate psychological mechanisms help us to make choices that are typically advantageous without having to engage in the costly effort of estimating future returns.

One of the most important proximate mechanisms for motivating cooperative behavior is empathy. The *empathy-altruism* hypothesis contends that *empathetic concern* is a human emotional response to taking the perspective of another person in need (Batson, Ahmad, & Lishner, 2011). Empathetic concern motivates humans to see that need relieved, often through cooperative helping. Across a large body of studies Batson and colleagues have showed that experimentally manipulating empathic concern increases cooperative behaviors, such as volunteering to take electric shocks instead of the target (Batson, Duncan, Ackerman, Buckley, & Birch, 1981) and cooperating with the target in economic games (Batson & Ahmad, 2001). Furthermore, evidence suggests that empathic individuals experience positive mood changes when they see a need get relieved, even if another agent caused the relief—suggesting that empathic concern reflects a genuine care for others (Batson et al., 1988).

The Social Heuristics Hypothesis

Any complete theory of cooperation must explain why we often observe cooperative behavior when no ultimate mechanisms appear to be operative, or they appear too weak to incentivize cooperation. For example, in January 2010, over 3 million people text-messaged the American Red Cross to donate $10 to disaster relief for the Haiti earthquake. It is hard to generate an ultimate explanation for such a behavior. Why would it pay off for individuals to give anonymous gifts to strangers in foreign countries? Likewise, in laboratory experiments, subjects routinely pay costs to benefit anonymous strangers in one-shot interactions in which they have nothing to gain (Camerer, 2003). Such "irrational" cooperation is quite common and is critical to the success of our societies and to our identities as moral actors.

The social heuristics hypothesis (SHH; Rand, Greene, & Nowak, 2012; Rand et al., 2014) provides one explanation for this intrinsically motivated cooperation. The SHH considers cooperative decisions from a *dual process* perspective, where decisions result from an interplay between two types of psychological processes: those that are fast, automatic, intuitive, and affective; and those that are slow, controlled, and deliberative (Kahneman, 2003; Sloman, 1996). The SHH contends that over time, strategies that are typically successful become automatized as default responses. As a result if cooperation typically pays off in the long run, individuals come to develop prosocial preferences and genuinely care for others. Individuals then carry these heuristics with them to other settings, such as one-shot anonymous laboratory experiments. Therefore, individuals who generally interact in environments where cooperation is advantageous should, as a result of spillover effects, be predisposed toward prosociality even when it will not actually pay off.

To provide evidence that cooperative behavior is the default response for many people, Rand et al. (2012) manipulated cognitive processing of subjects playing economic games. Across a series of studies PGG participants were randomly assigned to conditions that induced more intuitive decision making (via time pressure or writing about a time their intuitions had worked out well) or more deliberative decision making (via time delay or writing about a time that careful reasoning worked out well). The results showed significantly more cooperation among individuals induced to be intuitive than individuals induced to be deliberative. Thus, these results provided evidence that, on average, intuition favors cooperation in one-shot anonymous social dilemmas. It has also been found that impairing the function of the right lateral prefrontal cortex, a brain region associated with deliberation and control, increases giving in a unilateral money transfer, whereas amplifying this region decreases giving (Ruff, Ugazio, & Fehr, 2013), and that people by default project a cooperative frame onto neutrally framed prisoner's dilemma games (Engel & Rand, 2014).

Does this mean that cooperation is automatic for all people at all times? The answer is no. Social heuristics are flexible, and they change as we learn through experience whether cooperation tends to pay off. Although Rand et al. (2012) found that intuition favored cooperation overall, this pattern applied only to subjects who reported having cooperative daily-life interaction partners. In contrast, subjects who did *not* report having cooperative partners showed no evidence of defaulting toward cooperation. Further evidence of the moderating role of trust on intuitive cooperation comes from Rand and Kraft-Todd (In press). Whereas intuitive responses varied based on life experience, deliberation favored selfishness among both trusting and nontrusting subjects. Additional evidence moderators come from Tinghög et al. (2013), who find no effect of time pressure on PGG cooperation across several studies; and from Rand et al. (2014), who find a variety of time-pressure effect sizes across fifteen different studies that range from large and positive to null but never significantly negative. Similar results, where intuition sometimes favors prosociality and sometimes has no effect, have been found using cognitive load (Cornelissen, Dewitte, & Warlop, 2011; Hauge, Brekke, Johansson, Johansson-Stenman, & Svedsäter, 2009; Roch, Lane, Samuelson, Allison, & Dent, 2000; Schulz, Fischbacher, Thöni, & Utikal, 2014).

The SHH may help to explain why individuals from certain societies are more likely to be cooperative and trusting than those from others. Although mechanisms such as direct and indirect reciprocity are likely to be fairly universal across cultures, the quality of institutions varies dramatically from society to society (Ellingsen, Herrmann, Nowak, Rand, & Tarnita, 2012; Gächter & Herrmann, 2009; Henrich et al., 2010; Herrmann, Thoni, & Gächter, 2008). This variation creates environments where it is safer versus more dangerous to trust strangers. Thus, it may be that people living under effective institutions internalize cooperative norms to a greater extent than those living under poorly functioning institutions.

It is difficult to draw strong causal conclusions about cross-cultural variation in cooperation. However, correlational evidence suggests that institutions such as laws and markets play an important role in fostering cooperative norms. For example, people living in countries with less corruption and stronger rule of law were less likely to engage in antisocial punishment behavior in PGG games (Ellingsen et al., 2012; Gächter & Herrmann, 2009; Herrmann et al., 2008); furthermore, across a ranges of societies from hunter-gatherers to industrialized nations, greater market integration was associated with greater generosity in economic games (Henrich et al., 2010).

Direct experimental evidence also exists for the effect of social environment on heuristic-based cooperation in the form of experiments using repeated PDs as a model of the future consequences created by institutions. In these studies subjects were randomly assigned to play a series of repeated PDs either under a set of "good" rules that favored cooperation (large repeat probability) or a set of "bad" rules that favored defection (small repeat probability); afterward, all subjects played an identical battery of one-shot anonymous games (Peysakhovich & Rand, 2013). Subjects assigned to "good" environments were more cooperative, altruistic, trusting, and trustworthy in the subsequent one-shot games as well as more inclined to punish selfishness. Furthermore, this effect was especially strong among individuals who relied more on heuristic processing. These results provide direct evidence for spillover effects: the rules governing your interactions with others influence the cooperative heuristics you adopt.

Conclusion

Cooperation with unrelated individuals is a hallmark of humankind despite the temptation to behave selfishly. Mechanisms that create future consequences for present actions can make cooperation pay off in the longer term and allow cooperation to arise and be maintained. Some of the most powerful of these mechanisms are direct reciprocity, indirect reciprocity, and institutions. The cooperative behavior promoted by these mechanisms is then implemented with the help of various proximate psychological processes such as emotions like empathy. We argue that social heuristics form the bridge between ultimate and proximate explanations for cooperation: we internalize advantageous behavioral strategies as intuitive defaults, and thus mechanisms that promote cooperation may lead us to cooperate automatically, even in specific situations that are beyond the reach of any mechanism. Such "irrational" cooperation is central to the success of human civilization and forms a key component of morality.

We conclude this chapter by describing important open questions regarding human cooperative behavior. In this chapter we have proposed that intrinsic motivations for cooperative behavior arise as a result of extrinsic incentives that make cooperation advantageous. The evidence that we present for this hypothesis, however, runs counter to work showing that extrinsic incentives can *crowd out*, or undermine, intrinsic motivation (Titmuss, 1970). For example, in a classic study of extrinsic undermining, young children became less interested in drawing with markers when they were asked to draw in exchange for a "good player award" (Lepper, Greene, & Nisbett, 1973). Extrinsic undermining has also been shown to occur in social dilemma situations (e.g., Frey & Oberholzer-Gee, 1997; Gneezy & Rustichini, 2000). How can we reconcile these seemingly contradictory results? The answer likely concerns the *nature* of the incentives. For example, it may be that explicit financial incentives *crowd out* intrinsic motivation, whereas more implicit reputational incentives *crowd in* intrinsic motivation. Developing an understanding of how to apply incentives in a way that effectively creates cultures of cooperation with internalized prosocial norms is of great scientific interest as well as practical importance.

Another open question concerns how people *respond* to cooperative and selfish behaviors with behaviors such as reciprocity and norm enforcement. Not only do people engage in first-hand cooperative behaviors, but they also attend to and monitor the cooperative behaviors of others. Individuals respond to cooperative (and noncooperative) behavior both when the behavior has directly affected them (as *second parties*) and when it has not (as *third parties*). Evidence suggests that both second and third parties frequently reward cooperators and punish defectors, even at personal costs (Almenberg, Dreber, Apicella, & Rand, 2011; Fehr & Fischbacher, 2003, 2004). Reciprocity and norm enforcement are important behaviors to understand, as they create additional consequences for violating cooperative norms and thus help maintain cooperative societies.

As with cooperative behavior, the question arises as to why individuals engage in sanctioning behavior. Evidence suggests that second-party punishment is an intuitive and automatic response to unfair treatment (Grimm & Mengel, 2011; Rand & Nowak, 2013; Sanfey, Rilling, Aronson, Nystrom, & Cohen, 2003). This suggests that it may represent a generally

advantageous strategy—perhaps because it serves to deter future exploi-
tation. But why would a third-party observer pay costs to punish selfish
behavior when she has not been personally harmed? An understanding
of the ultimate and proximate mechanisms that support third-party norm
enforcement is needed. Research investigating these mechanisms will both
contribute to our theoretical understanding of human cooperation and
help us to understand how to best foster cultures of cooperation.

References

Almenberg, J., Dreber, A., Apicella, C. L., & Rand, D. G. (2011). *Third party reward and punishment: Group size, efficiency and public goods psychology and punishment.* Hauppauge, NY: Nova Science Publishers.

Andreoni, J., & Gee, L. (2012). Gun for hire: Delegated enforcement and peer punishment in public goods provision. *Journal of Public Economics, 96*(11), 1036-1046.

Axelrod, R. (1984). *The evolution of cooperation.* New York: Basic Books.

Baldassari, D., & Grossman, G. (2011). Centralized sanctioning and legitimate authority promote cooperation in humans. *Proceedings of the National Academy of Sciences USA, 108*(27), 11023–11027.

Barclay, P., & Willer, R. (2007). Partner choice creates competitive altruism in humans. *Proceedings. Biological Sciences, 274*(1610), 749–753.

Batson, C. D., & Ahmad, N. (2001). Empathy⊠induced altruism in a prisoner's dilemma II: What if the target of empathy has defected? *European Journal of Social Psychology, 31*(1), 25–36.

Batson, C. D., Ahmad, N., & Lishner, D. A. (2011). Empathy and altruism. In *The Oxford handbook of positive psychology,* 417-427. Oxford: Oxford University Press.

Batson, C. D., Duncan, B. D., Ackerman, P., Buckley, T., & Birch, K. (1981). Is empathic emotion a source of altruistic motivation? *Journal of Personality and Social Psychology, 40*(2), 290.

Batson, C. D., Dyck, J. L., Brandt, J. R., Batson, J. G., Powell, A. L., McMaster, M. R., et al. (1988). Five studies testing two new egoistic alternatives to the empathy-altruism hypothesis. *Journal of Personality and Social Psychology, 55*(1), 52.

Blonski, M., Ockenfels, P., & Spagnolo, G. (2011). Equilibrium selection in the repeated prisoner's dilemma: Axiomatic approach and experimental evidence. *American Economic Journal: Microeconomics, 3*(3), 164–192.

Bowles, S., Choi, J., & Hopfensitz, A. (2003). The co-evolution of individual behaviors and social institutions. *Journal of Theoretical Biology, 223*(2), 135–147.

Camerer, C. F. (2003). *Behavioral game theory: Experiments in strategic interaction.* Princeton, NJ: Princeton University Press.

Cornelissen, G., Dewitte, S., & Warlop, L. (2011). Are social value orientations expressed automatically? Decision making in the dictator game. *Personality and Social Psychology Bulletin, 37*(8), 1080–1090.

Dal Bó, P., & Fréchette, G. R. (2011). The evolution of cooperation in infinitely repeated games: Experimental evidence. *American Economic Review, 101*(1), 411–429.

Dreber, A., Fudenberg, D., & Rand, D. G. (2014). Who cooperates in repeated games? *Journal of Economic Behavior & Organization, 98*, 41–55.

Dreber, A., Rand, D. G., Fudenberg, D., & Nowak, M. A. (2008). Winners don't punish. *Nature, 452*(7185), 348–351.

Ellingsen, T., Herrmann, B., Nowak, M. A., Rand, D. G., & Tarnita, C. E. (2012). Civic capital in two cultures: The nature of cooperation in Romania and USA. Available at SSRN: http://papers.ssrn.com/sol3/papers.cfm?abstract_id=2179575.

Engel, C., & Rand, D. G. (2014). What does "clean" really mean? The implicit framing of decontextualized experiments. *Economics Letters, 122*(3), 386–389.

Fehr, E., & Fischbacher, U. (2003). The nature of human altruism. *Nature, 425*(6960), 785–791.

Fehr, E., & Fischbacher, U. (2004). Third-party punishment and social norms. *Evolution and Human Behavior, 25*(2), 63–87.

Feinberg, M., Willer, R., Stellar, J., & Keltner, D. (2012). The virtues of gossip: Reputational information sharing as prosocial behavior. *Journal of Personality and Social Psychology, 102*(5), 1015–1030.

Frey, B. S., & Oberholzer-Gee, F. (1997). The cost of price incentives: An empirical analysis of motivation crowding-out. *American Economic Review, 87*(4), 746–755.

Fudenberg, D., & Maskin, E. S. (1986). The Folk Theorem in repeated games with discounting or with incomplete information. *Econometrica, 54*(3), 533–554.

Fudenberg, D., Rand, D. G., & Dreber, A. (2012). Slow to anger and fast to forgive: Cooperation in an uncertain world. *American Economic Review, 102*(2), 720–749.

Gächter, S., & Herrmann, B. (2009). Reciprocity, culture and human cooperation: Previous insights and a new cross-cultural experiment. *Philosophical Transactions of the Royal Society of London. B, Biological Sciences, 364*(1518), 791–806.

Gneezy, U., & Rustichini, A. (2000). Pay enough or don't pay at all. *Quarterly Journal of Economics, 115*(3), 791–810.

Greif, A. (1993). Contract enforceability and economic institutions in early trade: The Maghribi traders' coalition. *American Economic Review, 83(3),* 525–548.

Grimm, V., & Mengel, F. (2011). Let me sleep on it: Delay reduces rejection rates in ultimatum games. *Economics Letters, 111*(2), 113–115.

Haley, K. J., & Fessler, D. M. T. (2005). Nobody's watching? Subtle cues affect generosity in an anonymous economic game. *Evolution and Human Behavior, 26,* 245–256.

Hauge, K. E., Brekke, K. A., Johansson, L.-O., Johansson-Stenman, O., & Svedsäter, H. (2009). *Are social preferences skin deep? Dictators under cognitive load.* University of Gothenburg Working Papers in Economics, (371).

Henrich, J., Ensminger, J., McElreath, R., Barr, A., Barrett, C., Bolyanatz, A., et al. (2010). Markets, religion, community size, and the evolution of fairness and punishment. *Science, 327*(5972), 1480–1484.

Herrmann, B., Thoni, C., & Gächter, S. (2008). Antisocial punishment across societies. *Science, 319*(5868), 1362–1367.

Kahneman, D. (2003). A perspective on judgment and choice: Mapping bounded rationality. *American Psychologist, 58*(9), 697–720.

Karlan, D., & McConnell, M. A. (2012). *Hey look at me: The effect of giving circles on giving.* National Bureau of Economic Research, Cambridge, MA.

Lacetera, N., & Macis, M. (2010). Social image concerns and prosocial behavior: Field evidence from a nonlinear incentive scheme. *Journal of Economic Behavior & Organization, 76*(2), 225–237.

Lepper, M. R., Greene, D., & Nisbett, R. E. (1973). Undermining children's intrinsic interest with extrinsic reward: A test of the "overjustification" hypothesis. *Journal of Personality and Social Psychology, 28*(1), 129.

Milgrom, P. R., & North, D. C. (1990). The role of institutions in the revival of trade: The law merchant, private judges, and the champagne fairs. *Economics and Politics, 2*(1), 1–23.

Milinski, M., Semmann, D., Bakker, T. C. M., & Krambeck, H. Jr. (2001). Cooperation through indirect reciprocity: Image scoring or standing strategy? *Proceedings of the Royal Society of London. B, Biological Sciences, 268*(1484), 2495–2501.

North, D. C. (1990). *Institutions, institutional change and economic performance.* Cambridge, UK: Cambridge University Press.

Nowak, M. A. (2006). Five rules for the evolution of cooperation. *Science, 314*(5805), 1560–1563.

Nowak, M. A., & Sigmund, K. (1992). Tit for tat in heterogeneous populations. *Nature, 355,* 250–253.

Nowak, M. A., & Sigmund, K. (2005). Evolution of indirect reciprocity. *Nature*, *437*(7063), 1291–1298.

Ostrom, E. (1990). *Governing the commons: The evolution of institutions for collective action.* Cambridge, UK: Cambridge University Press.

Ouss, A., & Peysakhovich, A. (2013). When punishment doesn't pay: "Cold Glow" and decisions to punish. Available at SSRN: http://papers.ssrn.com/sol3/papers.cfm?abstract_id=2247446.

Peysakhovich, A., & Rand, D. G. (2013). Habits of virtue: Creating norms of cooperation and defection in the laboratory. Available at SSRN: http://papers.ssrn.com/sol3/papers.cfm?abstract_id=2294242.

Pfeiffer, T., Tran, L., Krumme, C., & Rand, D. G. (2012). The value of reputation. *Journal of the Royal Society, Interface*, *9*(76), 2791–2797.

Rand, D. G., Fudenberg, D., & Dreber, A. (2013). It's the thought that counts: The role of intentions in reciprocal altruism. Available at SSRN: http://papers.ssrn.com/sol3/papers.cfm?abstract_id=2259407.

Rand, D. G., Greene, J. D., & Nowak, M. A. (2012). Spontaneous giving and calculated greed. *Nature*, *489*(7416), 427–430.

Rand, D. G., & Kraft-Todd, G. T. (in press). Reflection does not undermine self-interested prosociality. *Frontiers in Behavioral Neuroscience*.

Rand, D. G., & Nowak, M. A. (2013). Human cooperation. *Trends in Cognitive Sciences*, *17*(8), 413–425.

Rand, D. G., Ohtsuki, H., & Nowak, M. A. (2009). Direct reciprocity with costly punishment: Generous tit-for-tat prevails. *Journal of Theoretical Biology*, *256*(1), 45–57.

Rand, D. G., Peysakhovich, A., Kraft-Todd, G. T., Newman, G. E., Wurzbacher, O., Nowak, M. A., et al. (2014). Social heuristics shape intuitive cooperation. *Nature Communications*, *5*, 3677.

Roch, S. G., Lane, J. A. S., Samuelson, C. D., Allison, S. T., & Dent, J. L. (2000). Cognitive load and the equality heuristic: A two-stage model of resource overconsumption in small groups. *Organizational Behavior and Human Decision Processes*, *83*(2), 185–212.

Ruff, C. C., Ugazio, G., & Fehr, E. (2013). Changing social norm compliance with noninvasive brain stimulation. *Science*, *342*, 482–484.

Sanfey, A. G., Rilling, J. K., Aronson, J. A., Nystrom, L. E., & Cohen, J. D. (2003). The neural basis of economic decision-making in the ultimatum game. *Science*, *300*(5626), 1755–1758.

Schulz, J. F., Fischbacher, U., Thöni, C., & Utikal, V. (2014). Affect and fairness: Dictator games under cognitive load. *Journal of Economic Psychology 41*, 77–87.

Sloman, S. A. (1996). The empirical case for two systems of reasoning. *Psychological Bulletin, 119*(1), 3.

Sparks, A., & Barclay, P. (2013). Eye images increase generosity, but not for long: The limited effect of a false cue. *Evolution and Human Behavior, 34*(5), 317–322.

Tinghög, G., Andersson, D., Bonn, C., Böttiger, H., Josephson, C., Lundgren, G., et al. (2013). Intuition and cooperation reconsidered. *Nature, 497*(7452), E1–E2.

Titmuss, R. (1970). *The gift relationships*. London: George Allen.

Wedekind, C., & Milinski, M. (2000). Cooperation through image scoring in humans. *Science, 288*(5467), 850–852.

Yoeli, E., Hoffman, M., Rand, D. G., & Nowak, M. A. (2013). Powering up with indirect reciprocity in a large-scale field experiment. *Proceedings of the National Academy of Sciences USA, 110*(Supplement 2), 10424–10429.

III The Development of Morality

7 The Infantile Origins of Our Moral Brains

J. Kiley Hamlin

Historical Overview

The study of humans' moral origins is both old and new. A topic of philosophical inquiry since at least the fourth century B.C., in ancient Greece, Plato's *Meno* (380 B.C.) famously wonders to Socrates "whether virtue is acquired by teaching or by practice; or if neither by teaching nor practice, than whether it comes to man by nature, or in what other way?" Socrates replies that such a question requires understanding what virtue *is*, which in itself is tremendously difficult. In China during the same period Confucian philosopher Mencius stressed the importance of *both* nature and nurture in moral development. He claimed that all humans are innately endowed with four moral "sprouts" or "beginnings," predisposing humans to empathy, a sense of shame, respect for others and rules, and a sense of right and wrong. Mencius believed that the morality adults possess develops directly from these sprouts and would not be possible without them. Critically, Mencius viewed the moral sprouts only as predispositions setting men on the path toward virtue: just as bean sprouts need water and sunlight to become fully grown plants, moral sprouts require cultivation and nourishment to become full-fledged morality. He wrote, "Men have these Four Beginnings just as they have their four limbs. Having these Four Beginnings, but saying they cannot develop them is to destroy themselves" (Mencius, 2A:6, in Chan, 1963:65).

Fast-forward twenty-five centuries, and today's scholarly discussions regarding the origins of morality remain strikingly similar to those of Plato and Mencius: everyone wishes to know what morality is and how it develops; some are motivated to answer questions regarding human nature, others to understand atypical moral development, and others to create maximally virtuous individuals. Traditionally, these questions have been addressed by developmental psychologists who study children across the life span and who attempt to identify and account for both the beginnings of moral action and judgment in early childhood and the developmental change that occurs over time (see Killen & Smetana, 2006, 2013 for reviews).

Although this need not be the case (and there are some notable exceptions, e.g., Kagan, 1987), an implication of much of the moral developmental work to date has been that moral development is entirely a process of qualitative, rather than quantitative, change. That is, whether infants are conceived of as beginning as complete blank slates (innately nothing), begin with perceptual biases and domain-general learning skills (innately capable of learning and constructing knowledge), and/or begin already intensely oriented to the social world (innately social), there are no morality-specific tendencies or mental structures that emerge via biological maturation. Morality is entirely acquired, or constructed, after birth, as infants' amoral nature interacts with their experiences. Perhaps the most commonly hypothesized drivers of this moral change are (1) socialization regarding moral norms from caregivers and others and (2) the active construction of moral rules by children themselves as they navigate their own social relationships (see reviews in Killen & Smetana, 2006, 2013).

That both socialization and cognitive construction massively influence moral development is undeniable; such conclusions have been supported by decades of elegant research. However, although evidence of experience-based or cognitively constructed development does not in itself preclude an exploration of whether any aspects of morality develop in other ways, far less attention has been paid to the possibility that rather than as moral blank slates, infants enter the world equipped with motivations and/or cognitions with distinctly moral content. Perhaps akin to Mencius' "sprouts," this moral content could be the foundation on which socialization and construction processes build.

Although the possibility that some aspects of morality are built-in is a tenable and testable hypothesis, the lack of attention to whether morality has any innate components in most developmental literature is not terribly surprising. Indeed, there is not much about infants and toddlers that appears moral (and much that seems amoral or immoral), whereas the process of moral development that occurs across childhood is clear.

In addition, developmental psychologists have tended to accept a Kantian rationalist version of morality, whereby abstract notions of universal rights and justice are primary, holding that it is an essential component of moral action and judgment to be consciously aware both *that* one's actions and judgments are moral and *why* that is (Kant, 1785/1964; Kohlberg, 1969; Nucci & Turiel, 1978; Piaget, 1932; Rawls, 1971). Although it is clearly the case that some of the moral life of adults involves conscious reasoning (perhaps the best parts), this definition of morality has been challenged on both philosophical and empirical grounds for denying the role of intuitive emotional processes (Haidt, 2001; Hume, 1739/1978; Smith, 1759). More pressingly for the current purposes, the definition simply excludes infants from the study of morality: they clearly do not consciously ponder issues of universal rights.

This chapter overviews recent work suggesting that Mencius had it right: infants demonstrate striking moral capacities in the first year of life, ones that it would be difficult to claim are solely the result of socialization and constructivist processes. Critically, its purpose is not to propose that morality remains unchanged throughout the life span, nor that one's moral outcomes are not massively influenced by shared and unshared environmental inputs in combination with children's domain-general cognitive abilities. Indeed, such a proposal would be ridiculous. Rather, I suggest that this work provides evidence for "sprouts" or predispositions toward prosocial behavior, moral understanding, and moral evaluation within the first year of life, before children's experiences would be sufficient to create them from scratch; that is, that biological maturation has some foundational, although incomplete, role in moral development. I start by describing the theoretical basis for making nativist moral claims and then outline what is quickly becoming a large body of work on the moral lives of babies.

Theoretical Perspective: Morality Evolved

Humans are seriously nice: they cooperate with and help nonrelatives at cost to themselves at a rate that is unmatched in the animal kingdom (Tomasello, 2009; Warneken, 2013). Humans are also seriously judgmental: even though there is cross-cultural variation in exactly how certain actions are evaluated (see Haidt, 2012, for review), the tendency to view some actions as good, right, and praiseworthy and others as bad, wrong, and necessitating retaliation is a human universal (Brown, 1991). Questions of how human morality could have emerged—in particular, why are we so nice and why do we care how others act?—have led to marked increases in evolutionary theorizing regarding humans' moral origins (for reviews see Boehm, 2012; Hauser, 2006; Katz, 2000; Nowak, 2006).

In a nutshell the moral evolution perspective holds that humans' intensely cooperative nature is at the root of their success. By working together, sharing resources, caring about how others are doing, and helping those in need humans are stronger, safer, better fed, and smarter than they could ever be apart; by paying the relatively small costs of cooperating up front, individuals have the potential to be much better off in the long run. However, this "pay now, benefit later" structure is risky, leaving cooperative systems extremely vulnerable to cheaters who reap the benefits of others' cooperative acts but do not cooperate in return. This vulnerability to cheaters is hypothesized to be at the root of human moral judgments: if individuals are to have any hope of keeping cheaters at bay, they must be motivated to monitor others' cooperation-relevant acts, to judge them as good or bad, to punish misdeeds in the moment, and to remember their judgments for future encounters. From this perspective, humans (both individually and as a species) develop morality because it is required for cooperative systems to flourish.

Of course, whether morality is an evolutionarily relevant trait does not directly relate to how or when it emerges within individual life spans. It is perfectly consistent with an evolutionary account to assume that morality emerges during childhood as the product of (solely) parenting and socialization practices combined with children's ability to actively interpret and construct rules from their social worlds. Indeed, parenting practices and cognitive abilities were themselves shaped by evolution, and evolution only "cares" about whether a required behavior emerges—not about the specific way the behavior goes about emerging. By the time children are capable of contributing to cooperative endeavors and are at risk of being cheated, they may have already developed the morality they need to allow them to cooperate successfully. That is, moral animals need not start out as moral babies.

On the other hand, maybe moral animals do start as moral babies. Just as humans are born able to see and with ten fingers and toes, an evolutionary account of morality makes it reasonable to at least posit that some aspects of morality arise independently of experience; that is, maybe experience shapes morality, rather than creates it from scratch. This is of course an empirical question requiring empirical evidence to accept or reject; but the growing focus on the functional nature of morality suggests it is one worth asking. The remainder of this chapter reviews the growing body of evidence that supports the claim that human moral judgment and action are built on an innate moral core (for other recent reviews see Baillargeon et al., 2014; Bloom, 2013; Hamlin, 2013a; Wynn, 2008; Warneken, 2013).

Empirical Evidence: Do Infants Show Predispositions toward Moral Behavior?

Empathy, or the tendency to notice and share others' affective states, is considered a fundamental building block to moral behavior, driving individuals to care about others' needs (Decety, 2014; de Waal, 2008; Hastings, Miller, Kahle, & Zahn-Waxler, 2013). Indeed, psychopaths, the 1 percent of the general population who lack empathy, make up 20–30 percent of prison populations (Hare, 1999), highlighting the importance of empathy for promoting prosocial and inhibiting antisocial behaviors. Several studies now demonstrate that the beginnings of empathic responding may already be in place at birth: newborns cry when they hear recordings of other infants' cries, but they do not cry when hearing nonhuman sounds that are matched with crying for various perceptual features (Simner, 1971) and cry significantly less when hearing recordings of their own cries (Dondi, Simion, & Caltran, 1999). As a group these studies suggest that rather than resulting from an aversion to a particular perceptual stimulus, contagious crying in newborns represents a rudimentary form of empathy.

Empathic concern that is clearly directed at others (that is, where others' distress does more than cause an infant personal distress) has been documented from 8 to 16 months of age (Roth-Hanania, Davidov, & Zahn-Waxler, 2011). By 18 months toddlers show concern for others who have been harmed but do not show outward signs of distress, suggesting that empathic concern can be motivated solely by an understanding of how being a victim of antisocial acts makes others feel (Vaish, Carpenter, & Tomasello, 2009). Around this same time toddlers begin engaging in active comforting behaviors (hugging, patting, and providing comfort objects) (Svetlova, Nichols, & Brownell, 2010; Zahn-Waxler, Radke-Yarrow, Wagner,

& Chapman, 1992), such prosocial behaviors are correlated with toddlers' levels of concern (Eisenberg & Miller, 1987). Further highlighting the link between early concern for others and moral behavior, it was recently discovered that a failure to show empathic concern between 6 and 14 months of age is associated with increased levels of antisocial behavior in middle childhood and adolescence (Rhee et al., 2013).

Although one should be cautious not to lump all prosocial behaviors together into a single construct (Dunfield, Kuhlmeier, O'Connell, & Kelley, 2011), other forms of prosocial behavior also emerge early, as infants become physically capable of performing them. By 12 months of age infants are helpful communicators: they point to show the location of an item they know someone is searching for (Liszkowski, Carpenter, & Tomasello, 2008) and to warn an adult about negative action outcomes that they can foresee but the adult cannot (Knudsen & Liszkowski, 2013). By 14 to 18 months toddlers begin assisting with everyday household activities (Rheingold, 1982) and help adults to achieve their unfulfilled instrumental goals such as picking up dropped objects or stacking books (Warneken & Tomasello, 2006). Toddlers are often willing to share resources when prompted to do so, and spontaneous giving occurs, albeit rarely, in the second year of life (Rheingold, Hay, & West, 1976).

Why Are Infants Prosocial?

Although a variety of prosocial behaviors emerge early in life, so do the socialization processes and experiences that encourage them. By 18 months of age infants will surely have been actively encouraged to comfort, share with, and help others, and are likely praised and/or rewarded after doing so. That is, perhaps toddlers are simply motivated to please their parent (who is usually in the room) and/or anticipate external reinforcement. Arguing against this account are a growing number of empirical demonstrations that intrinsic, rather than extrinsic, motivations drive prosocial behavior in toddlers. For instance, 24-month-olds are equally likely to help an unfamiliar experimenter in need whether a parent is present or absent and whether or not they are directed to help (Warneken & Tomasello, 2013). At 20 months toddlers who are praised and rewarded after helping actually help *less* in the future than do nonreinforced toddlers (Warneken & Tomasello, 2008); reduction in a behavior following external reinforcement is one of the hallmarks of intrinsically motivated acts (Lepper, Greene, & Nisbett, 1973).

Perhaps they are not *prosocial* actions toddlers are motivated to perform but merely *social* ones, and someone in need simply provides toddlers with a good way to strike up an interaction. Surely toddlers are motivated to engage in all kinds of social interactions (Warneken, Gräfenhain, & Tomasello, 2012); however, several pieces of evidence suggest that early prosocial behaviors are more than this. For instance, Hepach, Vaish, and Tomasello (2012) showed 24-month-olds a person in need of instrumental help and then measured their pupil dilation, a measure of autonomic arousal. To explore whether the observed arousal was due to a desire that the needy individual receive help (a prosocial motivation) or a desire to provide help themselves (because they wish to engage in a social activity or anticipate being praised), Hepach and colleagues compared the subsequent decreases in arousal (pupil size) of toddlers who had either just observed someone else help the person or had just helped the person themselves. Consistent with the interpretation that toddlers' arousal was prosocially (as opposed to socially or extrinsically) motivated, the groups did not differ in rate of pupil attenuation. Similarly, in a study of the motivations underlying giving behaviors in toddlerhood, Aknin, Hamlin, and Dunn (2012) compared happiness in 20- to 23-month-olds (as rated by observers from videotape) when (1) engaging in social interactions that did or did not involve giving a toy back and forth and (2) giving away treats versus receiving treats themselves. In both cases toddlers were rated as happier after giving interactions than after nongiving interactions. In addition, toddlers were happier after giving away one of their own treats (real, costly giving) than after giving away one of the experimenter's treats (fake, noncostly giving); it is unclear from a brute socialization account why toddlers would distinguish costly from noncostly giving in this way. We have recently replicated these results with toddlers in our laboratory and with young children cross-culturally; all are consistent with the hypothesis that early prosociality is intrinsically motivated.

Empirical Evidence: Do Infants Show Predispositions toward Moral Judgments?

One of the hallmarks of moral judgments is that they are evaluations: identifying an act as immoral implies that one finds it negative in some way, whereas identifying an act as moral implies one finds it positive. Second, although we can evaluate our own actions and those others do to us using moral terms, quintessentially moral judgments apply to those actions that an evaluator is not involved in, which offer no immediate personal benefit or risk. With these two constraints in mind, the first studies relevant to the question of moral evaluation in infancy asked whether infants detect the difference between positive prosocial acts and negative antisocial acts between third parties. In the first of these Premack and Premack (1997) showed 12-month-old infants an animation in which a gray circle directed either a prosocial act (either helping to achieve a goal or affectionate touching) or an antisocial act (either preventing a goal or aggressive hitting) toward a black circle. Critically, although the circles engaged in self-propelled and goal-directed movements that are known to inspire agency attribution in both infants and adults (Heider & Simmel, 1944; Premack, 1990), they did not have faces or make any sounds, so there was nothing that might have clued infants in to the value of their behaviors other than the acts themselves. Infants were shown one of the possible four valenced acts ten times, at which point they were considered "habituated" to (bored with) the act. Critically, infants who were shown a prosocial act looked about equally to infants who were shown an antisocial act, suggesting that infants did not (for instance) find prosocial acts to be familiar or antisocial acts surprising. After habituation all infants were shown a new aggressive hitting action three times, and their attention was recorded. Results showed that infants' attention to hitting differed depending on the valence of the act they had seen in habituation. Infants who had been habituated to either helping or caressing dishabituated to hitting to a greater extent than did infants who had been habituated to hitting or hindering. That is, infants had habituated to the action's *valence*, and this influenced their attention to subsequent same- or oppositely valenced acts.

Kuhlmeier, Wynn, and Bloom (2003) advanced the Premacks' findings in several key ways. In their study 12-month-olds were shown instances of helping and hindering on a hill: a "climber" tried but failed to climb a hill and would be alternately pushed up the hill by a "helper" and pushed down the hill by a "hinderer." Critically, although once again having no eyes and making no sounds, each of the characters in the hill interaction had a distinct color and shape so that they could be identified across contexts, and each infant was habituated to both helping and hindering events. After habituating, infants saw two different test events. In the Consistent test event the climber approached the helper who had aided his goal; in the Inconsistent test event, the climber approached the hinderer who had prevented his goal. Infants reliably distinguished these test events, looking longer to the approach-helper events. In follow-up work in which the characters had eyes to facilitate agency detection, Kuhlmeier and colleagues found that 9-month-old infants looked longer to the inconsistent, approach-hinderer event (Kuhlmeier, Wynn, & Bloom, unpublished data).

These initial studies showed that by 9 to12 months of age, infants interpret the social valence of an event between unknown third parties, and this interpretation influences their attention to the third parties' subsequent interactions. The question remained, however, whether infants *themselves* evaluate the prosocial and antisocial actors: do infants care who does what to whom? It is this question that is most relevant to moral judgments. To address this question, Hamlin, Wynn, and Bloom (2007) recreated the hill scenario in 3-D, such that following habituation infants could choose between the helper and the hinderer. Strikingly, after both 6- and 10-month-olds watched puppet show versions of the hill scenarios, they robustly preferred (reached for) the helper over the hinderer. These results suggest they infants *do* care who does what to whom, at least enough for it to influence their social choices. In additional studies 6- and 10-month-olds preferred helpers over neutral characters but neutral characters over hinderers, suggesting that both positive and negative social evaluations are in place by age 6 months. Recent work using video displays and slight changes to the puppet shows (some infants saw the climber bouncing with joy after being helped, others did not; some infants saw the climber with unfixed eyes that (due to gravity) pointed *away* from his goal making it ambiguous) have replicated and extended these effects: infants only prefer helpers over hinderers when the climber's goal is clear, whether or not he bounces with joy (Hamlin, unpublished data). When the age group was lowered to 3 months, infants look longer (they are too young to reach) toward helpers than hinderers, and toward neutral characters than hinderers, but they do not yet distinguish helpers from neutral characters. These results suggest that an aversion to antisocial acts develops first (Hamlin, Wynn, & Bloom, 2010).

The same pattern of results is observed in studies where infants see different helping/hindering scenarios: from 3 months infants prefer a puppet who gives someone his ball back to a puppet who takes someone's ball away, and from 5 months they prefer a puppet who helps someone open a box to one that slams the box lid closed (Hamlin & Wynn, 2011). These studies also include control conditions, where the "helper" and "hinderer" perform exactly the same actions on an inanimate object unworthy of being helped and hindered; infants have not preferred helpers in any inanimate controls. To summarize data from studies using these three scenarios to date, 215 infants have been shown hill, box, and ball scenarios of various kinds in which there is a clear helper and hinderer, of which 179 chose the helper (binomial $p < .000000000001$); 130 infants have been shown control versions of these scenarios, of which only 56 chose the helper (binomial $p = .13$; χ^2 [df = 2] = 60.22, $p < .0001$ (Hamlin, Wynn, & Bloom, unpublished data). Infants' preference for helpers over hinderers, then, is present by just 3 months of age (we have not tested infants under 2.5 months due to methodological constraints), generalizes across distinct goal types (object retrievals, reaching a particular location), and is specific to the helpful and harmful aspects of the acts, rather than something perceptual about the displays (see Hamlin, unpublished data above). Finally, at no ages do infants show different attentional patterns to helper versus hinderer events, suggesting that helping and hindering are not differentially familiar or expected but rather are true evaluations: a critical aspect of moral judgments (but see Baillargeon et al., 2014, for a different interpretation).

Choice and preferential looking methodologies have been criticized as being unable to reveal anything more than simple preferences akin to liking ice cream better than cookies or rock star A better than rock star B (e.g., Killen & Rizzo, 2014). Although whole areas of moral philosophy (emotivism) would characterize moral evaluations in just this way, critics of a moral interpretation of this early work are correct that it is difficult to identify exactly what leads infants to choose one puppet over another. Although work with young infants is necessarily limited to simple methodologies, studies with toddlers have shown that 16-month-olds systematically avoid matching the food preferences of hinderers (Hamlin & Wynn, 2012), and 21-month-olds systematically give treats to helpers and take them away from hinderers. In addition, toddlers *avoid* taking treats from those whom *others* have hindered, suggesting they may feel reluctant to harm victims (Hamlin, Wynn, Bloom, & Mahajan, 2011). Basic social preferences may account for the helper and hinderer effects with toddlers, but the victim effect is harder to account for without appealing to morality (and is reminiscent of Vaish et al., 2009, in which toddlers showed concern for those who had been harmed but who were not showing distress).

In another study Dahl, Schuck, and Campos (2013) report that it is not until 27 months that infants preferentially help a prosocial over an antisocial *human*, suggesting that toddlers' "moral" behavior is at first specific to puppets (and therefore not actually moral). However, in this study the facial expressions/vocalizations of both the helpers and the hinderers were consistently positive. Because humans doing mean things while smiling mildly just does not look that mean (or looks sadistic), and smiling mildly after one's ball is taken away just does not make sense, the consistently positive expressions of the puppets may have confused things for toddler participants.

Returning to infant methodologies, there are several other ways in which infants' social evaluations may not be moral evaluations. Even without a rationalist definition of morality, there are many aspects of moral judgments not represented in the basic versions of helping and hindering studies described thus far, including notions of intentions and outcomes, the possibility of evaluating the same action as good, bad, or neither depending on context, and notions of deservingness. Clarifying just what infants' evaluations are like requires examining these more complex concepts.

A classic way to pit a moral versus a social interpretation of a particular action is via what has been referred to as the intention-outcome distinction (e.g., Piaget, 1932). Do infants prefer helpers because they have good intentions—at stake in moral questions—or because they caused a good outcome? In a recent study 8-month-olds were shown one puppet try but fail to help another puppet open a box (so the puppet did not get his toy, a bad outcome for the puppet) and a second try but fail to prevent the puppet from getting the box open (he did get his toy, a good outcome). When all the puppets were associated with the same good or bad outcome or when the valences of intentions and outcomes were pitted against each other, infants preferred those with helpful intentions over those with harmful intentions. In contrast, when all the puppets had the same helpful or harmful intention, infants did not prefer those who were associated with better outcomes (Hamlin, 2013b). In another study 10-month-olds only preferred helpers over hinderers when they could have known that what they were doing was helpful or not: if either the needy puppet's need was ambiguous or the helper and hinderer had not been present to see what it was, infants chose equally between the helper and hinderer despite all the actions and physical outcomes being identical (Hamlin, Ullman, Tenenbaum, Goodman, & Baker, 2013). In addition, we are currently in the process of replicating a study first reported by Le and Hamlin (2012) showing that 10-month-olds prefer those who help intentionally to those who help accidentally but prefer those who hinder accidentally versus those who hinder intentionally. Thus, just as mental states are critical to adults' and children's moral judgments (e.g., Baird & Moses, 2001; Cushman, 2008; Malle, 1999), they inform the social choices of 8- to 10-month-olds as well.

The tendency to discount accidental/ignorant acts and to focus on intentions is only part of the moral story: there are situations in which adults positively evaluate *intentionally harmful* acts, ones that are meant to prevent others' goals or hurt them in various ways. The paradigmatic case of this is punishment: for centuries public executions were treated as entertainment, and humans trust both punishers and punishing institutions (Barclay, 2006; Maurer, 1999). To test whether infants could ever positively evaluate an antisocial act, 5-, 8-, and 19-month-olds were shown two sets of puppet interactions (the box and ball shows described above). In the first an unknown needy puppet was helped and hindered in opening a box. In the second either the helper or the hinderer from the box show dropped his ball, which was either given back to him or taken away. Finally, infants chose between the giver and the taker. Both 8- and 19-month-olds showed a surprisingly robust flip in their choices depending on whether the giver and taker had acted on the former helper or the former hinderer: they preferred the puppet who gave to the helper, but they preferred the puppet who took from the hinderer (Hamlin et al., 2011). In this initial study 5-month-olds showed a consistent preference for the giver; however, in follow-up work in which they see more helper and hinderer box events during Phase 1 (perhaps boosting their memory for who did what) even 4.5-month-olds show prefer puppets who take from hinderers (Hamlin, 2014).

Additional studies extending this work suggest that this process may take the form of "the enemy of my enemy is my friend": infants also prefer those who harm a puppet who simply does not share their food preference (Hamlin, Mahajan, et al., 2013). Although this might be reason to exclude these findings from discussions of moral evaluations (and on some definitions of what counts as morality, it surely does; see Killen & Rizzo, 2014) it is unclear to what extent similar processes are not underlying most of *adults'* punitive sentiments. At the very least the work suggests that flexibly in interpreting prosocial and antisocial actions depending on context is in place remarkably early. Indeed, 3.5-month-olds (the youngest infants tested in Hamlin, 2014) who have no siblings at home to observe have probably *never even seen* an act of physical retaliation.

Finally, outside of the domain of helping and hindering, over the past 2 years there have been a number of empirical reports of toddlers showing sensitivity to fairness. For instance, Sloane, Baillargeon, and Premack (2012) reported that 20-month-olds look longer to unequal distributive acts than to equal ones, unless the recipients previously did unequal work. Schmidt and Sommerville (2011) demonstrated that those 15-month-olds who engage in prosocial giving (gave someone a nicer toy out of two) also looked longer to unequal distributions, suggesting there is a link between individual infants' expectations for fairness and prosocial behaviors. Finally, Geraci and Surian (2011) showed that, like their tendency to prefer helpers over hinderers, 16-month-olds prefer fair distributors to unfair distributors, providing evidence that the crucial evaluative component is present in toddlers' assessments of others' distributive acts.

Conclusions

Consistent with both philosophical theorizing in the fourth century B.C. and current interpretations of the evolutionary roots of moral action and judgment, human infants show signs of empathy, prosociality, and sociomoral evaluation at ages and in ways that suggest that experience, socialization, and cognitive construction are not alone in guiding moral development but, rather, aid in the development of initial moral predispositions. Indeed, an accurate understanding of human development requires *both* an account of the starting state and an accurate picture of what spurs change from there. Thus, rather than dismissing these findings as simply not moral, perhaps these findings can be used to guide theorizing about how such predispositions start children on the path toward optimal moral development and, perhaps most critically, help us to gain a better understanding of why and how moral development is sometimes suboptimal.

References

Aknin, L. B., Hamlin, J. K., & Dunn, E. W. (2012). Giving leads to happiness in young children. *PLoS ONE, 7*(6), e39211. doi:10.1371/journal.pone.0039211.

Baillargeon, R., Scott, R. M., He, Z., Sloane, S., Setoh, P., Jin, K., et al. (2014). Psychological and sociomoral reasoning in infancy. In P. Shaver & M. Mikulincer (Eds.-in-chief) & E. Borgida & J. Bargh (Vol. Eds.), *APA Handbook of Personality and Social Psychology:* Vol. 1. *Attitudes and Social Cognition.* Washington, DC: APA.

Baird, J. A., & Moses, L. J. (2001). Do preschoolers appreciate that identical actions may be motivated by different intentions? *Journal of Cognition and Development, 2*(4), 413–448.

Barclay, P. (2006). Reputational benefits for altruistic punishment. *Evolution and Human Behavior, 27*(5), 325–344.

Bloom, P. (2013). *Just babies: The origins of good and evil.* New York: Crown Publishers.

Boehm, C. (2012). *Moral origins: The evolution of virtue, altruism, and shame.* New York: Basic Books.

Brown, R. E. (1991). *Human universals.* New York: McGraw-Hill.

Chan, Wing-tsit (1963). *A source book in Chinese philosophy.* Princeton, NJ: Princeton University Press.

Cushman, F. (2008). Crime and punishment: Distinguishing the roles of causal and intentional analyses in moral judgment. *Cognition, 108*(2), 353–380.

Dahl, A., Schuck, R. K., & Campos, J. J. (2013). Do toddlers act on their social preferences? *Developmental Psychology, 49*(10), 1964–1970.

Decety, J. (2014). The neuroevolution of empathy and caring for others: Why it matters for morality. In J. Decety & Y. Christen (Eds.), *New frontiers in social neuroscience* (pp. 127–151). New York: Springer.

de Waal, F. B. (2008). Putting the altruism back into altruism: The evolution of empathy. *Annual Review of Psychology, 59*, 279–300.

Dondi, M., Simion, F., & Caltran, G. (1999). Can newborns discriminate between their own cry and the cry of another infant? *Developmental Psychology, 35*(2), 418–426.

Dunfield, K. A., Kuhlmeier, V. A., O'Connell, L., & Kelley, E. (2011). Examining the diversity of prosocial behavior: Helping, sharing, and comforting in infancy. *Infancy, 16*(3), 227–247.

Eisenberg, N., & Miller, P. A. (1987). The relation of empathy to prosocial and related behaviors. *Psychological Bulletin, 101*, 91–119.

Geraci, A., & Surian, L. (2011). The developmental roots of fairness: Infants' reactions to equal and unequal distributions of resources. *Developmental Science, 14*, 1012–1020.

Haidt, J. (2001). The emotional dog and its rational tail: A social intuitionist approach to moral judgment. *Psychological Review, 108*(4), 814–834.

Haidt, J. (2012). *The righteous mind: Why good people are divided by politics and religion.* New York: Vintage Books.

Hamlin, J. K. (2013a). Moral judgment and action in preverbal infants and toddlers: Evidence for an innate moral core. *Current Directions in Psychological Science, 22*, 186–193.

Hamlin, J. K. (2013b). Failed attempts to help and harm: Intention versus outcome in preverbal infants' social evaluations. *Cognition, 128*, 451–474.

Hamlin, J. K. (2014). Context-dependent social evaluation in 4.5-month-old human infants: The role of domain-general versus domain-specific processes in the development of social evaluation. *Frontiers in Psychology, 5*, 614.

Hamlin, J. K., Mahajan, N., Liberman, Z., & Wynn, K. (2013). Not like me = Bad infants prefer those who harm dissimilar others. *Psychological Science, 24*(4), 589–594.

Hamlin, J. K., Ullman, T., Tenenbaum, J., Goodman, N., & Baker, C. (2013). The mentalistic basis of core social cognition: Experiments in preverbal infants and a computational model. *Developmental Science, 16*(2), 209–226.

Hamlin, J. K., & Wynn, K. (2011). Young infants prefer prosocial to antisocial others. *Cognitive Development, 26*(1), 30–39.

Hamlin, J. K., & Wynn, K. (2012). Who knows what's good to eat? Infants fail to match the food preferences of antisocial others. *Cognitive Development, 27*(3), 227–239.

Hamlin, J. K., Wynn, K., & Bloom, P. (2007). Social evaluation by preverbal infants. *Nature, 450*(7169), 557–559.

Hamlin, J. K., Wynn, K., & Bloom, P. (2010). Three-month-olds show a negativity bias in their social evaluations. *Developmental Science, 13*(6), 923–929.

Hamlin, J. K., Wynn, K., Bloom, P., & Mahajan, N. (2011). How infants and toddlers react to antisocial others. *Proceedings of the National Academy of Sciences USA, 108*(50), 19931–19936.

Hare, R. (1999). *Without conscience: The disturbing world of the psychopaths among us.* New York: Guilford.

Hastings, P. D., Miller, J. G., Kahle, S., & Zahn-Waxler, C. (2013). The neurobiological bases of empathic concern for others. In M. Killen & J. Smetana (Eds.), *Handbook of moral development* (Vol. 2, pp. 818–864). New York: Psychology Press.

Hauser, M. D. (2006). *Moral minds: How nature designed our universal sense of right and wrong.* New York: Ecco.

Heider, F., & Simmel, M. (1944). An experimental study of apparent behavior. *American Journal of Psychology, 57*, 243–259.

Hepach, R., Vaish, A., & Tomasello, M. (2012). Young children are intrinsically motivated to see others helped. *Psychological Science, 23*(9), 967–972.

Hume, D. (1978). *A treatise of human nature.* London: John Noon. (Original work published 1739.)

Kagan, J. (1987). Introduction. In J. Kagan & S. Lamb (Eds.), *The emergence of morality in young children.* Chicago: University of Chicago Press.

Kant, I. (1964). *Groundwork of the metaphysic of morals.* New York: H.J. Paton, Harper Torchbooks. (Original work published 1785.)

Katz, L. D. (2000). *Evolutionary origins of morality: Cross-disciplinary perspectives.* Bowling Green, OH: Imprint Academic.

Killen, M., & Rizzo, M. T. (2014). Morality, intentionality, and intergroup attitudes. *Behavior, 151*, 337–359.

Killen, M., & Smetana, J. G. (2006). *Handbook of moral development.* Lawrence Erlbaum Associates.

Killen, M., & Smetana, J. G. (2013). *Handbook of moral development* (Vol. 2). New York: Psychology Press.

Knudsen, B., & Liszkowski, U. (2013). One-year-olds warn others about negative action outcomes. *Journal of Cognition and Development, 14*, 424–436.

Kohlberg, L. (1969). *Stage and sequence: The cognitive-developmental approach to socialization.* New York: Rand McNally.

Kuhlmeier, V., Wynn, K., & Bloom, P. (2003). Attribution of dispositional states by 12-month-olds. *Psychological Science, 14*(5), 402–408.

Le, D., & Hamlin, J. K. (2012). 10-month-olds' evaluations of accidental and intentional actions. Poster presented at the biannual meeting of the Society for Research in Child Development, Seattle, WA.

Lepper, M. R., Greene, D., & Nisbett, R. E. (1973). Undermining children's intrinsic interest with extrinsic reward: A test of the "overjustification" hypothesis. *Journal of Personality and Social Psychology, 28*, 129–137.

Liszkowski, U., Carpenter, M., & Tomasello, M. (2008). Twelve-month-olds communicate helpfully and appropriately for knowledgeable and ignorant partners. *Cognition, 108*(3), 732–739.

Malle, B. F. (1999). How people explain behavior: A new theoretical framework. *Personality and Social Psychology Review, 3*(1), 23–48.

Maurer, M. (1999). Why are tough on crime policies so popular? *Stanford Law & Policy Review, 11*, 9–22.

Nowak, M. A. (2006). Five rules for the evolution of cooperation. *Science, 314*, 1560–1563.

Nucci, L. P., & Turiel, E. (1978). Social interactions and the development of social concepts in preschool children. *Child Development, 49*(2), 400–407.

Piaget, J. (1932). *The moral judgment of the child.* London: Kegan Paul, Trench, Trubner and Co.

Plato. (380 B.C.). *Meno* (B. Jowett, Trans.) Available at http://classics.mit.edu/Plato/meno.html

Premack, D. (1990). The infant's theory of self-propelled objects. *Cognition, 36*(1), 1–16.

Premack, D., & Premack, A. J. (1997). Infants attribute value +/– to the goal- directed actions of self-propelled objects. *Journal of Cognitive Neuroscience, 9*(6), 848–856.

Rawls, J. (1971). *A theory of justice.* Cambridge, MA: Harvard University Press.

Rhee, S. H., Friedman, N. P., Boeldt, D. L., Corley, R. P., Hewitt, J. K., Knafo, A., et al. (2013). Early concern and disregard for others as predictors of antisocial behavior. *Journal of Child Psychology and Psychiatry, and Allied Disciplines, 54*, 157 166.

Rheingold, H. L. (1982). Little children's participation in the work of adults, a nascent prosocial behavior. *Child Development, 53*(1), 114–125.

Rheingold, H. L., Hay, D. F., & West, M. J. (1976). Sharing in the second year of life. *Child Development, 47*(4), 1148–1158.

Roth-Hanania, R., Davidov, M., & Zahn-Waxler, C. (2011). Empathy development from 8 to 16 months: Early signs of concern for others. *Infant Behavior and Development, 34*(3), 447–458.

Schmidt, M. F. H., & Sommerville, J. A. (2011). Fairness expectations and altruistic sharing in 15-month-old human infants. *PLoS ONE, 6*, e23223.

Simner, M. L. (1971). Newborns' response to the cry of another infant. *Developmental Psychology*, *5*, 136–150.

Sloane, S., Baillargeon, R., & Premack, D. (2012). Do infants have a sense of fairness? *Psychological Science*, *23*, 196–204.

Smith, A. (1759). *The theory of moral sentiments*. London: A. Millar.

Svetlova, M., Nichols, S. R., & Brownell, C. (2010). Toddlers' prosocial behavior: From instrumental to empathic to altruistic helping. *Child Development*, *81*(6), 1814–1827.

Tomasello, M. (2009). *Why we cooperate*. Cambridge, MA: MIT Press.

Vaish, A., Carpenter, M., & Tomasello, M. (2009). Sympathy through affective perspective-taking and its relation to prosocial behavior in toddlers. *Developmental Psychology*, *45*(2), 534–543.

Warneken, F. (2013). The origins of human cooperation from a developmental and comparative perspective. In G. Hatfield & H. Pittman (Eds.), *Evolution of mind, brain, and culture* (pp. 149–168). Philadelphia: Penn Museum Press.

Warneken, F., Gräfenhain, M., & Tomasello, M. (2012). Collaborative partner or social tool? New evidence for young children's understanding of joint intentions in collaborative activities. *Developmental Science*, *15*(1), 54–61.

Warneken, F., & Tomasello, M. (2006). Altruistic helping in human infants and young chimpanzees. *Science*, *311*(5765), 1301–1303.

Warneken, F., & Tomasello, M. (2008). Extrinsic rewards undermine altruistic tendencies in 20-month-olds. *Developmental Psychology*, *44*(6), 1785–1788.

Warneken, F., & Tomasello, M. (2013). Parental presence and encouragement do not influence helping in young children. *Infancy*, *18*(3), 345–368.

Wynn, K. (2008). Some innate foundations of social and moral cognition. In P. Carruthers, S. Laurence, & S. Stich (Eds.), *The innate mind:* Vol. 3. *Foundations and the future* (pp. 330–347). Oxford: Oxford University Press.

Zahn-Waxler, C., Radke-Yarrow, M., Wagner, E., & Chapman, M. (1992). Development of concern for others. *Developmental Psychology*, *28*, 126–136.

8 Mechanisms of Moral Development

Joshua Rottman and Liane Young

Moral evaluations constitute a fundamental aspect of human psychology. How does moral competence develop? For decades, this question has been addressed within cognitive-developmental frameworks (e.g., Killen & Smetana, in press; Kohlberg, 1971; Piaget, 1932), and the general answer has been that moral development is a *constructivist* process: children develop the ability to make increasingly sophisticated moral distinctions by actively reasoning about their social experiences. This proposal features prominently in "social domain theory" (Smetana, 1989, 2006; Turiel, 1983), which argues for coexisting domains of social understanding. Moral norms (characterized as governing actions with consequences for others' welfare) comprise one domain, and conventional norms (characterized as governing actions that affect social order) comprise another domain. A major claim of social domain theory is that young children construct different domains of social understanding at an early age by interacting with adults and peers and by attending to qualitatively distinct features of these social experiences. Does this account of moral development effectively explain the emergence of moral thought and the changes that occur throughout childhood?

We argue that, although social domain theory and other cognitive-developmentalist frameworks have provided powerful insights into moral development, an even more productive explanation of moral development comes from an analogy offered by Haidt (2012) in which the moral mind is likened to a set of "taste buds." These taste buds consist of receptors that respond innately to particular types of content, and there is a finite range over which these receptors can be adjusted by cultural factors. Thus, moral development may be understood by examining starting states and socially provided external input, rather than by focusing primarily on children's reasoning about their social experiences. This idea builds off a similar proposal by Mikhail (2007) and Hauser (2006), which applies linguistic theory to moral cognition, thus advocating the existence of a "universal moral grammar." Their principles-and-parameters theory also incorporates innate representations and environmentally induced diversification. However, we adopt the taste bud metaphor here because its proponents have emphasized a plurality of moral foundations (beyond harm and fairness) and an explicit focus on social communication as a mechanism of moral development.

In this chapter we present evidence from developmental psychology and other cognitive sciences to evaluate claims about the mechanisms of moral development. In particular, we first briefly review the constructivist process of moral development as presented in social domain theory. We then review evidence to assess the extent to which recent findings in moral psychology continue to support this model of moral acquisition and change. We find that, although social domain theory may be able to account for a subset of moral competence, other developmental mechanisms beyond constructivism are also crucial. In particular, we argue that moral development can be best explained by a theory focused on innate principles that are modified through social communication, which we refer to as the "taste bud theory."

The "Social Domain Theory" of Moral Development

Starting State
Social domain theorists assign little explanatory power to innate primitives and therefore pay minimal attention to the starting state. However, a few important biases are said to be present (Killen & Smetana, in press). First, babies are predisposed to be social; they are interested in other people and motivated to interact with them (see Killen & Rizzo, 2014). Second, beginning in the second year of life, perhaps with the acquisition of self-awareness, children begin to express empathic concern for others' pain and anguish and engage in reparative behaviors to alleviate this distress, regardless of whether they caused or merely observed the distress (Zahn-Waxler, Radke-Yarrow, Wagner, & Chapman, 1992). Third, children use their factual beliefs about the nature of reality to ascribe meanings to the social events they encounter (Turiel, Hildebrandt, & Wainryb, 1991; Turiel, Killen, & Helwig, 1987; Wainryb, 1991).

Constructivist Processes as the Motor of Moral Development

According to social domain theory, children's own agency is the catalyst of moral change, and the fundamental forces driving moral development are the interactional processes between children and their environments. Specifically, distinctions between different domains of social understanding (e.g., morals and conventions) are presumed to arise from children's reasoning about distinct kinds of social interactions. Therefore, ready-made features, both in the individual (i.e., innate intuitions) and in the environment (i.e., top-down socialization processes), are deemed insufficient for explaining the development of moral competence (Smetana, 2006; Turiel, 1983).

Social domain theory posits that children construct moral concepts from social interactions that involve violations of welfare or justice (Turiel, 1983). Certain actions (e.g., hitting) are characteristically linked to injury and pain, and these readily observed negative consequences lead children to assign moral status to such actions. In particular, children discover (either as observers or as victims themselves) that emotional distress commonly occurs as a result of harmful actions. Consistent with this account of how morals are constructed, research has shown that rowdy, extroverted children may grasp certain moral concepts precociously; active exuberance may relate to increased aggression and therefore greater direct experience with moral transgressions and their consequences (Smetana et al., 2012).

It is not the case that aggressive children have a consistent advantage in acquiring moral competence, however. Empathy and perspective taking also facilitate moral competence. Typically developing children spontaneously take the perspective of victims, allowing them to appreciate the pain that victims feel when they are harmed (Turiel, 1983). This empathic response also leads both children and adults to reprimand perpetrators by responding to the consequences of their actions (Nucci & Turiel, 1978). Therefore, children can construct conceptions of the moral domain both through their empathic reactions to harmful actions and the admonishments that follow such actions.

Not only can children use their own observations of suffering caused by harmful actions to construct an abstract understanding of morality, but they are also able to use their observations of suffering to construct specific moral concepts about unfamiliar actions. When preschoolers learn about a novel action that causes somebody to cry and are shown a picture of this crying victim, they form a belief that the action in question was immoral; however, a novel action that is prohibited but that does not result in distress tends not to be evaluated as immoral (Smetana, 1985). Similarly, when young children are told about and shown a pictorial depiction of a canonically innocuous action (petting) that causes pain in an unusual animal, they recognize the action as immoral. Conversely, if a canonically harmful action (hitting) causes pleasure in an unusual animal, the positive outcome outweighs the negative action for these children, and the action is judged morally permissible (Zelazo, Helwig, & Lau, 1996).

According to social domain theory, children construct not only moral concepts but also concepts of social convention (e.g., wearing black at a funeral). Children construct conventional concepts by participating in particular institutionalized systems, such as school and family life, and thereby witnessing the arbitrary regularities and expectations that facilitate orderly interactions through group consensus (Turiel, 1983, 2008). The features of these experiences are notably distinct from those in the moral domain. For example, emotional distress is uncommon in the aftermath of a conventional violation, and conventional transgressions primarily elicit responses from adult observers (rather than direct victims of the transgressions), who point out the importance of following rules and maintaining social order (Nucci & Turiel, 1978). Children can therefore recognize that conventions are created by rules and consensual agreement. Thus, conventions are distinguished from morals, which exist due to intrinsic features of actions and independent of the presence or absence of rules.

In sum, qualitatively different kinds of experiences are associated with moral and conventional transgressions. These qualitative differences are present even for toddlers, who are in the process of acquiring the capacity to distinguish between these domains (Smetana, 1989). According to social domain theorists, children as young as 3 years of age (Smetana & Braeges, 1990) actively extract featural regularities from their heterogeneous social interactions to construct a principled distinction between moral transgressions and other kinds of social transgressions (Smetana, 2006; Turiel, 1983).

Some Challenges to Social Domain Theory

Social domain theory has led to significant progress in understanding moral development (see Smetana, 2006). However, critiques have recently been mounted on several fronts. After reviewing some of these critiques, we turn to characterizing an alternate theory of moral development.

The Distinction between Moral and Conventional Domains Is Not Clear

Because social domain theory claims that the construction of different social domains is predicated on reasoning and universally widespread experiences, it follows that conceptions of morality and conventionality should be robust and invariant. However, this prediction has not been borne out. In particular, the moral/conventional distinction is less pronounced, and perhaps even absent, in some traditional or non-Western cultures. Shweder, Mahapatra, and Miller (1987) have found that there is no substantive domain distinction among religious Hindus in India. Instead individuals treat many norms as moral, even when they lack apparent consequences for the welfare of others (but see Turiel et al., 1987). Nisan (1987) has obtained convergent results with religious Muslims in Israel. Similarly, Brazilian children and adults view harmless but disgusting norm violations as possessing moral properties (Haidt, Koller, & Dias, 1993).

More radically, recent research has challenged the idea that the distinction between moral and conventional domains is psychologically meaningful even in Western cultures. Evidence suggests that the difference between these domains breaks down in cases that have not been chosen as exemplars of morals or conventions. For example, whipping derelict sailors is judged to have been more permissible in earlier historical eras, and spanking students is judged to be more permissible when a school principal says it is allowed, demonstrating that some harmful actions do not possess the typical "moral" properties of being wrong independent of contextual factors (e.g., historical time) or authority dictates (Kelly, Stich, Haley, Eng, & Fessler, 2007). Similarly, when harmful actions carry potential utilitarian benefits, the perceived wrongness of these actions becomes more dependent on the existence of rules (Piazza, Sousa, & Holbrook, 2013). Additionally, vegetarian children judge meat consumption to be wrong because of welfare considerations, but they do not believe their own "moral" judgments to be universally applicable (Hussar & Harris, 2010). Strikingly, even prototypical moral transgressions such as hurting another's feelings are not believed to be intrinsically wrong by 3- to 9-year-olds when outgroup members are the targets of the transgressions (Rhodes & Chalik, 2013).

Even social domain theorists have found that distinctions between moral and conventional domains are blurred. For example, this is the case with religious norms for Amish Mennonites and Orthodox Jews (Nucci & Turiel, 1993). A range of other "mixed-domain" actions, such as homosexuality and abortion, also elicit different patterns of judgments than prototypical moral or conventional transgressions, perhaps because they incorporate complex and disparate factual beliefs (Turiel et al., 1991). Finally, there is not a consistently sharp distinction even between prototypical moral and conventional norms; in one study, first- and second-grade children regarded wearing pajamas to school as wrong regardless of an authority figure's dictates, thus assigning "moral" weight to a canonically conventional violation (Tisak & Turiel, 1988).

It is nevertheless true that many studies conducted within the social domain framework have demonstrated a robust distinction between moral and conventional domains (Smetana, 2006). Indeed, these distinctions almost surely have some merit, and the growing divergent evidence may simply lead to friendly amendments to social domain theory. For example, it may be that children do not construct entirely discrete domains of social understanding but instead come to understand that social norms exist along different continuous dimensions, for example, generalizability and authority independence. Yet it is also the case that many other elements of moral competence cannot be easily encompassed within social domain theory, either in its current instantiation or in a modified form. For example, the constructivism posited by social domain theory cannot account for dissociations between judgments and justifications, moral concerns beyond harm and fairness, the innate moral concepts possessed by babies, or the role of intentions in moral judgments—phenomena we discuss below.

Moral Cognition Is Not Always Conscious, Effortful, and Reflective
A major paradigm shift in moral psychology has occurred in recent years; an ideal of rational judiciousness (Kohlberg, 1971; Turiel, 1983) has largely been overturned by findings that morality is often not grounded in consciously reasoned principles. People are sometimes unable to accurately justify their moral judgments, suggesting that certain moral evaluations result from gut intuitions (Haidt, 2001). Additionally, moral beliefs are often motivated by unconscious ideological biases (Uhlmann, Pizarro, Tannenbaum, & Ditto, 2009). These moral beliefs can in turn distort factual

beliefs (Alicke, 2000; Knobe, 2003; Leslie, Knobe, & Cohen, 2006). For instance, people who are persuaded to believe that capital punishment is wrong subsequently become more likely to believe that it does not effectively deter crime (Liu & Ditto, 2012). Further evidence that morality may not be entirely rooted in carefully reasoned assessments comes from studies demonstrating that irrelevant emotional primes can change moral judgments (e.g., Schnall, Haidt, Clore, & Jordan, 2008; Valdesolo & DeSteno, 2006; Wheatley & Haidt, 2005), and certain moral judgments recruit brain regions associated with emotional processing rather than controlled cognition (Decety, Michalska, & Kinzler, 2012; Greene & Haidt, 2002). Perhaps most strikingly, in a simple deception paradigm, people can be led to endorse reversals of their previously expressed moral attitudes, for example, to support government surveillance or illegal immigration one minute and then to argue against it the next (Hall, Johansson, & Strandberg, 2012). Notably, however, very little research of this nature has been conducted with children, and developmental evidence is needed to determine whether aspects of moral cognition are naturally emotional and automatic even in childhood rather than becoming affect-laden and automatized during development.

Personal Experience Cannot Explain Moral Judgments of Victimless Acts
As demonstrated by social domain theorists, children may learn harm norms by attending to negative consequences inherent in certain social interactions. However, folk concepts of morality encompass a multiplicity of concerns beyond harm and unfairness, including disloyalty and impurity (Graham, Haidt, & Nosek, 2009; Haidt, 2012; Shweder et al., 1987). For many of these victimless violations (e.g., religious or sexual taboos), no observable features indicate their moral wrongness (Edwards, 1987; Haidt, 2012; Shweder et al., 1987). Indeed, no obvious distress reactions can be used to infer that homosexuality or stem cell research is wrong. At least in these and other similar cases, children will not be able to autonomously construct moral concepts by attending to their own experiences. Even attending to internal emotional states will often be insufficient for robust moral acquisition, as the affect experienced in the wake of many victimless transgressions needs to be linked to these moral violations through sociocultural learning (Nichols, 2004; Rottman & Kelemen, 2012; Rozin, 1999).

The "Taste Bud Theory" of Moral Development

Given recent challenges to social domain theory, it is important to consider alternative accounts of moral development. The remainder of this chapter presents the fundamental features of our favored approach, which we call the "taste bud theory" (see Haidt, 2012). This theory proposes that, rather than needing to construct a moral sense through their own efforts, babies are born with certain prepared intuitions (the metaphorical taste receptors) that establish the boundaries for a mature moral sense. These intuitions are then modulated by cultural input, which adjusts the sensitivity of the receptors and the range of content to which they are responsive. This is therefore a form of nativism that allows for a large but finite degree of cultural variability, which can be provided by environmental input including social communication and which does not necessarily require reflective inferences about experienced social interactions (Haidt, 2012; Hauser, 2006; Mikhail, 2007).

There are several ways in which the "taste bud" analogy sheds light on moral development. In particular, taste receptors are biologically prepared to respond preferentially to particular stimulus categories. However, the sensitivity of these sets of receptors and their particular triggering stimuli differ based on cultural learning, and this process is especially flexible during early childhood. Additionally, the degree to which the range of diversity is constrained differs across receptors. For example, although children favorably respond to sugary and fatty foods with minimal experience or cultural input, children are likely to avoid bitter or irritating foods (such as coffee or chili) unless they receive cultural input to the opposite effect. Importantly, the constellation of positive and negative evaluations that a child develops in response to foods is not a product of logical reasoning or autodidactic constructivism but is rather an intuitive, emotionally laden set of responses derived from a complex interplay of innate biases and socially learned norms (Haidt, 2012).

Innate Primitives

As noted above, research in social psychology and neuroscience has increasingly demonstrated that many moral judgments are rooted in evolved, automatic, consciously inaccessible intuitions (see Greene, 2013; Haidt, 2012). These findings suggest that some fundamental aspects of moral cognition are not consciously constructed from experience. Instead, moral concepts have been found to emerge in the absence of relevant experiences and are even detectable in prelinguistic babies (see Baillargeon et al., 2014; Bloom, 2013; Hamlin, 2013b; Hamlin, chapter 7, this volume). Even as disinterested third parties, infants both expect and prefer others to act prosocially (e.g., Hamlin, Wynn, & Bloom, 2007; Kuhlmeier, Wynn, & Bloom, 2003; Premack & Premack, 1997). Relatedly, they also expect and prefer others to act fairly (e.g., Geraci & Surian, 2011; Schmidt & Sommerville, 2011; Sloane, Baillargeon, & Premack, 2012).

Other aspects of moral competence also show signatures of innateness in addition to arising surprisingly early in development. Research using the hypothetical "trolley problem," widely used for studying moral cognition (see Greene, 2013, for details), reveals innate moral principles. In two oft-compared variations of this moral dilemma, a train will imminently kill five people in its current path, but this outcome can be avoided either by pulling a lever that will divert the train onto a track where it will kill only one person (the "switch" case) or by pushing a heavy man off a footbridge that runs above the track, thus killing him and stopping the train (the "footbridge" case). Adults have repeatedly been found to make principled distinctions between the switch case and the footbridge case: the majority judgment is that it is permissible to flip the switch that harms one person but spares five others, but it is impermissible to push the man on the footbridge into harm's way to spare five others. People's stronger moral resistance to the footbridge case can be described by the "contact principle" (i.e., the notion that physical harm is worse than nonphysical harm) or the "doctrine of double effect" (i.e., the notion that it is permissible to cause harm as a side effect but not as a means to an end). However, adults are often unable to coherently articulate either principle when asked to defend their moral judgments (Cushman, Young, & Hauser, 2006; Hauser, Cushman, Young, Jin, & Mikhail, 2007). In age-appropriate versions of these two scenarios, 3- to 5-year-olds make the same distinctions as adults (Pellizzoni, Siegal, & Surian, 2010). Overall, these findings are suggestive of an underlying generative computational structure that is operative and intuitive from early in life (Dwyer, 2009; Mikhail, 2007).

Additionally, young children possess a domain-specific and potentially innate faculty for detecting violators of social prohibitions. In particular, when 3- to 4-year-olds are given the deontic rule that "all squeaky mice must stay in the house," and are presented with a toy house with squeaky and nonsqueaky toy mice, a majority of them understand that they must squeeze mice only outside the house to check whether the rule has been violated. A dramatic decrease in performance is found when children are instead given the nonsocial, nondeontic rule that "all squeaky mice are in the house." This pattern suggests that children's understanding of the deontic rule is guided not by a domain-general logical ability but by a capacity designed specifically to detect agents engaging in socially forbidden actions (Cummins, 1996). This advantage for detecting violations of prescriptive social rules over violations of descriptive nonsocial rules remains intact even when the rules are unusual and arbitrary, such as a requirement to put on a helmet before painting (Harris & Núñez, 1996). Although this competence has been studied only in relation to social conventions, detecting breaches of deontic rules is also a necessary component of moral cognition.

In general, the evidence reviewed in this section suggests that intuitions about "right" and "wrong" (or at least "good" and "bad") have evolved such that they emerge early in development and continue to play a substantial role in the moral judgments of adults. This pattern of early emergence may hold true especially for harm-based and perhaps also fairness-based moral beliefs. Moral intuitions about harm and fairness are characterized by typical properties of innate faculties, including rapid development despite a poverty of the stimulus (e.g., for the distinctions between trolley scenarios) and specialization for particular forms of input (e.g., for deontic rules or for actions causing distress). Therefore, the hypothesis that certain moral foundations are innate receives substantial support. This makes sense, as "value is not in the world" (Tooby, Cosmides, & Barrett, 2005), and thus it is difficult to determine how moral competence could be fully constructed from basic sensory and motivational primitives and content-independent, domain-general learning processes.

Sociocultural Learning as the Motor of Moral Change and Differentiation
According to the taste bud theory, innate intuitions form an initial foundation for moral psychology. On this theory certain types of norms will be more readily acquired than others—for example, those with affective resonance (Nichols, 2004) or those that relate to harm, fairness, loyalty, authority, or purity (Haidt, 2012). The taste bud theory suggests that the acquisition of cultural information through social learning is the paramount process through which mature moral competence becomes elaborated.

In some ways this mechanism of sociocultural change is consistent with social domain theory. Indeed, social domain theorists have acknowledged that children use adult testimony as an environmental input that they actively interpret and evaluate in their effort to construct moral competence (Smetana, 1989, 1999, 2006; Turiel et al., 1987; also see Grusec & Goodnow, 1994, for a similar perspective). As noted above, parental messages are qualitatively dissimilar in different interactional contexts, and these provide children with important information about whether transgressions are moral or not (Nucci & Turiel, 1978). Additionally, the quality of parental affect may influence children's motivational and attentional states (Smetana, 1999). However, although social domain theory acknowledges that parents and teachers can "facilitate" moral development as a component of the environment with which the child actively interacts, the locus of change is proposed to reside within the child, who heavily processes and reorganizes information from social communication. Therefore, although social domain theory and the taste bud theory intersect in assigning importance to social communication, the two theories differ in how much emphasis they place on children's interpretation, evaluation, and accommodation of the messages they receive. In particular, the taste bud theory allows for the possibility that children's moral beliefs are shaped through automatic and potentially unconscious processes (Haidt, 2001).

The taste bud theory suggests that communications from adults may influence moral development considerably more than social domain theorists acknowledge, such that socially transmitted information may even be the primary source of moral development (also see Edwards, 1987; Harris, 2012; Nichols, 2004; Shweder et al., 1987; Sripada & Stich, 2006). Support for this idea has often come from cross-cultural research. Recent research utilizing a series of dictator games has demonstrated that, although there is remarkable cross-cultural consistency in ideas about distributive justice in young children (likely due to shared innate principles), responses begin to diverge around 7 years of age to match the differences that can be observed in the adults of these cultures (House et al., 2013). Other research focusing on more local norms has found that cultural influences become pronounced even before middle childhood. For example, Shweder and colleagues (1987) uncovered substantial agreement between American 5-year-olds and adults, as well as between Indian 5-year-olds and adults, but very little agreement between matched age groups across these cultures. Both sets of evidence are consistent with the taste bud theory in suggesting that certain values have been emphasized or deemphasized through distinct patterns of cultural testimony.

Further evidence that adult testimony can drive moral acquisition comes in the form of studies showing that children do not need to experience or witness the negative consequences of an action in order to form a belief that it is wrong; merely hearing this information without seeing a bad outcome (as in Smetana, 1985, and Zelazo et al., 1996) can lead to the same effect (see Harris, 2012). Additionally, children's judgments about harm may not entirely result from inferences about their first-person experiences with distressed victims but may instead heavily depend on how their parents communicate to them in the aftermath of these situations of distress. In particular, mothers frequently use instances in which their children cause distress as moments for explicit teaching about morality. Crucially, mothers who explain the consequences of their child's harmful behaviors, using emotional language and absolute principles and rules, have children who are more likely to engage in reparative behaviors toward victims of their own actions (Zahn-Waxler, Radke-Yarrow, & King, 1979).

The taste bud theory proposes that both innate primitives and sociocultural learning will exert some degree of influence on all moral beliefs. However, the relative influence of these two factors is likely to vary, and a number of factors may lead social communication to become more or less effective in modulating moral intuitions. For instance, the impact of cultural input is likely to be moderated by the content of particular moral norms such that different kinds of moral beliefs exhibit distinct developmental trajectories (Jensen, 2008). As already described, some norms (e.g., those involving harm) may be more heavily prepared by innate principles, whereas others (e.g., those involving impurity) may be more heavily shaped by social communication. Additionally, certain routes of social transmission may be differentially effective. For example, some norms may be communicated most effectively through emotional narratives and exposure to moral exemplars (Haidt & Joseph, 2007), whereas other norms may be explicitly taught (Nichols, 2004).

Further research is needed to determine what kinds of cultural input are most influential and whether this changes across developmental time. Overall, however, current evidence suggests that sociocultural learning is responsible for important elements of moral competence and, in particular, for the incredible diversity of the moral domain. Therefore, adults not only have a facilitative role in moral development, but they are instead largely responsible for shaping children's initial predispositions and for specifying the detailed content of children's moral beliefs.

Additional Developmental Mechanisms: The Role of Intentions

According to the taste bud theory, conceptual change in the moral domain does not need to occur through the child's own rational efforts. Additionally, conceptual change is likely to be quantitative rather than qualitative, with certain principles being up-regulated or down-regulated over developmental time. Moral development may also involve an increasingly complex integration of various moral and nonmoral computations. This integration process has been demonstrated most fully for attributions of mental states (e.g., Cushman & Young, 2011; Decety et al., 2012; Young & Saxe, 2008).

Assessments of perpetrators' intentions and desires play a crucial role in mature evaluations of moral wrongness (see Young & Tsoi, 2013). Studies with adults have demonstrated that the right temporoparietal junction (RTPJ)—a brain region that supports thinking about agents' mental states (Saxe & Kanwisher, 2003)—is implicated in many moral judgments. This region is especially active for evaluations of attempted harm, which is judged to be immoral despite an absence of negative consequences (Young, Cushman, Hauser, & Saxe, 2007). Indeed, temporarily disrupting activity in this brain region causes people to judge attempted harms as more permissible (Young, Camprodon, Hauser, Pascual-Leone, & Saxe, 2010). The recruitment of mental state information is also necessary for absolving agents who cause harm accidentally, and there is an association between lenience toward accidental harm and enhanced activity in the RTPJ during moral judgment (Young & Saxe, 2009). Intentional and accidental harms are also distinguished by high-resolution spatial patterns of neural response within the RTPJ, as demonstrated by a recent study using multivoxel pattern analysis (Koster-Hale, Saxe, Dungan, & Young, 2013). Incredibly, this differentiation between intentional and accidental harms in the RTPJ occurs within less than one-tenth of a second after a stimulus is perceived (Decety & Cacioppo, 2012), demonstrating that adults automatically and immediately integrate information about intent into their harm-based moral judgments.

This integration of mental state information into moral judgments marks a major developmental milestone in children's moral competence.

As originally observed by Piaget (1932), young children focus more on outcomes than intentions when making moral judgments, whereas older children focus more on intentions. Although recent research has demonstrated even babies can take intentions into account in their moral evaluations (Hamlin, 2013a; Hamlin, chapter 7, this volume), it is still the case that a general shift toward explicitly integrating intentions into moral judgments occurs relatively late in development, by some accounts becoming reliable only around middle childhood (Cushman, Sheketoff, Wharton, & Carey, 2013; Helwig, Hildebrandt, & Turiel, 1995; Killen, Mulvey, Richardson, Jampol, & Woodward, 2011). Although children have acquired a robust understanding of intentions and beliefs by this point in development (see Wellman, Cross, & Watson, 2001), the integration of mental state information into explicit moral evaluations requires an additional step beyond the mere encoding of this information (Killen et al., 2011; Young & Saxe, 2008; Zelazo et al., 1996). This integration occurs more fully for some moral norms than for others; purity-based norms are more resistant to considerations of intentionality than harm-based norms (Chakroff, Dungan, & Young, 2013; Russell & Giner-Sorolla, 2011; Young & Saxe, 2011).

Conclusions

This chapter has examined the nature of the mechanisms that support moral development. A review of the research demonstrates that social domain theory can account for some of the evidence. However, moral cognition is incredibly rich and diverse, and some aspects of mature moral cognition are likely to have developed from innate intuitions in combination with sociocultural learning. We therefore argue that the taste bud theory accounts for more phenomena in moral development than social domain theory. There is surely room for the assimilation of these two theories, however. For example, research should examine whether logical inferences and peer interactions might account for some adjustments of innate principles and parameters.

Moral psychology is a quickly growing field, but additional attention needs to be paid to the process of moral psychological growth. In particular, more developmental studies must be conducted to test the veracity of the taste bud theory. Finally, as cognitive developmentalists have long realized, gaining insights into how morality emerges in children will provide scientists with a crucial key to discovering the architecture of the moral brain.

Acknowledgments

The authors were supported by NSF GRF DGE-1247312 to J.R. and by a John Templeton Foundation grant and an Alfred P. Sloan Foundation grant to L.Y. We are grateful to Lysa Adams, Peter Blake, Paul Harris, Deb Kelemen, Jon Lane, Sydney Levine, Samuel Ronfard, Judi Smetana, and Jen Cole Wright for their insightful comments.

References

Alicke, M. D. (2000). Culpable control and the psychology of blame. *Psychological Bulletin, 126*(4), 556–574.

Baillargeon, R., Scott, R. M., He, Z., Sloane, S., Setoh, P., Jin, K., et al. (2014). Psychological and sociomoral reasoning in infancy. In E. Borgida & J. Bargh (Eds.), *APA handbook of personality and social psychology, Vol. 1: Attitudes and social cognition* (pp. 79–150). Washington, DC: APA.

Bloom, P. (2013). *Just babies: The origins of good and evil.* New York: Crown.

Chakroff, A., Dungan, J., & Young, L. (2013). Harming ourselves and defiling others: What determines a moral domain? *PLoS ONE, 8*(9), e74434.

Cummins, D. D. (1996). Evidence of deontic reasoning in 3- and 4-year-old children. *Memory & Cognition, 24*(6), 823–829.

Cushman, F., Sheketoff, R., Wharton, S., & Carey, S. (2013). The development of intent-based moral judgment. *Cognition, 127*(1), 6–21.

Cushman, F., & Young, L. (2011). Patterns of moral judgment derive from nonmoral psychological representations. *Cognitive Science, 35*(6), 1052–1075.

Cushman, F., Young, L., & Hauser, M. (2006). The role of conscious reasoning and intuition in moral judgment: Testing three principles of harm. *Psychological Science, 17*(12), 1082–1089.

Decety, J., & Cacioppo, S. (2012). The speed of morality: A high-density electrical neuroimaging study. *Journal of Neurophysiology, 108*(11), 3068–3072.

Decety, J., Michalska, K. J., & Kinzler, K. D. (2012). The contribution of emotion and cognition to moral sensitivity: A neurodevelopmental study. *Cerebral Cortex, 22*(1), 209–220.

Dwyer, S. (2009). Moral dumbfounding and the linguistic analogy: Methodological implications for the study of moral judgment. *Mind & Language, 24*(3), 274–296.

Edwards, C. P. (1987). Culture and the construction of moral values: A comparative ethnography of moral encounters in two cultural settings. In J. Kagan & S. Lamb

(Eds.), *The emergence of morality in young children* (pp. 123–151). Chicago: University of Chicago Press.

Geraci, A., & Surian, L. (2011). The developmental roots of fairness: Infants' reactions to equal and unequal distributions of resources. *Developmental Science, 14*(5), 1012–1020.

Graham, J., Haidt, J., & Nosek, B. A. (2009). Liberals and conservatives rely on different sets of moral foundations. *Journal of Personality and Social Psychology, 96*(5), 1029–1046.

Greene, J. D. (2013). *Moral tribes: Emotion, reason, and the gap between us and them.* New York: Penguin.

Greene, J., & Haidt, J. (2002). How (and where) does moral judgment work? *Trends in Cognitive Sciences, 6*(12), 517–523.

Grusec, J. E., & Goodnow, J. J. (1994). Impact of parental discipline methods on the child's internalization of values: A reconceptualization of current points of view. *Developmental Psychology, 30*(1), 4–19.

Haidt, J. (2001). The emotional dog and its rational tail: A social intuitionist approach to moral judgment. *Psychological Review, 108*(4), 814–834.

Haidt, J. (2012). *The righteous mind: Why good people are divided by politics and religion.* New York: Pantheon.

Haidt, J., & Joseph, C. (2007). The moral mind: How five sets of innate intuitions guide the development of many culture-specific virtues, and perhaps even modules. In P. Carruthers, S. Laurence, & S. Stich (Eds.), *The innate mind, Vol. 3: Foundations and the future* (pp. 367–391). New York: Oxford University Press.

Haidt, J., Koller, S. H., & Dias, M. G. (1993). Affect, culture, and morality, or is it wrong to eat your dog? *Journal of Personality and Social Psychology, 65*(4), 613–628.

Hall, L., Johansson, P., & Strandberg, T. (2012). Lifting the veil of morality: Choice blindness and attitude reversals on a self-transforming survey. *PLoS ONE, 7*(9), e45457.

Hamlin, J. K. (2013a). Failed attempts to help and harm: Intention versus outcome in preverbal infants' social evaluations. *Cognition, 128*, 451–474.

Hamlin, J. K. (2013b). Moral judgment and action in preverbal infants and toddlers: Evidence for an innate moral core. *Current Directions in Psychological Science, 22*(3), 186–193.

Hamlin, J. K., Wynn, K., & Bloom, P. (2007). Social evaluation by preverbal infants. *Nature, 450*, 557–559.

Harris, P. L. (2012). *Trusting what you're told: How children learn from others*. Cambridge, MA: Harvard University Press.

Harris, P. L., & Núñez, M. (1996). Understanding of permission rules by preschool children. *Child Development, 67*(4), 1572–1591.

Hauser, M. D. (2006). *Moral minds: The nature of right and wrong*. New York: Harper.

Hauser, M., Cushman, F., Young, L., Jin, R. K., & Mikhail, J. (2007). A dissociation between moral judgments and justifications. *Mind & Language, 22*(1), 1–21.

Helwig, C. C., Hildebrandt, C., & Turiel, E. (1995). Children's judgments about psychological harm in social context. *Child Development, 66*(6), 1680–1693.

House, B. R., Silk, J. B., Henrich, J., Barrett, H. C., Scelza, B. A., Boyette, A. H., et al. (2013). Ontogeny of prosocial behavior across diverse societies. *Proceedings of the National Academy of Sciences USA, 110*(36), 14586–14591.

Hussar, K. M., & Harris, P. L. (2010). Children who choose not to eat meat: A study of early moral decision-making. *Social Development, 19*(3), 627–641.

Jensen, L. A. (2008). Through two lenses: A cultural–developmental approach to moral psychology. *Developmental Review, 28*(3), 289–315.

Kelly, D., Stich, S., Haley, K. J., Eng, S. J., & Fessler, D. M. T. (2007). Harm, affect, and the moral/conventional distinction. *Mind & Language, 22*(2), 117–131.

Killen, M., Mulvey, K. L., Richardson, C., Jampol, N., & Woodward, A. (2011). The accidental transgressor: Morally-relevant theory of mind. *Cognition, 119*(2), 197–215.

Killen, M., & Rizzo, M. T. (2014). Morality, intentionality and intergroup attitudes. *Behaviour, 151*, 337–359.

Killen, M., & Smetana, J. (in press). Morality: Origins and development. In M. Lamb & C. Garcia-Coll (Eds.), Handbook of child psychology (Vol. 3). *Social and emotional development* (7th ed.). New York: Wiley Blackwell.

Knobe, J. (2003). Intentional action in folk psychology: An experimental investigation. *Philosophical Psychology, 16*(2), 309–324.

Kohlberg, L. (1971). From is to ought: How to commit the naturalistic fallacy and get away with it in the study of moral development. In T. Mischel (Ed.), *Cognitive development and epistemology* (pp. 151–235). New York: Academic Press.

Koster-Hale, J., Saxe, R., Dungan, J., & Young, L. (2013). Decoding moral judgments from neural representations of intentions. *Proceedings of the National Academy of Sciences USA, 110*(14), 5648–5653.

Kuhlmeier, V., Wynn, K., & Bloom, P. (2003). Attributions of dispositional states by 12-month-olds. *Psychological Science, 14*(5), 402–408.

Leslie, A. M., Knobe, J., & Cohen, A. (2006). Acting intentionally and the side-effect effect: Theory of mind and moral judgment. *Psychological Science, 17*, 421–427.

Liu, B. S., & Ditto, P. H. (2012). What dilemma? Moral evaluation shapes factual belief. *Social Psychological and Personality Science, 4*(3), 316–323.

Mikhail, J. (2007). Universal moral grammar: Theory, evidence and the future. *Trends in Cognitive Sciences, 11*(4), 143–152.

Nichols, S. (2004). *Sentimental rules: On the natural foundations of moral judgment.* New York: Oxford University Press.

Nisan, M. (1987). Moral norms and social conventions: A cross-cultural comparison. *Developmental Psychology, 23*(5), 719–725.

Nucci, L. P., & Turiel, E. (1978). Social interactions and the development of social concepts in preschool children. *Child Development, 49*, 400–407.

Nucci, L., & Turiel, E. (1993). God's word, religious rules, and their relation to Christian and Jewish children's concepts of morality. *Child Development, 64*, 1475–1491.

Pellizzoni, S., Siegal, M., & Surian, L. (2010). The contact principle and utilitarian moral judgments in young children. *Developmental Science, 13*(2), 265–270.

Piaget, J. (1932). *The moral judgment of the child* (M. Gabain, Trans.). New York: Harcourt.

Piazza, J., Sousa, P., & Holbrook, C. (2013). Authority dependence and judgments of utilitarian harm. *Cognition, 128*(3), 261–270.

Premack, D., & Premack, A. J. (1997). Infants attribute value to the goal-directed actions of self-propelled objects. *Journal of Cognitive Neuroscience, 9*, 848–856.

Rhodes, M., & Chalik, L. (2013). Social categories as markers of intrinsic interpersonal obligations. *Psychological Science, 24*(6), 999–1006.

Rottman, J., & Kelemen, D. (2012). Aliens behaving badly: Children's acquisition of novel purity-based morals. *Cognition, 124*(3), 356–360.

Rozin, P. (1999). The process of moralization. *Psychological Science, 10*(3), 218–221.

Russell, P. S., & Giner-Sorolla, R. (2011). Moral anger, but not moral disgust, responds to intentionality. *Emotion, 11*(2), 233–240.

Saxe, R., & Kanwisher, N. (2003). People thinking about thinking people: The role of the temporo-parietal junction in "theory of mind." *NeuroImage, 19*(4), 1835–1842.

Schmidt, M. F., & Sommerville, J. A. (2011). Fairness expectations and altruistic sharing in 15-month-old human infants. *PLoS ONE, 6*(10), e23223.

Schnall, S., Haidt, J., Clore, G. L., & Jordan, A. H. (2008). Disgust as embodied moral judgment. *Personality and Social Psychology Bulletin, 34*(8), 1096–1109.

Shweder, R. A., Mahapatra, M., & Miller, J. G. (1987). Culture and moral development. In J. Kagan & S. Lamb (Eds.), *The emergence of morality in young children* (pp. 1–83). Chicago: University of Chicago Press.

Sloane, S., Baillargeon, R., & Premack, D. (2012). Do infants have a sense of fairness? *Psychological Science, 23*(2), 196–204.

Smetana, J. G. (1985). Preschool children's conceptions of transgressions: Effects of varying moral and conventional domain-related attributes. *Developmental Psychology, 21*(1), 18–29.

Smetana, J. G. (1989). Toddlers' social interactions in the context of moral and conventional transgressions in the home. *Developmental Psychology, 25*(4), 499–508.

Smetana, J. G. (1999). The role of parents in moral development: A social domain analysis. *Journal of Moral Education, 28*(3), 311–321.

Smetana, J. G. (2006). Social-cognitive domain theory: Consistencies and variations in children's moral and social judgments. In M. Killen & J. Smetana (Eds.), *Handbook of moral development* (pp. 119–153). Mahwah, NJ: Lawrence Erlbaum Associates.

Smetana, J. G., & Braeges, J. L. (1990). The development of toddlers' moral and conventional judgments. *Merrill-Palmer Quarterly, 36*(3), 329–346.

Smetana, J. G., Rote, W. M., Jambon, M., Tasopoulos-Chan, M., Villalobos, M., & Comer, J. (2012). Developmental changes and individual differences in young children's moral judgments. *Child Development, 83*(2), 683–696.

Sripada, C. S., & Stich, S. (2006). A framework for the psychology of norms. In P. Carruthers, S. Laurence, & S. Stich (Eds.), *The innate mind, Vol. 2: Culture and cognition* (pp. 280–301). New York: Oxford University Press.

Tisak, M. S., & Turiel, E. (1988). Variation in seriousness of transgressions and children's moral and conventional concepts. *Developmental Psychology, 24*(3), 352–357.

Tooby, J., Cosmides, L., & Barrett, H. C. (2005). Resolving the debate on innate ideas: Learnability constraints and the evolved interpenetration of motivational and conceptual functions. In P. Carruthers, S. Laurence, & S. Stich (Eds.), *The innate mind, Vol. 1: Structure and contents* (pp. 305–337). New York: Oxford University Press.

Turiel, E. (1983). *The development of social knowledge: Morality and convention.* Cambridge: Cambridge University Press.

Turiel, E. (2008). Thoughts about actions in social domains: Morality, social conventions, and social interactions. *Cognitive Development, 23*, 136–154.

Turiel, E., Hildebrandt, C., & Wainryb, C. (1991). Judging social issues: Difficulties, inconsistencies, and consistencies. *Monographs of the Society for Research in Child Development, 56*(2), 1–103.

Turiel, E., Killen, M., & Helwig, C. C. (1987). Morality: Its structure, functions, and vagaries. In J. Kagan & S. Lamb (Eds.), *The emergence of morality in young children* (pp. 155–243). Chicago: University of Chicago Press.

Uhlmann, E. L., Pizarro, D. A., Tannenbaum, D., & Ditto, P. H. (2009). The motivated use of moral principles. *Judgment and Decision Making, 4*(6), 476–491.

Valdesolo, P., & DeSteno, D. (2006). Manipulations of emotional context shape moral judgment. *Psychological Science, 17*(6), 476–477.

Wainryb, C. (1991). Understanding differences in moral judgments: The role of informational assumptions. *Child Development, 62*, 840–851.

Wellman, H. M., Cross, D., & Watson, J. (2001). Meta-analysis of theory-of-mind development: The truth about false belief. *Child Development, 72*(3), 655–684.

Wheatley, T., & Haidt, J. (2005). Hypnotic disgust makes moral judgments more severe. *Psychological Science, 16*(10), 780–784.

Young, L., Camprodon, J. A., Hauser, M., Pascual-Leone, A., & Saxe, R. (2010). Disruption of the right temporoparietal junction with transcranial magnetic stimulation reduces the role of beliefs in moral judgments. *Proceedings of the National Academy of Sciences USA, 107*(15), 6753–6758.

Young, L., Cushman, F., Hauser, M., & Saxe, R. (2007). The neural basis of the interaction between theory of mind and moral judgment. *Proceedings of the National Academy of Sciences USA, 104*(20), 8235–8240.

Young, L., & Saxe, R. (2008). The neural basis of belief encoding and integration in moral judgment. *NeuroImage, 40*(4), 1912–1920.

Young, L., & Saxe, R. (2009). Innocent intentions: A correlation between forgiveness for accidental harm and neural activity. *Neuropsychologia, 47*(10), 2065–2072.

Young, L., & Saxe, R. (2011). When ignorance is no excuse: Different roles for intent across moral domains. *Cognition, 120*(2), 202–214.

Young, L., & Tsoi, L. (2013). When mental states matter, when they don't, and what that means for morality. *Social and Personality Psychology Compass, 7*(8), 585–604.

Zahn-Waxler, C., Radke-Yarrow, M., & King, R. A. (1979). Child rearing and children's prosocial initiations toward victims of distress. *Child Development, 50*, 319–330.

Zahn-Waxler, C., Radke-Yarrow, M., Wagner, E., & Chapman, M. (1992). Development of concern for others. *Developmental Psychology, 28*(1), 126–136.

Zelazo, P. D., Helwig, C. C., & Lau, A. (1996). Intention, act, and outcome in behavioral prediction and moral judgment. *Child Development, 67*, 2478–2492.

9 The Neurocognitive Development of Moral Judgments: The Role of Executive Function

Ayelet Lahat

Over the past three decades a great deal of empirical work has been devoted to the study of children's moral development. These studies have examined children's understanding, conceptions, and reasoning about moral and social acts (Helwig & Turiel, 2011; Nucci, 1981; Smetana, 1981, 2006; Turiel, 1983, 2002). Most of this work has been conducted within *social domain theory*. According to this framework, adults, and even very young children, have different conceptions regarding violations of moral rules (prescriptions about fairness, others' welfare, and rights) and violations of social conventions (customs, traditions, and regulations that maintain social order). Thus, according to social domain theory, morality and social conventions are considered as two distinct domains of reasoning.

Recently, researchers in social cognitive neuroscience have begun to examine the neural correlates of moral judgments in adults (e.g., Greene, Nystrom, Engell, Darley, & Cohen, 2004; Greene, Sommerville, Nystrom, Darley, & Cohen, 2001; Luo et al., 2006; Moll et al., 2002). This chapter attempts to uncover the underlying mechanisms and processes that contribute to moral and social judgments. Only few studies have examined the neural correlates of the development of moral understanding in children and adolescents (e.g., Decety, Michalska, & Akitsuki, 2008; Decety, Michalska, & Kinzler, 2012; Eslinger et al., 2009; Lahat, Helwig, & Zelazo, 2013; Pujol et al., 2008). Research examining the neurocognitve development of moral and conventional judgments (Lahat, Helwig, & Zelazo, 2012; Lahat et al., 2013) has pointed to the role of executive function (EF), or the control over thought and action, in situations that require problem solving (Zelazo, Carlson, & Kesek, 2008).

This chapter provides an overview of recent work on the neurocognitive development of moral judgments. I argue that EF has a major contribution to moral development and provide evidence from developmental neuro-science to support this claim. First, I present a brief summary of empirical work based on social domain theory and follow this by a review of extant research examining the neural correlates of moral and social judgments in children and adolescents. Next, I present evidence suggesting that EF is linked to the neurocognitive development of moral judgments. To conclude I identify gaps that still exist in the literature and provide suggestions for future research.

Social Domain Theory

Different theoretical approaches in psychology have attempted to explain the acquisition of morality. These include behaviorism (Skinner, 1971), social learning (Aronfreed, 1968), and psychoanalytic approaches (Freud, 1930). An alternative approach to moral development stems from Jean Piaget's (1932) extensive study of children's moral judgments. This work was extended later by Lawrence Kohlberg (1981). According to Piaget's and Kohlberg's approach, moral development involves the construction of judgments about welfare, justice, and rights. Both Piaget and Kohlberg described a sequence for the development of moral judgments in which concepts of justice and rights are not constructed until late childhood or adolescence.

However, research from social domain theory has shown that children begin to develop moral judgments distinct from other domains of social judgments at a much earlier age (see Helwig & Turiel, 2011; Smetana, 2006). This line of inquiry has shown that children and adults do not reason in the same way about moral and social conventional acts (Nucci, 1981; Smetana, 2006; Turiel, 1983). Moral acts, such as hitting, lying, and stealing, are considered universal, unalterable, and independent from rules and authority. In contrast, social conventions, such as eating with one's fingers or wearing pajamas to school, can vary across different social systems, are contingent on societal rules, and can be altered by authority or social consensus (Nucci, 1981). For example, Nucci (1981) studied participants between seven and twenty years of age and found that at all ages children and adolescents thought that moral acts were wrong even if there was no social rule against these acts, whereas the vast majority of participants thought that conventional acts were acceptable if there was no rule prohibiting these behaviors.

In a different study (Turiel, 1983) the majority of children and adolescents (ages six to seventeen) thought that conventional game rules could be changed (86 percent said yes). However, they thought that a moral rule about stealing could not be changed (79 percent said no).

Children's distinction between moral and conventional acts is also evident from the justifications or reasons they provide for these judgments. Reasoning about moral acts is characterized by issues of harm, fairness, and rights, whereas reasoning about conventional acts is characterized by references to rules, customs, authority, and social organization (Helwig & Turiel, 2011; Nucci, 1981). When individuals make decisions about moral acts, they base these judgments on the intrinsic negative consequences or wrongness of the act, and social prohibitions are not relevant. In contrast, when making decisions regarding conventional violations, individuals consider social rules and evaluate the violation in light of these social prescriptions (Lahat & Zelazo, 2012). Social domain theory proposes that reasoning about each of these social acts constitutes a distinct organized system, or domain of social knowledge, that arises from children's experiences with different types of regularities in the social environment (Smetana, 2006). Accordingly, it is likely that the neurocognitive processing of judgments of moral and conventional violations is different.

Research has shown that the distinction between judgments of moral and conventional violations emerges by the preschool years (Smetana, 1981). In order to determine at what age children begin to distinguish moral and conventional rules, Smetana (1981) presented preschool children between the ages of three and five years with descriptions of conventional and moral violations. Children were asked to rate the seriousness of the violations. They were also asked whether the violations are acceptable in the absence of a rule and whether the violations are acceptable in different contexts. As expected, the moral violations were rated as more serious than the conventional violations. This finding was taken as evidence that the moral-conventional distinction emerges at an early age. In addition, the four- to five-year-olds treated the moral events as noncontingent on the presence of rules, and they also regarded them as generalizable across social contexts. The three- to four-year-olds also perceived the moral events as noncontingent on rules, but they did not regard them as generalizable. Therefore, it appears that for moral events, rule contingency may be an earlier-developing dimension than generalizabilty. In this study the results for the conventional events were not significant. Thus, it seems that children at this age have not yet formed stable understandings of the types of conventions examined in the study and that reasoning about moral violations may be an earlier-developing dimension than reasoning about conventional violations.

Developmental trends have also been found in children's justifications regarding moral and conventional judgments. For example, Davidson, Turiel, and Black (1983) presented six- to ten-year-old children with familiar and unfamiliar moral and conventional violations. The findings indicated that, for both familiar and unfamiliar violations, younger and older children were equally likely to explain that moral transgressions were wrong because they cause harm, but older children were more likely also to refer to conceptions of fairness in condemning these violations.

In sum, children's judgments about moral violations are distinct from their judgments about conventional violations (Nucci, 1981; Turiel, 1983). This distinction seems to develop as early as the preschool years (Smetana, 1981), and justifications regarding these two domains continue to develop into later childhood (Davidson et al., 1983). The conclusions of social domain theory have been drawn from research that used interview methods to examine individuals' judgments and reasoning about moral and conventional violations. If judgments of these violations do, in fact, correspond to two separate domains, it is likely that they may be processed differently at a neurocognitive level. In addition, the developmental trends described may be evident in neurocognitive processing as well. Examining the neurocognitive processing of these domains is important not only to better understand how children distinguish between these domains and how this distinction develops with age, but it may also have important implications for understanding moral (and immoral) behavior. The next section reviews studies examining the neural correlates of moral judgments in children and adolescents.

Neural Correlates of Moral Judgments in Children and Adolescents

Most of the work on the neural correlates of moral judgment has been conducted with adults. However, a few studies have been conducted with children and adolescents. For example, in a functional magnetic resonance imaging (fMRI) study, Pujol et al. (2008) scanned fourteen- to sixteen-year-old adolescents during judgments of moral dilemmas. The findings reveal increased focal activation in the posterior cingulate cortex during moral dilemmas as compared to a control condition, in which participants were asked to answer simple questions about facts presented in nonmoral scenarios. In a different study Eslinger et al. (2009) presented ten- to seventeen-year-old children with moral-straightforward, moral-ambiguous, and nonmoral scenarios. Results indicated at all ages a cluster of activation in the most rostral-medial (frontal polar) prefrontal region as well as the left lateral orbitofrontal, left temporoparietal junction, midline thalamus, and globus pallidus. Trials involving ambiguous-moral situations activated considerably more prefrontal and parietal regions than straight-forward-moral situations, suggesting the need for more neurocognitive resources in this condition (Eslinger et al., 2009).

Other developmental neuroimaging research examined the underlying circuits involved in the development of empathy, which refers to both sharing and understanding the emotional state of others in relation to one-self (Decety et al., 2008). In an fMRI study (Decety et al., 2008), typically developing children (ages seven to twelve years) were presented with short animated visual stimuli depicting painful and nonpainful situations, which involved pain caused either accidentally or intentionally. Following the fMRI scan, children rated how painful these situations appeared. The findings indicated that the perception of other people in pain was associated with increased activation in brain regions involved in the processing of first-hand experience of pain, including insula, somatosensory cortex, anterior midcingulate cortex, periaqueductal gray, and supplementary motor area. Additionally, when participants watched a person inflicting pain onto another, regions that are consistently engaged in representing social interaction and moral behavior (the temporoparietal junction, the paracingulate, orbital medial frontal cortices, amygdala) were additionally recruited and increased their connectivity with the frontoparietal attention network (Decety et al., 2008).

In a different study (Decety et al., 2012), participants between the ages of four and thirty-seven years viewed scenarios depicting intentional versus accidental actions that caused harm or damage to people and objects. The results indicated that although intentional harm was evaluated as equally wrong across all participants, ratings of deserved punishments and malevolent intent gradually became more differentiated with age. Furthermore, age-related increases in activation were detected in ventromedial prefrontal cortex in response to intentional harm to people as well as increased functional connectivity between this region and the amygdala (Decety et al., 2012).

Taken together, these studies show activation in posterior cingulate cortex, the temporoparietal junction, orbital and medial frontal cortices, as well as amygdala. Although this line of work has begun to examine brain regions involved in the development of moral judgments and empathy, it did not directly examine the underlying online cognitive processing that is involved in these judgments. Additionally, these studies did not examine the neurocognitive development of the distinction between judgments of morality and conventions. The next section presents evidence suggesting that the development of EF plays a role in children's and adolescent's moral and conventional judgments.

The Underlying Role of EF in Moral Development

Recent work on the development of moral and social understanding (Lahat et al., 2012, 2013; Beauchamp & Dooley, 2013) suggests that the development of these judgments relies on improvements in EF. According to Decety and Howard (2013), research in infancy and early childhood suggests that the development of EF is associated with increased moral understanding. For example, effortful control improves markedly during the preschool years (Kochanska, Murray, & Harlan, 2000), which is the same age that children begin to inhibit personal distress in favor of helping others (Decety & Howard, 2013). In preschool children, EF has been found to be highly correlated with an improved theory of mind, a skill often found necessary for mature moral understanding (Carlson, 2009).

The link between moral judgments and EF has been directly examined in a behavioral study with adolescents (Beauchamp & Dooley, 2013). Participants between the ages of thirteen and twenty completed a battery of EF tasks. These participants were also administered a moral reasoning computer task in which they were presented with age-appropriate moral dilemmas. Participants were asked to make a decision regarding the dilemma and also provide a justification for their choice. The results suggest that four key EFs were related to moral reasoning maturity: conceptual reasoning, cognitive flexibility, verbal fluency, and feedback utilization (Beauchamp & Dooley, 2013).

Neuroimaging work examining moral judgments in adults also suggests that EF is involved in judgments of moral dilemmas. For example, Greene, Nystrom, Engell, Darley, and Cohen (2004) found that judgments of difficult personal dilemmas, as compared to easy personal dilemmas, involved increased activity bilaterally in both the dorsolateral prefrontal cortex and anterior cingulate cortex. These findings have been taken as evidence that judgments of difficult dilemmas engage brain areas associated with the detection of cognitive conflict and cognitive control not only to reflect on the dilemma but also potentially to overcome any prepotent reactions to such dilemmas (Greene et al., 2004).

From the perspective of social domain theory, moral judgments are based on the intrinsic wrongness of the act, they are independent from rules and authority, and they do not require consideration of the context or social prohibitions against the act (Nucci, 1981; Turiel, 1983). Thus, prototypical moral violations are often uncomplicated and may be judged on the basis of the negative consequences of the act (Lahat & Zelazo, 2012; Richardson, Mulvey, & Killen, 2012). In contrast, however, conventional judgments require consideration of the circumstances and social prohibitions surrounding the violation, as well as the particular social context in which the violation occurs. It seems likely, therefore, that conventional judgments would require conscious reflection about the context in which the act occurs and the rules within this social context.

To illustrate, in judging a moral act such as hitting another person, societal prohibitions do not necessarily need to be considered; rather, the judgment may often be made in a relatively unreflective fashion, according to the intrinsic wrongness of the act. In contrast, judging whether it is acceptable or not to chew gum during class (conventional) more typically depends on whether a certain teacher and/or school has a rule about this. This type of conventional judgment requires more conscious reflection and consideration of the context in which the violation occurs.

Recent work (Lahat et al., 2012, 2013) examining behavioral and electrophysiological correlates of the development of moral and conventional judgments provides evidence for this suggestion. In order to assess the cognitive processes involved in the development of moral and conventional judgments, we (Lahat et al., 2012) developed a paradigm in which participants were asked to read scenarios presented on a computer screen that had one of three possible endings: (1) moral violations, (2) conventional violations, and (3) neutral acts. Children (aged ten years), adolescents (aged thirteen years), and undergraduates judged whether the act was acceptable or unacceptable in a condition in which social rules were assumed or in a condition in which they imagined the absence of rules (rule-removed condition). Results indicated that, at all ages, reaction times (RTs) were faster for moral than conventional violations when a rule was assumed. These findings suggest that judgments of conventional violations require increased cognitive processing relative to moral judgments. Furthermore, RTs in the rule-removed condition were slower than in the rule-assumed condition only for adults' moral judgments. In addition to this age difference adolescents made more normative judgments than children. These findings extend previous studies (e.g., Davidson et al., 1983; Nucci, 1981; Smetana, 1981) in social domain theory by showing different time courses of processing conventional versus moral violations and revealing age-related differences in the tendency to make these judgments (Lahat et al., 2012).

In order to better understand the processes underlying these judgments and their development, we conducted a second study (Lahat et al., 2013) in which we used the same paradigm with adolescents (ages twelve to fourteen years) and undergraduates while event-related potentials (ERPs) were recorded. We focused on the N2 component of event-related potential, which is a negativity in the ERP waveform usually observed between 200 and 400 ms poststimulus and is thought to be an index for the detection of cognitive conflict (Nieuwenhuis, Yeung, van den Wildenberg, & Ridderinkhof, 2003). Our findings indicated that adolescents had larger (i.e., more negative) N2 amplitudes than adults for moral and neutral, but not conventional, acts. N2 amplitudes were larger when a rule was removed than assumed for moral, but not conventional, violations. Taken together these findings suggest that the neurocognitive processing involved in moral and conventional judgments continues to develop between early adolescence and young adulthood, consistent with the development of prefrontal cortical networks. Furthermore, this set of studies points to differences in the recruitment of EF between moral and conventional judgments.

In sum, behavioral research with adolescents has shown links between EF and moral judgments (Beauchamp & Dooley, 2013). Neuroimaging work with adults (Greene et al., 2004) and ERP work with children and adolescents (Lahat et al., 2013) point to neural circuits associated with EF, specifically cognitive conflict. Finally, although studies from social domain theory have shown that even preschoolers can distinguish between moral and conventional violations (Smetana, 1981), ERP evidence suggests that the understating of these types of violations continues to develop between adolescence and young adulthood (Lahat et al., 2013).

Future Directions

Research on the neurocognitive development of moral and social judgments suggests that improvements in EF are linked to age-related changes in these judgments (Lahat et al., 2012, 2013). However, this work focuses on a particular type of EF—cognitive conflict. Given behavioral research on the link between EF and the development of moral reasoning (Beauchamp & Dooley, 2013), other EFs may be involved in moral decision making, and future research on the neural correlates involved in the development of moral judgments should examine EF abilities other than cognitive conflict. Additionally, the moral and conventional violations described in this line of work (Lahat et al., 2012, 2013) are straightforward prototypical situations. However, in reality these violations are much more complex, and future research should examine the link between EF and the neurocognitive development of multifaceted violations, which include a combination of both moral and conventional concerns.

Understanding the online processing of judgments of moral and conventional violations has important implications for behavior, especially for atypical populations. For example, research shows that psychopaths have often failed to make the moral-conventional distinction (Blair, 1995). Additionally, children with behavior problems have more difficulty with this distinction than children with fewer behavior problems (Blair, Monson, & Frederickson, 2001).

In a recent study (Lahat, Gummerum, Mackay, & Hanoch, in press), we examined the cognitive processes related to offenders' moral and conventional judgments in order to better understand their criminal behavior. Offenders, undergraduates, and control participants were administered the moral-conventional judgments computer task in which participants read scenarios and were asked to judge whether the act was acceptable or unacceptable when rules were either assumed or removed. Additionally, participants completed an EF task so we could examine the relation between EF and moral and social judgments. The findings indicated that controls and undergraduates had faster RTs and a higher percentage of normative judgments than offenders. Additionally, RTs of moral and conventional judgments in most conditions were related to EF among students but not controls or offenders. Thus, offenders, as compared to controls and students, may rely more on rule-oriented responding and may rely less on EF when making moral and social judgments (Lahat et al., in press). Future research might usefully examine the ERP correlates for these individuals, and such findings may have important implications for offenders' and psychopaths' deviant behavior. Additionally, future research might examine the underlying neourocognitve processing among antisocial children as well as children with conduct disorder.

Another population that may benefit from the implications of behavioral and ERP research on moral judgments is children with autism. These individuals have been found to be able to make the morality-convention distinction (Blair, 1996), but other research (Grant, Boucher, Riggs, & Grayson, 2005) shows that they have problems providing reasons for their judgments. These findings could suggest that children with autism may be processing these violations differently than controls, and this may be reflected in ERPs.

Conclusion

Children acquire an understanding of moral rules at an early age, and they are able to distinguish between moral and conventional violations (Helwig & Turiel, 2011; Nucci, 1981; Smetana, 1981, 2006; Turiel, 1983). This distinction seems to be established by the preschool years (Smetana, 1981), and justifications regarding these two domains continue to develop into later childhood (Davidson et al., 1983). Moral acts are based on the intrinsic negative consequences or wrongness of the act, and social prohibitions are not relevant. In contrast, conventional violations are judged according to social rules and prescriptions (Lahat & Zelazo, 2012). Based on this claim it is likely that moral and conventional judgments are processed differently at a neurocognitive level.

Most research on the neural correlates of moral judgment has been conducted with adults (e.g., Greene et al., 2001, 2004; Luo et al., 2006; Moll et al., 2002). A few studies have investigated the neural basis of moral judgments (Eslinger et al., 2009; Pujol et al., 2008) as well as empathy (Decety et al., 2008, 2012) among children and adolescents. These studies show activation in posterior cingulate cortex, the temporoparietal junction, orbital and medial frontal cortices, as well as amygdala.

Recently, behavioral and ERP studies have begun examining the role of EF in the development of moral judgments (Beauchamp & Dooley, 2013; Lahat et al., 2012, 2013). These studies show that EF is associated with the development of moral judgments. Importantly, this line of work extends social domain theory by suggesting that understating of moral and conventional violations continues to develop between adolescence and young adulthood (Lahat et al., 2013). Finally, research on the neurocognitive development of moral judgments may have important implications for special populations, such as children with antisocial behavior as well as children with autism.

References

Aronfreed, J. (1968). *Conduct and conscience: The socialization of internalized control over behavior*. New York: Academic Press.

Beauchamp, M. H., & Dooley, J. J. (2013). Exploring the cognitive and affective predictors of moral reasoning in adolescence using the So-Moral task. Paper presented at the biennial meeting of the Society for Research in Child Development, Seattle.

Blair, R. J. R. (1995). A cognitive developmental approach to morality: Investigating the psychopath. *Cognition, 57*(1), 1–29. doi:10.1016/0010-0277(95)00676-p.

Blair, R. J. R. (1996). Brief report: Morality in the autistic child. *Journal of Autism and Developmental Disorders, 26*(5), 571–579.

Blair, R. J. R., Monson, J., & Frederickson, N. (2001). Moral reasoning and conduct problems in children with emotional and behavioural difficulties. *Personality and Individual Differences, 31*(5), 799–811. doi:10.1016/s0191-8869(00)00181-1.

Carlson, S. M. (2009). Social origins of executive function development. *New Directions for Child and Adolescent Development, 123*, 87–98.

Davidson, P., Turiel, E., & Black, A. (1983). The effects of stimulus familiarity on the use of criteria and justifications in children's social reasoning. *British Journal of Developmental Psychology, 1*, 49–65.

Decety, J., & Howard, L. H. (2013). The role of affect in the neurodevelopment of morality. *Child Development Perspectives, 7*(1), 49–54.

Decety, J., Michalska, K. J., & Akitsuki, Y. (2008). Who caused the pain? An fMRI investigation of empathy and intentionality in children. *Neuropsychologia, 46*(11), 2607–2614.

Decety, J., Michalska, K. J., & Kinzler, K. D. (2012). The contribution of emotion and cognition to moral sensitivity: A neurodevelopmental study. *Cerebral Cortex*, *22*(1), 209–220.

Eslinger, P. J., Robinson-Long, M., Realmuto, J., Moll, J., de Oliveira-Souza, R., Tovar-Moll, F., et al. (2009). Developmental frontal lobe imaging in moral judgment: Arthur Benton's enduring influence 60 years later. *Journal of Clinical and Experimental Neuropsychology*, *31*(2), 158–169.

Freud, S. (1930). *Civilization and its discontents*. London: Hogarth Press.

Grant, C. M., Boucher, J., Riggs, K. J., & Grayson, A. (2005). Moral understanding in children with autism. *Autism*, *9*(3), 317–331.

Greene, J. D., Nystrom, L. E., Engell, A. D., Darley, J. M., & Cohen, J. D. (2004). The neural bases of cognitive conflict and control in moral judgment. *Neuron*, *44*(2), 389–400.

Greene, J. D., Sommerville, R. B., Nystrom, L. E., Darley, J. M., & Cohen, J. D. (2001). An fMRI investigation of emotional engagement in moral judgment. *Science*, *293*(5537), 2105–2108.

Helwig, C. C., & Turiel, E. (2011). Children's social and moral reasoning. In C. Hart & P. Smith (Eds.), *Wiley-Blackwell handbook of childhood social development* (Rev. 2nd ed., pp. 567–583). Hoboken, NJ: Wiley-Blackwell.

Kochanska, G., Murray, K. T., & Harlan, E. T. (2000). Effortful control in early childhood: Continuity and change, antecedents, and implications for social development. *Developmental Psychology*, *36*(2), 220–232.

Kohlberg, L. (1981). *Essays on moral development* (Vol. 1). *The philosophy of moral development*. San Francisco: Harper & Row.

Lahat, A., Gummerum, M., Mackay, L., & Hanoch, Y. (in press). Cognitive processing of moral and social judgments: A comparison of offenders, students, and control participants. *Quarterly Journal of Experimental Psychology*. doi:10.1080/17470218.2014.944918.

Lahat, A., Helwig, C. C., & Zelazo, P. D. (2012). Age-related changes in cognitive processing of moral and social conventional violations. *Cognitive Development*, *27*(2), 181–194.

Lahat, A., Helwig, C. C., & Zelazo, P. D. (2013). An ERP study of adolescents' and young adults' judgments of moral and social conventional violations. *Child Development*, *84*, 955–969.

Lahat, A., & Zelazo, P. D. (2012). Towards a process model of children's reasoning about social domains. *Human Development*, *55*, 26–29.

Luo, Q., Nakic, M., Wheatley, T., Richell, R., Martin, A., & Blair, R. J. R. (2006). The neural basis of implicit moral attitude: An IAT study using event-related fMRI. *NeuroImage, 30*(4), 1449–1457.

Moll, J., de Oliveira-Souza, R., Eslinger, P. J., Bramati, I. E., Mourão-Miranda, J., Andreiuolo, P. A., et al. (2002). The neural correlates of moral sensitivity: A functional magnetic resonance imaging investigation of basic and moral emotions. *Journal of Neuroscience, 22*(7), 2730–2736.

Nieuwenhuis, S., Yeung, N., van den Wildenberg, W., & Ridderinkhof, K. (2003). Electrophysiological correlates of anterior cingulate function in a go/no-go task: Effects of response conflict and trial type frequency. *Cognitive, Affective & Behavioral Neuroscience, 3*(1), 17–26. doi:10.3758/cabn.3.1.17.

Nucci, L. (1981). Conceptions of personal issues: A domain distinct from moral or societal concepts. *Child Development, 52*, 114–121.

Piaget, J. (1932). *The moral judgment of the child*. London: Routledge & Kegan Paul.

Pujol, J., Reixach, J., Harrison, B. J., Timoneda-Gallart, C., Vilanova, J. C., & Pérez-Alvarez, F. (2008). Posterior cingulate activation during moral dilemma in adolescents. *Human Brain Mapping, 29*(8), 910–921.

Richardson, C. B., Mulvey, K. L., & Killen, M. (2012). Extending social domain theory with a process-based account of moral judgments. *Human Development, 55*(1), 4–25.

Skinner, B. F. (1971). *Beyond freedom and dignity*. New York: Knopf.

Smetana, J. G. (1981). Preschool children's conceptions of moral and social rules. *Child Development, 52*, 1333–1336.

Smetana, J. G. (2006). Social-cognitive domain theory: Consistencies and variations in children's moral and social judgments. In M. Killen & J. G. Smetana (Eds.), *Handbook of moral development* (pp. 119–153). Mahwah, NJ: Lawrence Erlbaum Associates.

Turiel, E. (1983). *The development of social knowledge: Morality and convention*. Cambridge: Cambridge University Press.

Turiel, E. (2002). *The culture of morality: Social development, context, and conflict*. Cambridge: Cambridge University Press.

Zelazo, P. D., Carlson, S. M., & Kesek, A. (2008). Development of executive function in childhood. In C. A. Nelson & M. Luciana (Eds.), *Handbook of developmental cognitive neuroscience* (2nd ed., pp. 553–574). Cambridge, MA: MIT Press.

10   Girl Uninterrupted: The Neural Basis of Moral Development among Adolescent Females

Abigail A. Baird and Emma V. Roellke

The nature of our moral selves has intrigued and eluded scholars in religion, philosophy, and psychology for centuries. Advances in modern science have given us many reasons to think that we are increasingly close to uncovering the substrates of what makes a person "moral." Morality can be described as an intricate system of beliefs, values, and ideas that ultimately influences how an individual distinguishes between right and wrong and acts upon these judgments (Kalsoom, Behlol, Kayani, & Kaini, 2012). In fact, our evolved morality is thought to be one of the things that, along with complex language, sets us apart from other species. It is interesting that both language and morality appear to be uniquely human as they share critically important attributes. Moral and linguistic development are not hard-wired abilities but, rather, intricate capacities that are shaped by the context of our experience (Caravita, Gini, & Pozzoli, 2012; Kalsoom et al., 2012; Nelson & Buchholz, 2003). The plasticity observed in both moral and linguistic development is at once the most fascinating and mysterious feature of both abilities. Like language, morality has a developmental course (see Baird, 2007 for a review) that is shaped by an idiosyncratic interaction of nature and nurture; this interaction has made it nearly impossible to construct a single model for how humans come to be moral creatures. Ironically, those whose moral standards we often question the most, namely adolescents, may have the most to teach us about how we acquire our moral reasoning as adults. Although children as young as three years old have been shown to exhibit an understanding of moral versus immoral behaviors (Caravita et al., 2012), morality develops throughout the life span, with the most profound period of moral development occurring during adolescence (Walker, 1989). Adolescence is a particularly crucial period of development given the significant changes that occur during this time. Moral development during this period is aided by a complex combination of biopsychosocial factors, including neuronal maturation, changes in cognition, and a shift from a parent-centered to peer-centered social world.

People are never more certain about what is right or wrong as they are during adolescence (independent of the accuracy of these thoughts). Unprecedented gains in abstract thinking that follow the neural matura-tion of puberty enable adolescents to think about their own thoughts, as well as the thoughts of others, from a third-party (or non-egocentric) point of view for the first time in their lives. The ability to contemplate their own thinking, as well as the thoughts of those around them, fills most adolescents with a great sense of accomplishment, often misperceived as self-righteousness or egocentrism. What young adolescents have yet to real-ize, however, is that this ability is just the beginning of a learning process, driven by context, that will provide them with the skills needed to engage in adult levels of moral reasoning and, in most cases, moral behavior. The means by which an individual comes to integrate his/her own beliefs with those of the people and larger society around her/him is precisely what pre-pares an adolescent to enter the adult social world. This process is so critical to the survival of the human race that it may in fact be the best lens through which to understand how our brains contribute to moral reasoning.

The very nature of adolescence makes a compelling case for using it as a vantage point from which to study the brain bases of moral thought because it brings two critically important factors to the forefront of the discussion. The first of these resides in the functional purpose of adoles-cence—namely to differentiate the sexes from one another in order to even-tually enable procreation. Although sex differences have been studied a bit in the extant literature, individual differences also need to be considered in terms of gender—the social construct that accompanies the presentation of biological sex. Second, although few would argue that morality is contextu-ally bound, at no other point in one's life is the context more influential to brain development than during the transition into adult society. For most adolescents, context will consist largely of the community and culture in which they reside, mixed with a healthy dose of the contemporary culture within their peer group. The discussion that follows explores these ideas by first briefly reviewing the most recent literature on the brain bases of moral reasoning in adults and the sex differences that have been reported. Next we turn to the importance of sexual dimorphism during adolescence and how this manifests in unique aspects of female peer relationships. Finally, we attempt to integrate these ideas and suggest possible directions for future research on the neural substrates of moral thought.

The Moral Brain circa 2015

Although recent years have produced a number of exciting findings from the study of the neurophysiological underpinnings of adult moral reasoning, most investigations have referenced, albeit to different extents, the same "characters" (i.e., neuroanatomical regions). It is worth a short review before we think about how we might revise our conceptions of moral development. As previously discussed, moral behavior among most adults is the result of the coordination of a number of neural regions and networks. Each component of morality, from emotional recognition and empathy to the evaluation of outcomes and decision making, involves a number of brain regions, all of which work together in various combinations to eventually allow an individual to engage in moral behavior. Although a complete review of these regions is beyond our scope here, the most fundamental are worth mentioning. It is possible to divide the neural regions associated with moral reasoning in the adult brain into three groups on the basis of their function: emotional experience, mentalizing, and behavioral regulation.

The *emotional experience* group consists of the amygdala, the anterior insula, and the dorsal anterior cingulate cortex—all of which contribute to the experience and memory of emotion. The *amygdala* is particularly important in the process of evaluating potential rewards or punishments of a given situation. It is typically associated with the "fight or flight" response and is most reactive to visual and potentially threatening stimuli (Adolphs, 1999; Blair, 2007). The amygdala is a structure that is active, in some form, in nearly every study of human emotion. The experience of emotion relies most heavily on interoception. *Interoception* refers to the dynamic processing of afferent homeostatic sensory information and the ability for that information to reach conscious awareness. This also includes the creation of abstracted feeling states (cold, hunger, pain) from the diverse set of discrete sensations that arrive from multiple internal sensory systems.

The *anterior insula*, which is home to the primary cortical representation of the body's internal state, is consistently referred to as the critical hub (along with the anterior cingulate) for interoceptive (emotional) experience.

It seems that a major function of the anterior insula is to assemble diverse sensory information into coherent feeling states and to assess the salience of those states in service of executive control (Craig, 1996, 2004; Critchley, Wiens, Rotshtein, Ohman, & Dolan, 2004). Another critical aspect of interoception is the experience of physical pain. A number of studies have validated both the anterior insula (Ostrowsky et al., 2002) and the posterior portion of the anterior cingulate cortex (see Price, 2000, for a review) as being critical to the experience of physical pain.

What is most relevant about the function of these two regions is that human beings have overlaid prosocial emotions onto the primary sensory function of these regions, co-opting their function for the underpinnings of moral behavior. In the same way that the posterior portion of the anterior cingulate aids in the experience of physical pain, a series of elegant studies led by Eisenberger have demonstrated that this same tissue also processes social pain and exclusion (see Eisenberger, 2012, for a review). In addition to physical pain, guilt may be the most powerful emotion in terms of moral motivation. Studies have reliably shown that the anterior insula is highly active during the experience of interpersonal guilt, as a derivative of the more primitive interoceptive state of disgust (Phillips et al., 1997; Shin et al., 2000). Disgust is a universal human emotion tied to a set of nonverbal behaviors that clearly and rapidly convey a visceral, repulsive sensation. Together, the evidence described above suggests that the relatively complex emotion of guilt has, over the course of evolution, co-opted the neural hardware that enables enduring avoidance of noxious stimuli, following just a single experience. This primary response has been expanded in the social realm, where the very same social signals that reduce the probability of an individual ingesting toxic substances are also able to significantly reduce the chances of an individual violating important moral standards. The insula and anterior cingulate cortex produce aversive visceral responses associated with witnessing or engaging in immoral behaviors (Bechara, 2001; Krach et al., 2011; Vogt, Finch, & Olson, 1992), such as experiencing nausea accompanied by feelings of guilt or anxiety, which are likely to reduce future immoral behavior. Finally, insula-driven guilt has been associated with a desire to compensate others and engage in self-punishment (Berthoz, Grèzes, & Armony, 2006; Yu, Hu, Hu, & Zhou, 2013).

The *mentalizing group* is comprised of the posterior cingulate, precuneus, retrosplenial cortex, as well as the dorsolateral portion of the parietal cortex (which includes the supramarginal and angular gyri). Collectively, these regions support processes that require understanding the perspective of others and integrating it with one's own experience. These regions have also been shown to participate in the creation of the individual's socioemotional "narrative" (Greene & Haidt, 2002) through the integration of emotion, mental imagery, and contextually specific memory (Fletcher et al., 1995; Moll, Eslinger, & de Oliveira-Souza, 2001). More specifically, the posterior cingulate cortex (PCC) is consistently described as the central node in the default mode network (see Buckner et al., 2008, for a review as well as Fair et al., 2009, for a review of its developmental course). The PCC has been shown to play a prominent role in the processing of both pain and contextually relevant episodic memory (Maddock, Garrett, & Buonocore, 2001; Nielsen, Balslev, & Hansen, 2005). Additionally, previous results have suggested that the PCC inhibits the parietal cortices to avoid distractions and simultaneously activates the medial prefrontal cortex to redirect attention so the individual can internally generate mental strategies (Small et al., 2003). In sum, the PCC subserves a constellation of functions that relate, most fundamentally, to intrinsic experience and the flexible nature of thoughtful and self-preserving behavior (Pearson, Heilbronner, Barack, Hayden, & Platt, 2011).

The precuneus is known to contribute to processes of reflective self-awareness and autobiographical recall (Kjaer, Nowak, & Lou, 2002; Lundstrom, Ingvar, & Petersson, 2005). In close collaboration with the precuneus, the retrosplenial cortex also contributes to self-relevant aspects of memory. Specifically, it has been shown to support processes related to planning and hypothetical reasoning (Vann, Aggleton, & Maguire, 2009). Both regions contribute to moral reasoning by recognizing socially significant visual cues, aiding in the theory of mind process, and reflecting on complex conceptions of "humanness" (Allison, Puce, & McCarthy, 2000; Brothers & Ring, 1992; Frith, 2001). Together, the supramarginal and angular gyri are often referred to as the *inferior parietal cortex*, the *parietal operculum*, or the *temporoparietal junction*. This area has been shown to correspond with tasks that require an individual to make inferences about the mental states of others, especially when compared to physical qualities about people. It is among the most commonly observed areas of activity during tasks that require theory of mind strategies (Saxe & Kanwisher, 2003) and/or the ability to distinguish between one's own thoughts and the thoughts of others (Decety & Sommerville, 2003).

Finally, the medial prefrontal cortex makes up the *behavioral regulation group* of regions. The group includes the ventral anterior cingulate cortex, the dorsomedial and ventromedial prefrontal regions, and the hippocampus. Collectively, these regions contribute to the attentional, organizational, and regulatory aspects of emotional information as it pertains to the individual. Generally, the *medial prefrontal cortex* enables individuals to integrate their emotions with decision-making processes. Additionally, this region is critical to the development of conscious moral planning (Damasio, 1994; Reiman, 1997). The ventral portion of the ACC is most reliably engaged during conditions in which, under high arousal, it is important to direct attention to the processing of emotional information. In these instances, the ventral ACC seems to work closely with the ventromedial prefrontal cortex in terms of attention to potentially rewarding information (see Phillips, Drevets, Rauch, & Lane, 2003, for a review). The ventromedial prefrontal cortex is primarily responsible for evaluating rewards and punishments and for providing individuals with the ability to control and inhibit potentially disadvantageous behaviors (Blair, 2001; Damasio, 1994; Damasio, Grabowski, Frank, Galaburda, & Damasio, 1994; O'Doherty, Kringelbach, Rolls, Hornack, & Andrews, 2001). The dorsomedial prefrontal cortex has been shown to be critical for making highly adaptive, very rapid real-world social decisions (Cooper, Dunne, Furey, & O'Doherty, 2012). It has been consistently linked with cognitions about the "self," making distinctions between "self" and "other" (Mitchell, Banaji, & Macrae, 2005; Pfeifer, Lieberman, & Dapretto, 2007), and contributing to the manipulation of information related to the "self" (Ochsner et al., 2004).

Finally, our ability to experience emotion, organize and perform cognitive functions on acquired emotional information, and regulate our behavior using previous experience all critically rely on the functionality of the hippocampus. Among the innumerable functions the hippocampus subserves, it is critical for the consolidation of cohesive experiences in which all of the functions described above (and more) are meaningfully integrated and stored in long-term memory (see Murray & Kensinger, 2013, for a review).

Theories of Moral Development

There has always been a great deal of variance in how morality is defined, both theoretically and operationally. The variations in the most basic ideas of moral theory are critical to understanding why modeling moral development is such a herculean (or possibly Sisyphean) task. In the psychological literature, the work of Lawrence Kohlberg forms the backbone of how we understand moral development. Kohlberg's (1974) theories of moral development describe the progression from an obedience and punishment orientation, to the native hedonistic and instrumental orientation, to the good boy/girl orientation, and, finally, to the law and authority orientation. Although Kohlberg's theories set a framework for psychologists to gauge developmental milestones in relation to morality, other theorists (Baumrind, 1986; Gilligan, 1982) have suggested that his orientations contain a bias that favors traditionally "male" reasoning and, in doing so, fail to address decision-making processes and morality formation as typically experienced by females.

Carol Gilligan (1982) has offered a more feminist perspective of moral development. Kohlberg theorized that females rarely progress past the third (good girl/boy) stage of moral development, whereas males consistently exhibit morality development through the fourth (law and authority) stage (Muuss, 1988). Gilligan argues that females are just as morally developed as males but that they simply approach morality from a different perspective. According to her theoretical framework, females employ an interpersonal outlook that fosters interdependent relationships, emphasizes responsibility toward others, and focuses on sensitivity toward humanity. Males, on the other hand, employ a justice-oriented approach that focuses on upholding rules, engaging in logical thinking, and preserving autonomy (Kalsoom et al., 2012; Muuss, 1988; Silfver & Helkama, 2007; Walker, 1989). Gilligan is careful to explain that the two approaches are neither hierarchical nor mutually exclusive. That is, both approaches are equally valid, and individuals of both sexes tend to engage in a combination of interpersonal- and justice-oriented moral development. Thus, although women and men engage in morality formations that bring them to the same conclusions of "right" and "wrong," they often prefer different routes of processing, which urge them to rely more heavily on one orientation than the other.

Although Gilligan's morality theories have been widely referenced throughout the field of psychology, most researchers have been unsuccessful in their attempts to find empirical validation for her ideas. One possible explanation for this lack of evidence results from the fact that tests of moral reasoning have traditionally compared participants on the basis of biological sex, as opposed to gender. Although it is beyond the scope of the present discussion to explore this idea entirely, it is worth consideration.

"Sex" and "gender" are different despite the fact that there is most often a great deal of overlap between the two. Whereas sex refers to the chromosomes and biology that make one male or female, gender is more difficult to define. Gender is often influenced by both context and the individual's beliefs about her or his own identity. When a baby is born (with a small number of exceptions), the child is immediately identified as male or female based on its visible genitalia, but it will be a few years before the child's gender develops (see Bussey & Bandura, 1999, for a thorough review). If we are to responsibly seek out models of moral development, the effects of both neurobiological factors associated with sex and the sociocultural influences of gender must be prominently considered.

Young children do not possess secondary sex characteristics, such as facial hair or full breasts, physical signs meant to signal their biological sex to others. These traits do not emerge until adolescence simply because they are physical manifestations of the increases in sex hormones that accompany puberty. Evolutionarily, these differences have evolved because they enable people of reproductive age to recognize each other with greater ease and speed. It was once believed that the way in which these traits appeared determined how masculine or feminine a person was, but it is now understood that feelings and perceptions of gender are much more complex and nuanced ideas. Simply, as societies evolve, so do their gender norms. At present, males still show more "traditionally male," and females "traditionally female" behavior, but there are increasing numbers of individuals who show a mix of the two (Gilligan, 1982; Muuss, 1988). In terms of understanding individual differences in moral reasoning, adding gender (in addition to biological sex) might enable us to forge a more parsimonious model of moral development by integrating the work of both Kohlberg and Gilligan.

Although there are few consistently reported sex differences in moral reasoning, research has indicated a number of small, albeit important, distinctions on the basis of gender. For example, girls have been shown to exhibit an increased tendency toward guilt, and they often employ a more implicit and empathic processing route in order to reach these conclusions. Boys, on the other hand, exhibit guilt less frequently, and they usually rely more on cognition and reasoning to form these feelings (Silfver & Helkama, 2007). Additionally, girls tend to prefer evaluations of social desirability (i.e., being considered kind or well-liked), as opposed to boys who more readily strive for social status (i.e., being considered popular or socially influential) (Caravita et al., 2012).

The following sections focus on a female perspective. This point of view is presented not because moral dilemmas faced by boys are less important or complex; rather, we explore feminine experiences with morality because it is important to approach the topic from a rarely considered perspective. It will be equally important for future work to explore moral development from a male (sex), or masculine (gender), vantage point. For the purpose of this chapter, we refer only to the differences between self-identifying males and females, although future researchers may consider incorporating gender identifications that do not fit the typical binary model. Additionally, it should be noted that although "relational" people tend to be female, gender and relational personality are not direct correlates of one another and thus may produce variation in measured outcomes. Thus, "females" and "relational people" usually, although not universally, fall into Gilligan's model of interdependence/caring.

Female Puberty and Peer Relationships

The learning and organization of social behavior that takes place during adolescence occurs within a sensitive period, when the biology is uniquely attuned to socially relevant information that is able to be acquired at a particularly astounding rate (Nelson & Guyer, 2011). The remarkable developmental plasticity of the human brain enables adolescents to learn seemingly endless information about their unique and highly variable social contexts. As a result, it makes practical sense to think of the adolescent brain as a primarily social organ with the capacity to acquire knowledge and behavior that is essential for thriving in a highly complex social environment (Burnett, Thompson, Bird, & Blakemore, 2011; Sebastian, Viding, Williams, & Blakemore, 2010).

The myriad changes that occur in adolescence require substantial revision to the systems responsible for understanding one's experience. At puberty, females acquire the capacity to give birth and care for an infant; therefore, it is only logical that they would be predisposed to emotional, empathic, and sociocognitive processes that are unique. Taylor's (2006) "tend and befriend" model posits that women may be more likely to resort to forming interdependent relationships in a time of physical or moral crisis. Rooted in an evolutionary perspective, her theory explains females' tendencies to engage in relationship formation during stressful situations as indicative of what was, historically, a necessary survival mechanism for families faced with a threat. In other words, a woman's act of engaging in relationship formation serves as a remnant of a mother's responsibility to protect her offspring in the face of danger. It follows logically, then, that this strategy would emerge following brain changes that occur during puberty. Humans universally agree that hurting children is a deplorable act, but few respond with the ferocity and tenacity of the mother of a child who has been injured. This development begins early in adolescence wherein adolescents become more prosocial than younger children (Fabes & Kupanoff, 1999) and friendships become increasingly important (Berndt, 1982; Brown, 2004; Larson & Richards, 1991; Richards, Crowe, Larson, & Swarr, 1998).

A large body of behavioral evidence has underscored the importance of same-sex peer relationships, especially among girls, during this time (Ma & Huebner, 2008; Prinstein, Cheah, Borelli, Simon, & Aikins, 2005; Rudolph, 2002). In general, girls have a greater propensity than boys to form close, intimate, self-disclosing friendships (Claes, 1992; Ma & Huebner, 2008). From an evolutionary perspective, greater affiliation among women is advantageous because it ensures group survival. Similarly, from a middle school perspective, a "tend and befriend" pattern (Taylor et al., 2000) among adolescent girls would seem to be advantageous because tightly knit friend groups tend to outline and adhere to common norms of behavior (which is often the bedrock of moral reasoning). These strategies serve to reduce individuals' uncertainty about how to "survive" within the larger school environment (Zwolinski, 2008). The socialization style observed among girls likely reflects the nature of both the biological changes initiated by adolescence and the sociocognitive transformations that accompany this maturation. For females, exploring and learning about interpersonal difficulties and moral dilemmas involving friends (i.e., engaging the interdependent orientation) may be an important step in shaping subsequent ideals of morality and determining future actions. Although there are certainly difficult, even painful, lessons to be learned during adolescence, this may be somewhat adaptive. It is important to remember that social pain (like physical pain) can facilitate memory of unpleasant or immoral actions and, in doing so, decrease the likelihood of their reoccurrence.

Within the context of their intense interpersonal relationships, gender roles and societal values often create a paradox for developing girls. For example, they are expected, as women, to be caring and relational, and yet, at the same time, patriarchal, individualistic Western culture rewards individuals who are less inclined to place central importance on relational thinking and behavior (Kalsoom et al., 2012). Again, it is likely that these opposing forces contribute to the lack of coherence in models of how moral reasoning develops.

Female Brain Development and Moral Reasoning

Returning to the descriptions of the neural substrates related to moral reasoning in adults, it makes sense to examine the development of these networks during adolescence and to highlight the ways in which females may differ from males. It is important to recognize that, as differences in brain structure and/or function are reviewed, it is often impossible to know the precise relationship between the two. It is equally likely that observed sex differences in neural networks are the result of idiosyncrasies in behavioral strategies or that sex differences in neural networks are responsible for the variations in reported experience and/or behavioral strategies. The increasing presence of techniques that allow scientists to model developing networks of functional connectivity shows great promise with regard to understanding how the brain functionally develops. Although still in its methodological infancy, exciting work from Power and colleagues (see Power, Fair, Schlaggar, & Petersen, 2010) has reliably demonstrated that although some neural networks in the developing brain show increased distribution across regions, with the "beefing up" of long-range connections (e.g., the frontoparietal networks known to mature in late adolescence), there are other networks that become more locally coordinated (e.g., the improved coordination among functionally distinct prefrontal regions). Ultimately, the integration of behavioral data with structural, functional, and network data from human imaging holds great promise for improving our understanding of moral development.

Emotional Experience

Earlier in this review the brain regions described as being most closely related to the emotional experience component of moral development included the amygdala, anterior insula, and the dorsal anterior cingulate cortex. All three of these regions have demonstrated relevant structural and functional differences as a product of sex across development. The amygdala is of particular interest as it is known to be a structure critical for social and emotional learning. In terms of human development it has also been implicated in the understanding of emotional "reactivity" or basic temperament (see Kagan, Snidman, Kahn, & Towsley, 2007, for a review). This could easily predispose an individual to a certain propensity toward a "fight or flight" response that is shaped by the dense connections among the amygdala, insula, and anterior cingulate cortex that are known to emerge during adolescence. The fluid coordination of these regions is what enables adolescents to use interoceptive information to inform moral reasoning.

Evidence from previous developmental research suggests these changes would likely be seen in areas that integrate internal sensory information with higher cognitive processes. Early sensory areas are typically mature before the end of childhood, reaching adult levels of cortical thickness by roughly the age of eight (Gogtay et al., 2004; Shaw et al., 2008). In contrast, many higher-order cortical regions, such as the dorsolateral prefrontal cortex and anterior cingulate, are not fully mature until the mid-twenties (Bennett & Baird, 2006; Gogtay et al., 2004).

One interoceptive structure that possesses a similar protracted maturation is the anterior insula (Shaw et al., 2008). This subregion of the insula is considered by many to bridge the interoceptive sensory system with higher-order cognitive processes (Craig, 2002; Critchley et al., 2004). If the anterior insula is involved in interoceptive imagery, it is likely to show developmental differences in activity between adults and adolescents. This would represent a kind of functional disconnection between the construction of abstract interoceptive feeling states and higher-order executive control.

Among the many functions related to moral reasoning, the emotional perception of empathy relies heavily on the functions subsumed by the anterior insula. It has been shown that females show greater activity in the anterior insula (relative to males) while witnessing others being treated unfairly. It has also been shown that changes to this experimental paradigm, such as the recipient of inequity treating others unfairly or an unfair individual being subjected to physical pain, resulted in diminished activity in the anterior insula in male participants; however, insula activity remained heightened among females (Bernhardt & Singer, 2012). Given that the insula does not fully mature until relatively late in human development, it is reasonable to assume that the observed sex differences likely emerge as a result of complex social learning during adolescence. This idea is supported by the work of Decety and Michalska, who reported a positive correlation between age and activity in the anterior insula and a negative correlation between age and activity in the amygdala during a task in which participants observed other individuals in pain that has been inflicted by another (compared with pain inflicted on oneself). It is also noteworthy that pain inflicted by others was perceived as more painful by younger subjects than by adults; and further, individuals' ratings of pain correlated positively with amygdala activity. Given the rich connections between the amygdala and anterior insula, the authors posit a developmental transition from the more primitive, survival-based response of the amygdala to a more nuanced and integrated moral response (Decety & Michalska, 2010). Decety and colleagues have also demonstrated a developmental progression whereby greater emphasis is placed on intentionality with regard to interpersonal harm, which is the cornerstone of moral reasoning. The developmental transition from an immature amygdala-based affective response to a more mature approach to moral reasoning that relies on judgments of intent was reflected in a positive correlation between age and greater functional connectivity between the amygdala and ventromedial prefrontal cortex (Decety, Michalska, & Zinzler, 2011).

Mentalizing

In terms of moral development, *mentalizing* is understood as the ability to understand the social and emotional perspective of another, to hold that person's emotional perspective in mind while keeping it separate from your own. Generally speaking, previous work has shown that children who perform better on mentalizing tasks are more sophisticated in their moral judgments as adults. This is likely because mentalizing lays the foundation for incorporating more diverse perspectives in adulthood, resulting in more complex and nuanced moral reasoning (Lane, Wellman, Olson, LaBounty, & Kerr, 2010). A simpler version of this process, theory of mind, emerges in early childhood. Saxe and Kanwisher (2003) provide elegant evidence for how the temporoparietal region becomes increasingly specialized for social information. Prior to about nine years of age, this region is shown to be highly responsive to general social information about others. However, as individuals move closer to puberty, it becomes increasingly specialized in its responsiveness, becoming preferentially more active in response to the social and emotional states of others (Saxe, Whitfield-Gabrieli, Scholz, & Pelphrey, 2009). This transition is critical for moral reasoning based on the experience and beliefs of socially relevant others. It is reasonable to speculate that these maturational improvements are the result of improvements in connectivity between the temporoparietal region and frontal regions, as it is well established that areas engaged in higher-order information processing take the longest amount of time to fully mature (Giedd et al., 1999; Gogtay et al., 2004).

In terms of gender differences, Harenski and colleagues (2008) have reported that adult females who responded to pictures of unpleasant moral violations showed a strong modulatory interaction between activity in the posterior cingulate cortex and anterior insula. Additionally, it was revealed that this activity was proportional to the intensity of their ratings of the degree of moral violation. Unlike females, male participants showed a greater response in the temporoparietal region that tracked with their ratings of moral transgression. Importantly, the ratings between males and females did not differ significantly, which indicates that while engaging in moral reasoning, female participants may rely more heavily on neural structures that support both emotional experience and mentalizing, whereas males may rely solely on mentalizing brain regions. This distinction is consistent with models of moral reasoning that suggest females may approach moral dilemmas from a more care-based or empathic strategy relative to males.

Behavioral Regulation
In adults, it is thought that the ventral anterior cingulate cortex, the dorsomedial and ventromedial prefrontal regions, as well as the hippocampus all contribute to the organizational and regulatory aspects of emotion as it pertains to socially appropriate (morally thoughtful) behavior. This grouping of regions has demonstrated some interesting sex differences among adolescents. For example, because the orbitofrontal cortex (OFC) and ventrolateral prefrontal cortex (vPFC) structurally mature earlier in females, females seem to be able to learn new social rules and engage in social inhibition at an earlier age than males (Nelson & Guyer, 2011). Nelson and Guyer further speculate that the interaction of pubertal hormones and gendered behavior also supports the notion that young adolescent females are likely to

be (relative to age-matched males) more adept at learning new social rules (e.g., rules related to peer relationships) and to possess better inhibitory control, especially in the presence of socially salient circumstances (e.g., moral transgressions).

Estrogen is the primary organizer when it comes to female adolescence. The deluge of estrogen circulating in the body during female puberty also affects the brain. Increased amounts of estrogen in the brain have major effects on both cognition and emotion. There are two brain structures in particular that are significantly shaped by female adolescence: the hippo-campus and the prefrontal cortex. Studies have shown that both of these structures contain large numbers of estrogen receptors, and they vary in size and function as a result of estrogen's influence (Campbell, 2008; Giedd et al., 1996; Taylor, 2006). The increased maturation of these regions means that they are making new and more efficient connections with both local and distant areas of the brain (Power et al., 2010). The hippocampus is often referred to as the "seat of memory" in human beings, as it is known to be a key player in many aspects of human memory. Hu and colleagues (Hu, Prussner, Coupe, & Collins, 2013) found that controlling for pubertal change produced relative increases in hippocampal volume in females as a function of increased "puberty score" (while producing relative decreases in volume in males). Given the increasing complexity of adolescent girls' social lives, as well as the potential for child rearing, it makes sense to see a great deal of development in the hippocampus, a brain structure known for integrating and consolidating different aspects of memory to form cohesive personal narratives (Casebeer & Churchland, 2003).

Together, these findings underscore that although the outcomes of many tasks and behaviors may appear similar, there may be significant differences in the manner (or strategy) and concomitant neural regions that accom-pany similar performance. Differences in behavioral strategy should not be minimized, especially when reliably observed, as they likely are reflective of developmental commonalities among those being studied. Therefore, as quantifiable as some of these sex differences are, it is critical to appreciate that many of them are influenced by gender, a relationship that neurosci-ence is just beginning to explore. It is also the case that both sex and gender are only two of many "systems" along with culture, peers, temperament, and others that influence how adolescents develop (Mills, Lalonde, Clasen, Giedd, & Blakemore, 2012).

Final Thoughts

Cognitively, adolescents undergo a transformation in which they begin to engage in more logical, abstract, and idealistic thinking. Additionally, they develop the ability to reflect on past events and integrate them into the present. These changes in cognition allow individuals to approach moral dilemmas logically, apply moral codes of conduct in an abstract manner, consider ideal outcomes, and recall past outcomes in order to influence decision making in the present (see Blakemore, 2008 for a review). Changes in social environment during adolescence such as the shift from a parent-dominated to peer-dominated world must also be considered in the study of moral development, as individuals become more likely to learn through peer observations, punishments, and reinforcements than through experiences associated with parental figures (Caravita, Gini, & Pozzoli, 2012). The internalization of social norms—particularly gender norms—during adolescence can also lead to a deviation from dispositional traits and parental lessons and to a prioritization of external, as opposed to internal, values (Gilligan, 1982; Nelson & Buchholz, 2003). The work of the developing individual is to integrate her or his own inter- and intrapersonal characteristics with the vast number of cultural contexts in which they exist. Learning to balance and blend familial expectations, gender norms, peer demands, and broader cultural expectations is probably one of the many reasons that few adults would voice a desire to return to adolescence, and this is undoubtedly more true among the female population.

In terms of acquiring moral reasoning, adolescence and early adulthood represent an extended "practice time," during which adolescents begin to regulate their own frontal networks. Later, emerging adults become increasingly adept at self-regulation (by gradually wresting control away from external sources of behavioral regulation, namely parents and peers) and begin to more fluidly integrate their own cognitive and emotional processes and learn from the consequences of their actions.

As adolescents continue to mature, their frontal systems become increasingly coordinated and able to effectively regulate or communicate with more posterior regions in the brain. As young adults emerge from adolescence, their decision-making processes begin to approximate those observed among mature individuals. This highly functional and personally tailored process results from a great deal of experience in an infinite variety of contexts.

The neural maturation described above sets the stage for the integration of emotion and cognition, which is fundamental not only to adult decision making but also to the formation of adult-like social attachments. The adolescent has the capacity to discern future feelings and to make subtle distinctions regarding expressed emotion. Moreover, affective states become integrated with formal thought operations. The maturation that takes place during adolescence, namely the initial integration of visceral emotion (largely from the networks that support emotional and interoceptive experience) and social cognition, is essential for fully developed moral reasoning that functions intuitively and automatically (i.e., requiring minimal cognitive effort). This integration is the principal goal of adolescent development and comes about largely as a function of learning about both the self and the self in context. The integration of visceral emotion and social cognition is an elemental aspect of moral development (Hinson, Jameson, & Whitney, 2002). During adolescence, emotional experience as well as mentalizing and behavioral regulation are all translocated from the self to self-in-relationship, a domain in which the complex unfolding of visceral, mental, and behavioral states in both the self and other can be mutually recognized, integrated, and learned from. All of these critical processes are undoubtedly influenced by measurable individual differences in sensitivity to sex hormones and gender norms; and until we as scientists are able to start teasing apart these various factors, we are likely to continue to feel befuddled by the emergence of moral reasoning in adulthood.

References

Adolphs, R. (1999). Social cognition and the human brain. *Trends in Cognitive Sciences, 3,* 469 479.

Allison, T., Puce, A., & McCarthy, G. (2000). Social perception from visual cues: Role of the STS region. *Trends in Cognitive Sciences, 4,* 267–278.

Baird, A. A. (2007). Moral reasoning in adolescence: The integration of emotion and cognition. In W. Sinnott-Armstrong (Ed.), *Moral psychology* Vol. 3: *The neuroscience of morality: Emotion, disease, and development* (pp. 323–342). Cambridge, MA: MIT Press.

Baumrind, D. (1986). Sex differences in moral reasoning: Response to Walker's (1984) conclusion that there are none. *Child Development, 57,* 511–521.

Bechara, A. (2001). Neurobiology of decision-making: Risk and reward. *Seminars in Clinical Neuropsychiatry, 6*(3), 205–216.

Bennett, C. M., & Baird, A. A. (2006). Anatomical changes in the emerging adult brain: A voxel-based morphometry study. *Human Brain Mapping, 27*(9), 766–777.

Berndt, T. (1982). The features and effects of friendship in early adolescence. *Child Development, 53*(6), 1447–1460.

Bernhardt, B. C., & Singer, T. (2012). The neural basis of empathy. *Annual Review of Neuroscience, 35*, 1–23.

Berthoz, S., Grèzes, J., & Armony, J. L. (2006). Affective response to one's own moral violations. *NeuroImage, 31*(2), 945–950.

Blair, R. J. (2001). Neurocognitive models of aggression, the antisocial personality disorders, and psychopathy. *Journal of Neurology, Neurosurgery, and Psychiatry, 71*, 727–731.

Blair, R. J. R. (2007). The amygdala and ventromedial prefrontal cortex in morality and psychopathy. *Trends in Cognitive Sciences, 11*(9), 387–392.

Blakemore, S.-J. (2008). The social brain in adolescence [review]. *Nature Reviews. Neuroscience, 9*(4), 267–277.

Brothers, L., & Ring, B. (1992). A neuroethological framework for the representation of other minds. *Journal of Cognitive Neuroscience, 4*, 107–118.

Brown, B. (2004). Adolescents' relationships with peers. In R. Lerner & L. Steinberg (Eds.), *Handbook of adolescent psychology* (2nd ed., pp. 363–394). New York: Wiley.

Buckner, R. L., Andrews-Hanna, J. R., & Schacter, D. L. (2008). The brain's default network: Anatomy, function, and relevance to disease. *Annals of the New York Academy of Sciences, 1124*, 1–38.

Burnett, S., Thompson, S., Bird, G., & Blakemore, S.-J. (2011). Pubertal development of the understanding of social emotions: Implications for education. *Learning and Individual Differences, 21*(6), 681–689.

Bussey, K., & Bandura, A. (1999). Social cognitive theory of gender development and differentiation. *Psychological Review, 106*(4), 676–713.

Campbell, A. (2008). Attachment, aggression, and affiliation: The role of oxytocin in female social behavior. *Biological Psychology, 77*(1), 1-10.

Caravita, S. C. S., Gini, G., & Pozzoli, T. (2012). Main and moderated effects of moral cognition and status on bullying and defending. *Aggressive Behavior, 38*, 456–468.

Casebeer, W. D., & Churchland, P. S. (2003). The neural mechanisms of moral cognition: A multiple-aspect approach to moral judgment and decision-making. *Biology and Philosophy, 18*, 169–194.

Claes, M. E. (1992). Friendship and personal adjustment during adolescence. *Journal of Adolescence, 15*(1), 39–55.

Cooper, J. C., Dunne, S., Furey, T., & O'Doherty, J. P. (2012). Dorsomedial prefrontal cortex mediates rapid evaluations predicting the outcome of romantic interactions. *Journal of Neuroscience, 32*, 15647–15656.

Craig, A. D. (1996). An ascending general homeostatic afferent pathway originating in lamina I. *Progress in Brain Research, 107*, 225–242.

Craig, A. D. (2002). How do you feel? Interoception: The sense of the physiological condition of the body. *Nature Reviews. Neuroscience, 3*(8), 655–666.

Craig, A. D. (2004). Human feelings: Why are some more aware than others? *Trends in Cognitive Sciences, 8*(6), 239–241.

Critchley, H. D., Wiens, S., Rotshtein, P., Ohman, A., & Dolan, R. J. (2004). Neural systems supporting interoceptive awareness. *Nature Neuroscience, 7*(2), 189–195.

Damasio, A. R. (1994). *Descartes' error: Emotion, reason, and the human brain*. New York: Putnam.

Damasio, H., Grabowski, T., Frank, R., Galaburda, A. M., & Damasio, A. R. (1994). The return of Phineas Gage: Clues about the brain from the skull of a famous patient. *Science, 264*, 1102–1105.

Decety, J., & Michalska, K. J. (2010). Neurodevelopmental changes in the circuits underlying empathy and sympathy from childhood to adulthood. *Developmental Science, 13*, 886–899.

Decety, J., Michalska, K. J., & Zinzler, K. D. (2011). The contribution of emotion and cognition to moral sensitivity: A neurodevelopmental study. *Cerebral Cortex, 22*(1), 209–220.

Decety, J., & Sommerville, J. A. (2003). Shared representations between self and other: A social cognitive neuroscience view. *Trends in Cognitive Sciences, 7*(12), 527–533.

Eisenberger, N. I. (2012). Broken hearts and broken bones: A neural perspective on the similarities between social and physical pain. *Current Directions in Psychological Science, 21*(1), 42–47.

Fabes, C., & Kupanoff, L. (1999). Early adolescence and prosocial/moral behavior: The role of individual processes. *Journal of Early Adolescence, 19*(1), 5–16.

Fair, D. A., Cohen, A. L., Power, J. D., Dosenbach, N. U. F., Church, J. A., Miezin, F. M., et al. (2009). Functional brain networks develop from a "local to distributed" organization. *PLoS Computational Biology, 5*(5), e1000381.

Fletcher, P. C., Frith, C. D., Baker, S. C., Shallice, T., Frackowiak, R. S., & Dolan, R. J. (1995). The mind's eye—precuneus activation in memory-related imagery. *NeuroImage, 2*, 195–200.

Frith, U. (2001). Mind blindness and the brain in autism. *Neuron, 32*, 969–979.

Giedd, J. N., Vaituzis, A. C., Hamburger, S. D., Lange, N., Rajapakse, J. C., Kaysen, D., et al. (1996). Quantitative MRI of the temporal lobe, amygdala, and hippocampus in normal human development: Ages 4–18 years. *Journal of Comparative Neurology, 366*(2), 223–230.

Giedd, J. N., Blumenthal, J., Jeffries, N. O., Castellanos, F. X., Liu, H., Zijedenbos, A., et al. (1999). Brain development during childhood and adolescence: A longitudinal MRI study. *Nature Neuroscience, 2*(10), 861–863.

Gilligan, C. (1982). *In a different voice: Psychological theory and women's development.* Cambridge, MA: Harvard University Press.

Gogtay, N., Giedd, J. N., Lusk, L., Hayashi, K. M., Greenstein, D., Vaituzis, A. C., et al. (2004). Dynamic mapping of human cortical development during childhood through early adulthood. *Proceedings of the National Academy of Sciences USA, 101*(21), 8174–8179.

Greene, J., & Haidt, J. (2002). How (and where) does moral judgment work? *Trends in Cognitive Sciences, 6*, 517–523.

Harenski, C. L., Antonenko, O., Shane, M. S., & Kiehl, K. A. (2008). Gender differences in neural mechanisms underlying moral sensitivity. *Social Cognitive and Affective Neuroscience, 3*(4), 313–321.

Hinson, J. M., Jameson, T. L., & Whitney, P. (2002). Somatic markers, working memory, and decision making. *Cognitive, Affective & Behavioral Neuroscience, 2*(4), 341–353.

Hu, S., Pruessner, J. C., Coupe, P., & Collins, D. L. (2013). Volumetric analysis of medial temporal lobe structures in brain development from childhood to adolescence. *NeuroImage, 74*, 276–287.

Kagan J, Snidman N, Kahn V, & Towsley S. (2007) The preservation of two infant temperaments into adolescence. *Monographs of the Society for Research in Child Development, 72*(2):1–75, vii; discussion 76–91.

Kalsoom, F., Behlol, M. G., Kayani, M. M., & Kaini, A. (2012). The moral reasoning of boys and girls in light of Gilligan's theory. *International Education Studies, 5*(3), 15–23.

Kjaer, T. W., Nowak, M., & Lou, H. C. (2002). Reflective self-awareness and conscious states: PET evidence for a common midline parietofrontal core. *NeuroImage, 17*(2), 1080–1086.

Kohlberg, L. (1974). Education, moral development and faith. *Journal of Moral Education, 4*(1), 5–16.

Krach, S., Cohrs, J. C., Cruz de Echeverria Loebell, N., Kircher, T., Sommer, J., Jansen, A., et al. (2011). Your flaws are my pain: Linking empathy to vicarious embarrassment. *PLoS ONE, 6*(4), e18675.

Lane, J. D., Wellman, H. M., Olson, S. L., LaBounty, J., & Kerr, D. C. (2010). Theory of mind and emotion understanding predict moral development in early childhood. *British Journal of Developmental Psychology, 28*, 871–889.

Larson, R., & Richards, M. H. (1991). Daily companionship in late childhood and early adolescence: Changing developmental contexts. *Child Development, 62*(2), 284–300.

Lundstrom, B. N., Ingvar, M., & Petersson, K. M. (2005). The role of precuneus and left inferior frontal cortex during source memory episodic retrieval. *NeuroImage, 27*(4), 824–834.

Ma, C. Q., & Huebner, E. S. (2008). Attachment relationships and adolescents' life satisfaction: Some relationships matter more to girls than boys. *Psychology in the Schools, 45*, 177–190.

Maddock, R. J., Garrett, A. S., & Buonocore, M. H. (2001). Remembering familiar people: The posterior cingulate cortex and autobiographical memory retrieval. *Neuroscience, 104*(3), 667–676.

Mills, K. L., Lalonde, F., Clasen, L. S., Giedd, J. N., & Blakemore, S.-J. (2012). Developmental changes in the structure of the social brain in late childhood and adolescence. *Social Cognitive and Affective Neuroscience, 9*, 123–131.

Mitchell, J. P., Banaji, M. R., & Macrae, C. N. (2005). The link between social cognition and self-referential thought in the medial prefrontal cortex. *Journal of Cognitive Neuroscience, 17*, 1306–1315.

Moll, J., Eslinger, P. J., & de Oliveira-Souza, R. (2001). Frontopolar and anterior temporal cortex activation in a moral judgment task: Preliminary functional MRI results in normal subjects. *Arquivos de Neuro-Psiquiatria, 59*, 657–664.

Murray, B. D., & Kensinger, E. A. (2013). A review of the neural and behavioral consequences for unitizing emotional and neutral information. *Frontiers in Behavorial Neuroscience, 7*, 1–42.

Muuss, R. E. (1988). Carol Gilligan's theory of sex differences in the development of moral reasoning during adolescence. *Adolescence, 23*(89), 229–243.

Nelson, A. K., & Buchholz, S. (2003). Adolescent girls' perceptions of goodness and badness and the role of will in their behavioral decisions. *Adolescence, 38*(151), 421–440.

Nelson, E. E., & Guyer, A. E. (2011). The development of the ventral prefrontal cortex and social flexibility. *Developmental Cognitive Neuroscience, 1*, 233–245.

Nielsen, F. A., Balslev, D., & Hansen, L. K. (2005). Mining the posterior cingulate: Segregation between memory and pain components. *NeuroImage, 27*(3), 520–532.

Ochsner, K. N., Knierim, K., Ludlow, D. H., Hanelin, J., Ramachandran, T., Glover, G., et al. (2004). Reflecting upon feelings: An fMRI study of neural systems supporting the attribution of emotion to self and other. *Journal of Cognitive Neuroscience, 16,* 1746–1772.

O'Doherty, J., Kringelbach, M. L., Rolls, E. T., Hornak, J., & Andrews, C. (2001). Abstract reward and punishment representations in the human orbitofrontal cortex. *Nature Neuroscience, 4,* 95–102.

Ostrowsky, K., Magnin, M., Ryvlin, P., Isnard, J., Guenot, M., & Mauguiere, F. (2002). Representation of pain and somatic sensation in the human insula: A study of responses to direct electrical cortical stimulation. *Cerebral Cortex, 12*(4), 376–385.

Pearson, J. M., Heilbronner, S. R., Barack, D. L., Hayden, B. Y., & Platt, M. L. (2011). Posterior cingulate cortex: Adapting behavior to a changing world. *Trends in Cognitive Sciences, 15*(4), 143–151.

Pfeifer, J. H., Lieberman, M. D., & Dapretto, M. (2007). I know you are but what am I: Neural bases of self- and social knowledge retrieval in children and adults. *Journal of Cognitive Neuroscience, 19,* 1323–1337.

Phillips, M. L., Young, A. W., Senior, C., Brammer, M., Andrew, C., Calder, A. J., et al. (1997). A specific neural substrate for perceiving facial expressions of disgust. *Nature, 389*(6650), 495–498.

Phillips, M. L., Drevets, W. C., Rauch, S. L., & Lane, R. (2003). Neurobiology of emotion perception I: The neural basis of normal emotion perception. *Biological Psychiatry, 54,* 504–514.

Power, J. D., Fair, D. A., Schlaggar, B. L., & Petersen, S. E. (2010). The development of human functional brain networks. *Neuron, 67,* 735–748.

Price, D. D. (2000). Psychological and neural mechanisms of the affective dimension of pain. *Science, 288*(5472), 1769–1772.

Prinstein, M. J., Cheah, C. S. L., Borelli, J. L., Simon, V. A., & Aikins, J. W. (2005). Adolescent girls' interpersonal vulnerability to depressive symptoms: A longitudinal examination of reassurance seeking and peer relationships. *Journal of Abnormal Psychology, 114*(4), 676–688.

Reiman, E. M. (1997). The application of positron emission tomography to the study of normal and pathologic emotions. *Journal of Clinical Psychiatry, 58*(Suppl. 16), 4–12.

Richards, M. H., Crowe, P. A., Larson, R., & Swarr, A. (1998). Developmental patterns and gender differences in the experience of peer companionship during adolescence. *Child Development, 69*(1), 154–163.

Rudolph, K. D. (2002). Gender differences in emotional responses to interpersonal stress during adolescence. *Journal of Adolescent Health, 30*(4), 3–13.

Saxe, R., & Kanwisher, N. (2003). People thinking about thinking people. The role of the temporo-parietal junction in "theory of mind". *NeuroImage, 19*, 1835–1842.

Saxe, R. R., Whitfield-Gabrieli, S., Scholz, J., & Pelphrey, K. A. (2009). Brain regions for perceiving and reasoning about other people in school-aged children. *Child Development, 80*, 1197–1209.

Sebastian, C., Viding, E., Williams, K. D., & Blakemore, S.-J. (2010). Social brain development and the affective consequences of ostracism in adolescence. *Brain and Cognition, 72*, 134–145.

Shaw, P., Kabani, N. J., Lerch, J. P., Eckstrand, K., Lenroot, R., Gogtay, N., et al. (2008). Neurodevelopmental trajectories of the human cerebral cortex. *Journal of Neuroscience, 28*(14), 3586–3594.

Shin, L. M., Dougherty, D. D., Orr, S. P., Pitman, R. K., Lasko, M., MacKlin, M. L., et al. (2000). Activation of anterior paralimbic structures during guilt-related script-driven imagery. *Biological Psychiatry, 48*(1), 43–50.

Silfver, M., & Helkama, K. (2007). Empathy, guilt, and gender: A comparison of two measures of guilt. *Scandinavian Journal of Psychology, 48*, 239–246.

Small, D. M., Gitelman, D. R., Gregory, M. D., Nobre, A. C., Parrish, T. B., & Mesulam, M. M. (2003). The posterior cingulate and medial prefrontal cortex mediate the anticipatory allocation of spatial attention. *NeuroImage, 18*(3), 633–641.

Taylor, S. E. (2006). Tend and befriend: Biobehavioral bases of affiliation under stress. *Current Directions in Psychological Science, 15*(6), 273–277.

Taylor, S. E., Klein, L. C., Lewis, B. P., Gruenewald, T. L., Gurung, R. A., & Updegraff, J. A. (2000). Biobehavioral responses to stress in females: Tend-and-befriend, not fight-or-flight. *Psychological Review, 107*(3), 411–429.

Vann, S. D., Aggleton, J. P., & Maguire, E. A. (2009). What does the retrosplenial cortex do? *Nature Reviews. Neuroscience, 10*(11), 792–802.

Vogt, B. A., Finch, D. M., & Olson, C. R. (1992). Functional heterogeneity in cingulate cortex: The anterior executive and posterior evaluative regions. *Cerebral Cortex, 2*(6), 435–443.

Walker, L. J. (1989). A longitudinal study of moral reasoning. *Child Development, 60*(1), 157–166.

Yu, H., Hu, J., Hu, L., & Zhou, X. (2013). The voice of conscience: Neural bases of interpersonal guilt and compensation. *Social Cognitive and Affective Neuroscience.* (Aug 19. Epub ahead of print).

Zwolinski, J. (2008). Biopsychosocial responses to social rejection in targets of relational aggression. *Biological Psychology, 79*(2), 260–267.

IV   The Affective and Social Neuroscience of Morality

11   Neural Correlates of Human Morality: An Overview

Ricardo de Oliveira-Souza, Roland Zahn, and Jorge Moll

Morality as a Phenomenologically Distinctive Experience

The ability to know what is morally "right" and "wrong" and to act on this knowledge has developed in humans to a degree of sophistication that has not been demonstrated in any other species (Nowak & Sigmund, 2005). Human moral abilities may have resulted from evolutionary pressures acting not only on the survival of the individual but also on small groups for whom altruistic members may have conferred a competitive advantage to the whole community (Gintis, Henrich, Bowles, Boyd, & Fehr, 2008). From a phenomenological perspective, moral conduct is the outward expression of a qualitatively discrete state of mind figuratively described by philosophers and theologians as "moral sense" and "moral compass" (Marazziti et al., 2013). The moral sense acts as a device that continuously screens our environment and thoughts for entities and events with moral salience. Deviations from rightness are detected as in a compass that always points northward, providing the individual with an intuition on how much he has drifted from the right path. In this chapter we use the expressions "moral stance" or simply "morality" as short labels for the experience of the moral sense and that compass (Pfaff, 2007). Lewis (1952) put down one of the clearest modern accounts of morality. Departing from the physical laws, which cannot be broken by man, to the Law of Nature, which is intrinsic to human nature, he states:

Now this Law or Rule about Right and Wrong used to be called the Law of Nature. Nowadays, when we talk of "the laws of nature" we usually mean things like gravitation ... or the laws of chemistry. But when the older thinkers called the Law of Right and Wrong "the Law of Nature," they really meant the Law of Human Nature. The idea was that, just as all bodies are governed by the law of gravitation, and organisms by biological laws, so the creature called man also had his law—with this great difference, that a body could not choose whether it obeyed the law of gravitation or not, but a man could choose either to obey the Law of Human Nature or to disobey it. (p. 4)

These, then, are the two points I wanted to make. First, that human beings, all over the earth, have this curious idea that they ought to behave in a certain way, and cannot really get rid of it. Secondly, that they do not in fact behave in that way. They know the Law of Nature; they break it. These two facts are the foundation of all clear thinking about ourselves and the universe we live in. (p. 8)

Neuroscientific efforts over the past three decades have shown that these phenomenological constructs can experimentally be reduced to subordinate mental and behavioral entities with relatively independent neural correlates. The puzzle remains, however, that none of these subordinate neuropsychological constructs have so far been shown to be exclusive to morality or, conversely, that the moral stance is exclusively dependent on any one of them to take place. For example, although empathy is certainly a core ingredient of morality (Decety & Jackson, 2004), the experience of empathy need not be linked to moral experience (Lenhoff, Wang, Greenberg, & Bellugi, 1997), and, conversely, the experience of morality may occur independently from empathy (Haidt & Joseph, 2007). Therefore, the neuroscientific investigation of morality will be completed only when an account in neural terms is provided for the experience of the moral sense and compass as they appear to the conscious self, namely, as immediate unitary objects of consciousness. At least two broad alternatives are equally plausible in this regard. Either (1) the experience of morality is the product of an as yet unidentified dedicated neural system [the "moral center" of classical authors (Maudsley, 1876)], or (2) such a discrete "moral system" does not actually exist, and moral experience is an emergent property of the synchronized activity of subordinate neuropsychological components (Damasio, 1989; Moll et al., 2005), many of which have been dissected out by modern research, as indicated in figure 11.1.

Morality as a Product of the Social Brain

The social brain concept has been formulated by neurologists (Gazzaniga, 1985) and neuroscientists (Brothers, 1991) to describe an ensemble of brain structures and pathways primarily engaged in the organization of the social life and behavior of social species (figure 11.2).

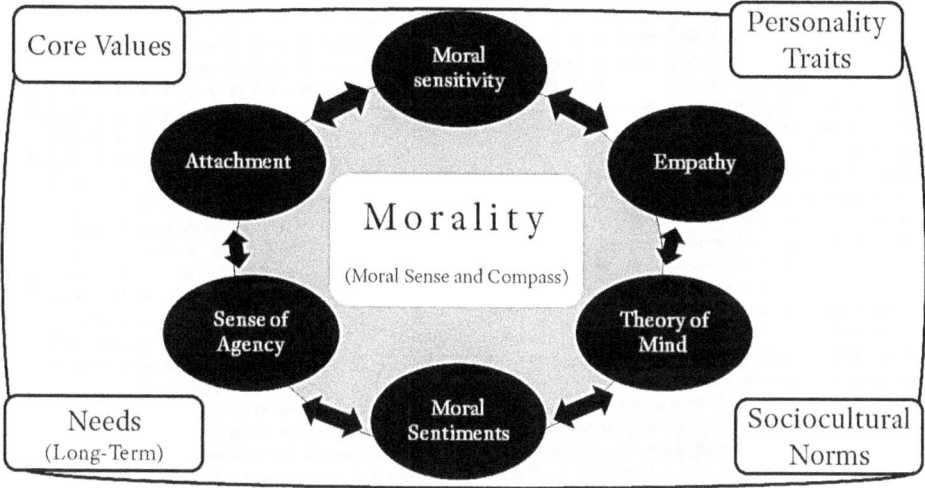

Figure 11.1
The multidimensional structure of morality. As discussed in the text, moral experience and conduct are conditioned by two tiers, each composed of subordinate
components. An inner tier (black boxes) represents short-term, contextual, episodic
mental states, whereas the outer tier (white boxes) represents stable long-term, non-
contextual (transsituational), modifiers of moral experience and conduct. Interactions among components are profuse and dynamic.

One of the tenets of the social brain concept is that the relative increase
in brain volume and its concurrent internal reorganization were primarily driven by the increase in social complexity (Holloway, 2008). However, despite its appealing explanatory power, the social brain hypothesis
had to be validated against competing alternatives, particularly the epi-
phenomenal, ecological, and developmental brain hypotheses (Dunbar,
1998). Research over the past decade has given plain support to the social
brain hypothesis, especially when human culture is envisaged as the most
characteristic expression of the neurobiological nature of man (Adolphs,
2009). Research has also shown that the social brain is a multidimensional
construct with subordinate domains that may experimentally be studied
as neuropsychological components. Studies on normal volunteers and
patients with strategic brain lesions suggest that the neural structures that
underpin these components are distributed within partly overlapping and
partly segregated corticosubcortical loops that functionally interact in specific ways depending on the prevailing component at work (Park & Friston,

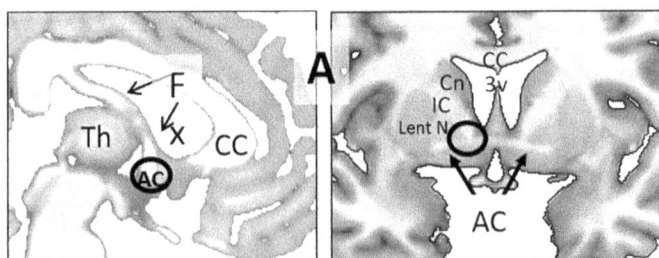

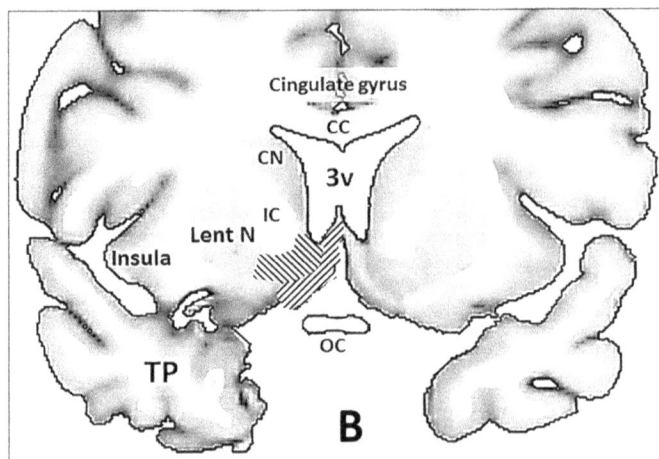

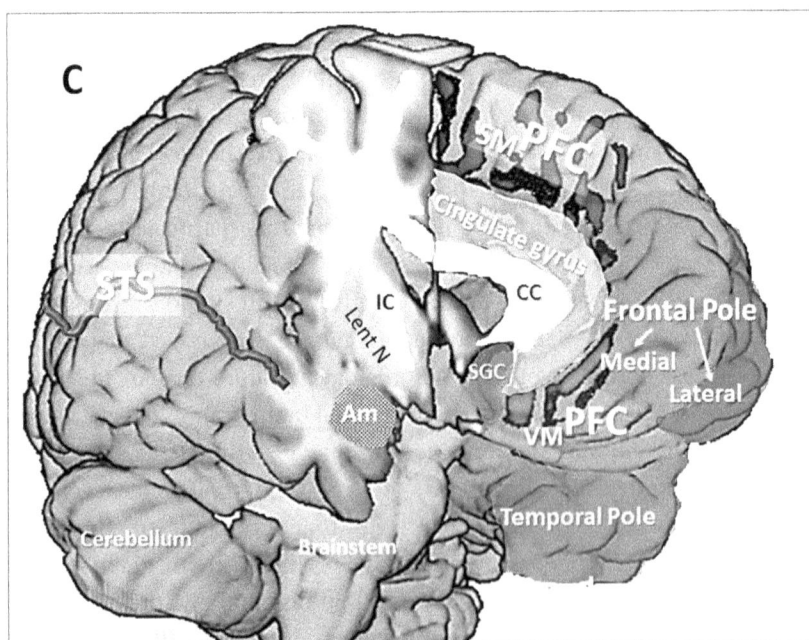

2013). In the remainder of this chapter we put forth some of the main cognitive constituents of morality and their purported neural underpinnings.
In the end we advance some possible applications of the neuroscience of
morality to practical issues.

Neural Components of Morality

Pathways to Moral Decision Making

Neuroscientific research has shown that moral decisions—decisions that
eventually come up with a judgment of right or wrong—can assume two
general forms in normal adults. The ordinary form is an automatic and
effortless type of judgment that relies on moral intuitions; justifications
for such judgments are often post hoc and contradictory (Haidt & Joseph,
2007). The second and less common type of judgment is far from automatic
and intuitive; on the contrary, it consists of a laborious process that relies
on conscious logicoverbal musings (Kahneman, 2011). This type of moral
reasoning is typically elicited by moral dilemmas, which consist in hypothetical or real-life situations for which no readymade right-versus-wrong
decision is obvious (Crockett, 2013). These dilemmas typically arise by pitting two or more decision outcomes that are equivalent in utility, thereby
generating a decision conflict. The conflict in itself tells very little about
underlying psychological and neural representations of values associated
with the alternatives, an issue that has been overlooked in many accounts
of moral judgment (Moll & de Oliveira-Souza, 2007).

Figure 11.2
Schematic depiction of the main anatomical landmarks implicated in morality as
discussed in the text. (A) The perifornical region (circle) is a reference landmark for
a heterogeneous array of structures that belong to functional systems that underpin
attachment and social motivation. It comprises a small volume of neural tissue at the
crossroads of the anterior commissure and the fornix (Fx), at which point the fornix
bifurcates into a precommissural and a postcommissural bundle. (B) Coronal section
of the brain rostral to the anterior commissure (AC) showing two major macrostruc-
tural forebrain systems, the nucleus accumbens (\\\\) and the septal nuclei (////).
(C) Three-dimensional view of the cerebral hemispheres showing the temporal (TP)
and frontal poles, the ventromedial prefrontal cortex (vMPFC), the subgenual cortex (SGC), and the superior temporal sulcus (STS). The right temporal pole is cut in
the coronal plane to expose the amygdala (Am). CC, corpus callosum; CN, caudate
nucleus; IC, internal capsule; Lent N, lenticular nucleus; OC, optic chiasma; sMPFC,
superomedial prefrontal cortex; 3v, third ventricle.

The Fundamental Neurocognitive Structure of Morality

Humans continuously and automatically scan thoughts, ideas and ongoing perceptions for moral salience. Indeed, many of our casual assertions and opinions about ordinary matters that are part of everyday life have some moral tinge. The neural basis of such moral sensitivity has been investigated with functional neuroimaging (fMRI) in a study in which adult volunteers were asked to pay attention to a series of visual scenarios depicting moral situations. They were not warned in advance on the content of the pictures, moral or other (Moll et al., 2002). This "passive," nonjudgmental condition differentially engaged the right medial frontopolar cortex [Brodmann area (BA) 10], ventromedial prefrontal cortex (vmPFC, BA 11), the superior temporal sulcus (STS, BA 21, 39), and the left posterior midtemporal (BA 22, 19) and lateral occipital (BA 18, 19) gyri. This experimental setup is probably the closest we have come to imaging the neural substrates of the moral sense. In a related experiment participants were asked to judge the moral content of social scenarios as either right or wrong. In this condition a partially overlapping pattern of activation emerged. Critically, the bilateral frontopolar cortex (BA 10) and the right temporal pole (BA 38) were engaged when a moral decision had to be made concerning the rightness or wrongness of each scenario (Moll et al., 2002).

The third fundamental neurocognitive component of morality is agency, or the inner conviction that we own our actions and are directly responsible for them (Milgram, 1974). The attribution of intentionality to our own behavior and to the behavior of others is a critical element of morality. In the criminal justice system, for example, the intent to harm (*mens rea*), regardless of whether harm is actually inflicted, is critical for court decisions. Dracon's legal code (ca. 624 BCE) was the first to distinguish between murder (killing with intent) and manslaughter (killing without intent), thus incorporating a psychological distinction into law that has since been decisive for making individuals criminally accountable for their acts (Green & Groff, 2003).

In a recent study on moral agency Moll et al. (2007) evoked the most basic form of agency (i.e., emotionally neutral agency) with scripts such as "You heard your neighbors talking next door; they were chatting about business." In this condition activation was elicited in the mid-STS and middle temporal gyrus, lateral orbitofrontal cortex (OFC), medial OFC-subgenual cortex, dorsal anterior cingulate cortex, and anterior insula. Scripts evocative of moral compared to neutral agency elicited additional activation in other corticobasal forebrain regions. The loci of these activations varied depending on the moral emotion recruited by the different agency-evoking scripts. For example, in comparison to the other-critical emotion of indignation (elicited by scripts such as "Your colleague said that her boss has abused her, but she's afraid of reporting him to the police"), the prosocial emotion of guilt (elicited by scripts such as "You took your nephew to the park, and he got lost in the crowd") differentially engaged the right temporopolar cortex (BA 38) and the left medial OFC (BA 11). These findings suggest that the fundamental components of morality are neither monolithic nor static but are modified by cognitive-emotional representations and contextual elements as well as by the structure of personality, as discussed below.

Moral Motivations as the Steam of Moral Conduct

A common thread permeates most formulations on morality, namely, the ability to behave in prosocial, or altruistic, ways, especially when such actions oppose our concurrent selfish inclinations. In fact, the emergence of lofty forms of altruism in humans may have played a decisive role in hominid phylogenesis (Isaac, 1978). In the eighteenth century Adam Smith provided a clear account on a particular class of sentiments—the "social passions" or "benevolent affections"—which are experienced as sympathy and expressed through behaviors that ultimately promote the welfare of others (Smith, 1790/2002). He further contrasted these prosocial sentiments with the selfish, and therefore unsocial, emotions such as hatred and resentment. The influence of Smith's views can be felt today in the formulation of the concept of "moral emotions," or emotions that are linked to the interests or welfare either of society as a whole or at least of persons other than the judge or agent (Haidt, 2003).

Moral sentiments interact swiftly with moral motivations to compel behavior toward goals and decisions depending on the prevailing moral sentiment, such as pity/sympathy (when observing the suffering of others) or guilt for being the agent of harming another person. Moral sentiments are often tied to abstract values, virtues (e.g., "honesty," "generosity"), or belief systems, which transform moral knowledge into goals that guide behavior in specific situations or into general principles that transcend particular situations (Schwartz, 2006).

Research over the past two decades has shown that the neural organization of prosocial behaviors is far more intricate than that which underpins
self-centered behaviors. In large part this complexity results from the combination of subtle and often fleeting mental states that are recruited by
moral experience and that have just begun being characterized. The neural
circuits involved in morality share extensive regions of the PFC, anterior
temporal lobes, insula, and the temporoparietal segment of the STS. These
regions are profusely interconnected with each other as well as with focal
areas of the basal forebrain macrosystems with which long and short corticosubcortical loops are sustained (Zahm, 2006). Investigations with fMRI
have shown that different aspects of morality engage correspondingly discrete sectors of this broad neural framework. Likewise, small lesions within
this general region may impair different aspects of morality with concurrent changes in social conduct and personality.

Lesion analysis and fMRI studies have provided complementary information on the engagement of discrete sectors of this broad fronto-temporo-
insular region in morality. For example, damage to certain sectors of the
PFC usually gives rise to a severe change in personality that represents the
overall effect of impairments in different social cognitive domains (Barrash
et al., 2011). The "acquired sociopathy" of these patients in fact results from
a mix of impairments among which the most significant are a shift toward
the selfish pole on the selfish-prosocial continuum (Moll et al., 2011), an
impairment in the ability to form and sustain interpersonal bonds (Rankin
et al., 2006), and a decrease of emotional empathy (Shamay-Tsoory et al.,
2005) and of the sense of guilt (Krajbich, Adolphs, Tranel, Denburg, & Camerer, 2009). Besides such dramatic personality changes, a less well-studied
change in the structure of individual values has been reported. Patients
with degeneration primarily affecting the right prefrontal cortex (Miller
et al., 2001) or with temporolimbic epilepsy (Bear, 1979) may undergo a
profound change in religious, political, or moral values (Saver & Rabin,
1997). These cases, however, are relatively rare because as a rule they occur
as minor manifestations within complex associations of symptoms that
include dementia, rage outbursts, sexual disinhibition, somnolence, and
hyperphagia (Alpers, 1937; Poeck & Pilleri, 1965; Reeves & Plum, 1969).
Functional neuroimaging studies have attempted to fill this gap by revealing discrete functional systems within the broad regions indicated by lesion
studies. For example, altruistic behaviors selectively engage the subgenual

cortex (Moll & de Oliveira-Souza, 2009), whereas interpersonal attachment selectively engages a small perifornical region that includes the diagonal band of Broca and the ventromedial hypothalamus/preoptic area (Moll et al., 2011). These small regions and their connections are usually damaged in the aforementioned cases, and the symptoms of their destruction naturally go unnoticed.

There is growing evidence that certain frontotemporal-basal forebrain subregions within the social cognition network are more strongly activated for some moral sentiments relative to others. The frontopolar cortex (FPC) is one of the regions most consistently engaged by moral sentiments in general (Moll et al., 2005). Direct comparisons between prosocial (guilt, embarrassment, compassion) and other-blaming moral sentiments (indignation, disgust) have demonstrated stronger FPC activation for prosocial and lateral orbitofrontal and anterior insular activations for other-blaming sentiments (Moll et al., 2007). Furthermore, guilt and compassion (which involve empathic feelings but not embarrassment) recruit activity in the aforementioned basal forebrain structures. Patients with frontotemporal dementia provide causal evidence for a role of the FPC and septal region in the experience of guilt and compassion (Moll et al., 2011).

Morality as a Core Dimension of the Inner Self

There are scant data on the neural correlates of morality as an integral part of the self. Psychologists have long demonstrated that individual predispositions to act in selfless ways are strongly modulated by the structure of personality, personal systems of values and beliefs, and the strength of immediate and long-term needs (Moffitt et al., 2011). Thus, depending on the particular interactions among prebehavioral determinants, actual behavior in specific situations will largely vary among individuals. The task for neuroscientists is to delve into these processes, striving to find the *individual* correlates of morality in *individual* brains. To date we have only scratched the surface of this complex organization. Returning to the central issue posed at the beginning of this chapter, there is still a long way to run before we understand if the sensing of rightness and wrongness is a product of the workings of a dedicated neural system. So far, we can feel confident that several subordinate components that contribute qualitatively distinct information to moral conscience have strong correlates in the neural substance (Moll & Schulkin, 2009; Pujol et al., 2012; Young & Dungan, 2012).

192 Ricardo de Oliveira-Souza, Roland Zahn, and Jorge Moll

The emphasis on individual (in contrast to average or transcultural) in the present context has to do with the practical implications of studies on the neurological organization of morality. For both practicing physicians and psychologists, diagnosis, treatment, and prognosis are made on an individual basis. Diagnostic neuroimaging would be of critical help if it could be used as an ancillary exam in the diagnosis of neurologic and neuropsychiatric disorders that manifest as abnormalities in moral cognition and behavior. Moreover, depending on the subordinate component primarily affected in different disorders (e.g., antisocial behaviors in psychopathy versus autism), treatments should ideally be tailored to individual patterns and degrees of dysfunction.

Concluding Remarks

1. The preceding considerations lead us to the following general conclusions concerning the behavioral-anatomical correlates of morality. Integrity of...

... the vmPFC is critical for selfless behaviors as well as for the experience of prosocial sentiments in specific contexts.

... the FPC is critical for the mental simulation of the possible social outcomes of current moral judgments.

... the right anterior temporal lobe is critical for the storage and retrieval of social conceptual knowledge and for the experience of moral sentiments.

... hypothalamus, septum, and adjoining basal forebrain macrosystems are critical for moral motivation and for the experience and expression of affiliative behaviors.

... the subgenual cortex is critical for the sense of agency and social belongingness.

2. There is a need for studies on the prebehavioral determinants of actual behavior in what concerns personal core values, long-term needs, and personality traits as well as the modulating role of sociocultural norms on actual moral experience and decision making.

References

Adolphs, R. (2009). The social brain: Neural basis of social knowledge. *Annual Review of Psychology, 60,* 693–716.

Alpers, B. J. (1937). Relation of the hypothalamus to disorders of personality: A case report. *Archives of Neurology and Psychiatry, 38,* 291–303.

Barrash, J., Asp, E., Markon, K., Manzel, K., Anderson, S. W., & Tranel, D. (2011). Dimensions of personality disturbance after focal brain damage: Investigation with the Iowa Scales of Personality Change. *Journal of Clinical and Experimental Neuropsychology, 33,* 833–852.

Bear, D. M. (1979). Temporal lobe epilepsy: A syndrome of sensory-limbic hyperconnection. *Cortex, 15,* 357–384.

Brothers, L. (1991). The social brain: A project for integrating primate behavior and neurophysiology in a new domain. *Concepts in Neuroscience, 1,* 27–51.

Crockett, M. J. (2013). Models of morality. *Trends in Cognitive Sciences, 17,* 363–366.

Damasio, A. R. (1989). The brain binds entities and events by multiregional activation from convergence zones. *Neural Computation, 1,* 123–132.

Decety, J., & Jackson, P. L. (2004). The functional architecture of human empathy. *Behavioral and Cognitive Neuroscience Reviews, 3,* 71–100.

Dunbar, R. I. M. (1998). The social brain hypothesis. *Evolutionary Anthropology, 6,* 178–190.

Gazzaniga, M. S. (1985). The social brain. *Psychology Today, 19,* 29–38.

Gintis, H., Henrich, J., Bowles, S., Boyd, R., & Fehr, E. (2008). Strong reciprocity and the roots of human morality. *Social Justice Research, 21,* 241–253.

Green, C. D., & Groff, P. R. (2003). *Early psychological thought. Ancient accounts of mind and soul.* Westport, CT: Praeger.

Haidt, J. (2003). The moral emotions. In R. J. Davidson, K. R. Scherer, & H. H. Goldsmith (Eds.), *Handbook of Affective Sciences* (pp. 852–870). Oxford: Oxford University Press.

Haidt, J., & Joseph, C. (2007). The moral mind: How 5 sets of innate intuitions guide the development of many culture-specific virtues, and perhaps even modules. In P. Carruthers, S. Laurence, & S. Stich (Eds.), *The innate mind* (Vol. 3, pp. 852–870). New York: Oxford University Press.

Holloway, R. L. (2008). The human brain evolving: A personal retrospective. *Annual Review of Anthropology, 37,* 1–19.

Isaac, G. (1978). The food-sharing behavior of protohuman hominids. *Scientific American, 238*, 90–108.

Kahneman, D. (2011). *Thinking, fast and slow*. New York: Farrar, Strauss & Giroux.

Krajbich, I., Adolphs, R., Tranel, D., Denburg, N. L., & Camerer, C. F. (2009). Economic games quantify diminished sense of guilt in patients with damage to the prefrontal cortex. *Journal of Neuroscience, 29*, 2188–2192.

Lenhoff, H. M., Wang, P. P., Greenberg, F., & Bellugi, U. (1997). Williams syndrome and the brain. *Scientific American, 277*, 68–73.

Lewis, C. S. (1952). *Mere Christianity*. New York: HarperCollins.

Marazziti, D., Baroni, S., Landi, P., Ceresoli, D., & Dell'Osso, L. (2013). The neurobiology of moral sense: Facts or hypotheses? *Annals of General Psychiatry, 12*, 6.

Maudsley, H. (1876). *Responsibility in mental disease* (3rd ed.). London: Henry S. King & Co.

Miller, B. L., Seeley, W. W., Mychack, P., Rosen, H. J., Mena, I., & Boone, K. (2001). Neuroanatomy of the self. Evidence from patients with frontotemporal dementia. *Neurology, 57*, 817–821.

Milgram, S. (1974). *Obedience to authority. An experimental view*. New York: Harper Perennial.

Moll, J., & de Oliveira-Souza, R. (2007). Moral sentiments and reason: Friends or foes? *Trends in Cognitive Sciences, 11*, 323–324.

Moll, J., & de Oliveira-Souza, R. (2009). "Extended attachment" and the human brain: Internalized cultural values and evolutionary implications. In J. Verplaetse, J. De Schrijver, S. Vanneste, & J. Braeckman (Eds.), *The moral brain: Essays on the evolutionary and neuroscientific aspects of morality* (pp. 69–85). New York: Springer.

Moll, J., & Schulkin, J. (2009). Social attachment and aversion in human moral cognition. *Neuroscience and Biobehavioral Reviews, 33*, 456–465.

Moll, J., de Oliveira-Souza, R., Bramati, I. E., & Grafman, J. (2002). Networks in emotional moral and nonmoral social judgments. *NeuroImage, 16*, 696–703.

Moll, J., de Oliveira-Souza, R., Garrido, G. J., Bramati, I. E., Caparelli, E. M. A., Paiva, M. L. M. F., et al. (2007). The self as a moral agent: Linking the neural bases of social agency and moral sensitivity. *Social Neuroscience, 2*, 336–352.

Moll, J., Zahn, R., de Oliveira-Souza, R., Bramati, I. E., Krueger, F., Tura, B., et al. (2011). Impairment of prosocial sentiments is associated with frontopolar and septal damage in frontotemporal dementia. *NeuroImage, 54*, 1735–1742.

Moll, J., Zahn, R., de Oliveira-Souza, R., Krueger, F., & Grafman, J. (2005). The neural basis of human moral cognition. *Nature Reviews. Neuroscience, 6*, 799–809.

Moffitt, T. E., Arseneault, L., Belsky, D., Dicksonc, N., Hancox, R. J., Harrington, H. L., et al. (2011). A gradient of childhood self-control predicts health, wealth, and public safety. *Proceedings of the National Academy of Sciences USA, 108,* 2693–2698.

Nowak, M. A., & Sigmund, K. (2005). Evolution of indirect reciprocity. *Nature, 437,* 1291–1298.

Park, H.-J., & Friston, K. (2013). Structural and functional brain networks: From connections to cognition. *Science, 342,* 1238411.

Pfaff, D. W. (2007). *The neuroscience of fair play: Why we (usually) follow the Golden Rule.* New York: Dana Press.

Poeck, K., & Pilleri, G. (1965). Release of hypersexual behavior due to lesion in the limbic system. *Acta Neurologica Scandinavica, 41,* 233–244.

Pujol, J., Batalla, I., Contreras-Rodríguez, O., Harrison, B. J., Pera, V., Hernández-Ribas, R., et al. (2012). Breakdown in the brain network subserving moral judgment in criminal psychopathy. *Social Cognitive and Affective Neuroscience, 7,* 917–923.

Rankin, K. P., Gorno-Tempini, M. L., Allison, S. C., Stanley, C. M., Glenn, S., Weiner, M. W., et al. (2006). Structural anatomy of empathy in neurodegenerative disease. *Brain, 129,* 2945–2956.

Reeves, A. G., & Plum, F. (1969). Hyperphagia, rage, and dementia accompanying a ventromedial hypothalamic neoplasm. *Archives of Neurology, 20,* 616–624.

Saver, J. L., & Rabin, J. (1997). The neural substrates of religious experience. *Journal of Neuropsychiatry and Clinical Neurosciences, 9,* 498–510.

Schwartz, S. H. (2006). Les valeurs de base de la personne: Théorie, mesures et applications. *Revue Francaise de Sociologie, 42,* 249–288.

Shamay-Tsoory, S. G., Tomer, R., Berger, B. D., Goldsher, D., & Aharon-Peretz, J. (2005). Impaired "affective theory of mind" is associated with right ventromedial prefrontal damage. *Cognitive and Behavioral Neurology, 18,* 55–67.

Smith, A. (2002). *The Theory of Moral Sentiments* (6th ed.), K. Haasonssen (Ed.). Cambridge: Cambridge University Press. (Original work published 1790.)

Young, L., & Dungan, J. (2012). Where in the brain is morality? Everywhere and maybe nowhere. *Social Neuroscience, 7,* 1–10.

Zahm, D. S. (2006). The evolving theory of basal forebrain functional-anatomical "macrosystems." *Neuroscience and Biobehavioral Reviews, 30,* 148–172.

12   The Cognitive Neuroscience of Moral Judgment and Decision Making

Joshua D. Greene

Cognitive neuroscience aims to understand the mind in physical terms. This endeavor assumes that the mind *can* be understood in physical terms, and, insofar as it is successful, validates that assumption. Against this philosophical backdrop, the cognitive neuroscience of moral judgment takes on special significance. Moral judgment is, for many, the quintessential operation of the mind beyond the body, the earthly signature of the soul (Greene, 2011). (In many religious traditions it is, after all, the quality of a soul's moral judgment that determines where it ends up.) Thus, the prospect of understanding moral judgment in physical terms is especially alluring, or unsettling, depending on your point of view. In this brief review I provide a progress report on our attempts to understand how the human brain makes moral judgments and decisions.

The Paradox of the "Moral Brain"

The fundamental problem with the "moral brain" is that it threatens to take over the entire brain and thus ceases to be a meaningful neuroscientific topic. This is not because morality is meaningless but, rather, because neuroscience is centrally concerned with physical mechanisms, and it is increasingly clear that morality has few, if any, neural mechanisms of its own (Greene & Haidt, 2002; Parkinson et al., 2011; Young & Dungan, 2012).

By way of analogy consider the concept of a *vehicle*. Motorcycles and sailboats are vehicles. Lawnmowers and kites are not. But, mechanically speaking, motorcycles have more in common with (gas-powered) lawnmowers than with sailboats, and sailboats operate more like kites than motorcycles. This does not mean that the concept of a vehicle is meaningless. Rather, the world's vehicles are united, not by their internal mechanisms, but at a more abstract functional level. So, too, with morality. More specifically, I (Greene, 2013), like many others (Darwin, 1871/2004; Frank, 1988; Gintis, Bowles, Boyd, & Fehr, 2005; Haidt, 2012), believe that morality is a suite of cognitive mechanisms that enable otherwise selfish individuals to reap the benefits of cooperation. That is, we have psychological features that are straightforwardly moral (such as empathy, righteous indignation, and an aversion to harming innocent people) and others that are not (such as gossip, embarrassment, vengefulness, and ingroup favoritism) because they enable us to achieve goals that we cannot achieve through collective self-ishness. I will not defend this controversial thesis here. Instead, my point is that *if* this unified theory of morality is correct, it does not bode well for a unified theory of moral neuroscience. What is more, as we shall see, the data increasingly bear out this skepticism. In the early days of moral neuroscience it was thought, perhaps not unreasonably, that one might isolate the distinctive neural mechanisms of moral thought (Moll, Eslinger, & de Oliveira-Souza, 2001) and that the human brain might house a dedicated "moral organ" (Hauser, 2006). These views, however, are no longer tenable. It is now clear that the "moral brain" is, more or less, the whole brain applying its computational powers to problems that we, on nonneuroscientific grounds, identify as "moral."

Understanding this is, itself, a kind of progress, but it leaves the cognitive neuroscience of morality—and the author of a chapter that would summarize it—in an awkward position. To truly understand the neuroscience of morality, we must understand the many neural systems that shape moral thinking, none of which, so far, appears to be specifically moral. These include systems that enable the representation of value and that motivate its pursuit (Knutson, Taylor, Kaufman, Peterson, & Glover, 2005; Pessoa, 2010; Rangel, Camerer, & Montague, 2008; Schultz, Dayan, & Montague, 1997), systems that orchestrate thought and action in accordance with internal goals (Miller & Cohen, 2001), systems that enable the imagination of complex distal events (Buckner, Andrews-Hanna, & Schacter, 2008; Raichle et al., 2001), and systems that enable the representation of people's hidden mental states (Frith & Frith, 2006; Mitchell, 2009), among others. In short, if you want to understand the neuroscience of morality, you might start by working your way through this weighty volume.[1]

Of course, some neuroscientific topics bear more directly on morality than others, as indicated by my nonrandom list of relevant neural systems.

This suggests that the present task is not hopeless, that we can make some useful generalizations about the cognitive neuroscience of morality, even while acknowledging that the moral brain is not a distinct entity. This field, properly understood, will not isolate and describe the mechanisms essential for morality while the rest of cognitive neuroscience goes about its business. Instead, it provides a set of useful *entry points* into the broader problems of complex cognition and decision making (cf. Buckholtz & Meyer-Linden-berg, 2012 for a parallel view of psychopathology). More specifically, we can study the brains of people who reliably commit basic moral transgressions, the reactions of healthy brains to such transgressions, and the ways in which our brains handle more complex moral problems. Along the way we encounter some recurring themes that point the way toward a more encompassing account of moral, and nonmoral, cognition.

Bad Brains

In the 1990s Damasio and colleagues published a series of path-breaking studies of decision making in patients with damage to ventromedial pre-frontal cortex (VMPFC), one of the regions damaged in the famous case of Phineas Gage (Damasio, 1994). VMPFC patients were mysterious because their real-life decision making was clearly impaired, but their deficits typically evaded detection using standard neurological measures of executive function (Saver & Damasio, 1991) and moral reasoning (Anderson, Bechara, Damasio, Tranel, & Damasio, 1999). Using a game designed to simulate real-world risky decision making (the Iowa Gambling Task), Bechara, Tranel, Damasio, and Damasio (1996) documented these behavioral deficits and demonstrated, using autonomic measures, that these deficits are emotional. It seems that such patients make poor decisions because they are unable to generate the feelings that guide adaptive decision making in healthy individuals. These early studies, while identifying a key biological substrate for moral choice, also underscore the critical role of learning in moral development. Late-onset VMPFC damage typically results in poor decision making and a deterioration of "moral character" (Damasio, 1994), but children with early-onset VMPFC damage are likely to develop into "sociopathic" adults who, in addition to being reckless and irresponsible, are duplicitous, aggressive, and strikingly lacking in empathy (Anderson et al., 1999; Grattan & Eslinger, 1992).

Studies of psychopaths and other individuals with antisocial personality disorder (APD) underscore the importance of emotion in moral decision making. APD is a catchall diagnosis for individuals whose behavior is unusually antisocial. Psychopathy, in contrast, is a more specific, somewhat heritable disorder (Viding, Blair, Moffitt, & Plomin, 2005) whereby individuals exhibit a pathological degree of callousness, lack of empathy or emotional depth, and lack of genuine remorse for their antisocial actions (Hare, 1991). Psychopaths tend to engage in instrumental aggression, whereas other individuals with APD are characterized by reactive aggression (Blair, 2001).

Psychopathy is characterized by profound but selective emotional deficits. Psychopaths exhibit normal electrodermal responses to threat cues (e.g., a picture of a shark's open mouth) but reduced responses to distress cues (e.g., a picture of a crying child; Blair, Jones, Clark, & Smith, 1997). In a classic study Blair (1995) provided evidence that psychopaths fail to distinguish between rules that authorities cannot legitimately change ("moral" rules, e.g., a classroom rule against hitting) from rules that authorities can legitimately change ("conventional" rules, e.g., a rule prohibiting talking out of turn). According to Blair (1995), psychopaths see all rules as *mere* rules because they lack the emotional responses that lead ordinary people to imbue moral rules with genuine, authority-independent moral legitimacy. Although this is consistent with what is generally known about psychopathic psychology, a more recent study challenges the original finding that psychopaths do not draw the moral/conventional distinction (Aharoni, Sinnott-Armstrong, & Kiehl, 2012).

Studies of psychopathy and APD implicate a wide range of brain regions including the insula, posterior cingulate cortex (PCC), parahippocampal gyrus, and superior temporal gyrus (Kiehl, 2006; Raine & Yang, 2006). However, as emphasized by Blair (2007), two interconnected structures take center stage: the amygdala and the VMPFC. These regions, along with subregions of subgenual anterior cingulate cortex (ACC) and lateral prefrontal cortex, form a network that is essential for generating and regulating responses to salient stimuli (Pessoa, 2010). Blair (2007) has proposed that psychopathy arises primarily from amygdala dysfunction, which is crucial for stimulus-reinforcement learning (Davis & Whalen, 2001) and thus for normal moral socialization (Oxford, Cavell, & Hughes, 2003). In psychopaths (or individuals with psychopathic traits) the amygdala exhibits weaker responses to fearful faces (Marsh et al., 2008), to emotional words (Kiehl et al., 2001), to pictures indicating moral violations (Harenski, Harenski, Shane, & Kiehl, 2010; Harenski, Kim, & Hamann, 2009), and to dilemmas involving harmful actions (Glenn, Raine, & Schug, 2009). As noted above, the amygdala operates in tight conjunction with the VMPFC, and, consistent with this, psychopathic individuals also exhibit reduced VMPFC responses to morally salient stimuli (Harenski et al., 2010). Beyond the amygdala-VMPFC circuit, psychopaths also exhibit hypoactivity in the default mode network (DMN) (Buckner et al., 2008; Raichle et al., 2001) during moral judgment (Pujol et al., 2012), consistent with this network's heightened response to emotionally engaging moral dilemmas in healthy people (Greene, Sommerville, Nystrom, Darley, & Cohen, 2001). (Note that some of the paraticipants in this study failed to meet standard criteria for psychopathy. See Schaich Borg & Sinnott-Armstrong, 2013.)

Psychopaths, in addition to their weak affective responses to harm, are known for their impulsive behavior (Hare, 1991). The VMPFC serves as part of the frontostriatal pathway, responsible for representing the values of outcomes and actions based on past experience (Knutson et al., 2005; Rangel et al., 2008). Individuals with psychopathic traits (specifically, impulsive antisocial behavior) exhibit heightened responses to reward within this system (Buckholtz et al., 2010) along with increased striatal volume (Glenn, Raine, Yaralian, & Yang, 2010). Finally, their emotional deficits may sometimes cause them to rely more heavily on explicit reasoning, dependent on the frontoparietal control network (Glenn, Raine, Schug, Young, & Hauser, 2009; Koenigs, Kruepke, Zeier, & Newman, 2012). Thus, while the origins of psychopathy may lie in one or more discrete neural abnormalities, their influence is felt throughout the brain.

Good Brains

Studies of healthy individuals responding to moral transgressions are generally consistent with studies of psychopaths and others with APD. They, too, highlight the importance of the amygdala and VMPFC (Blair, 2007; Decety & Porges, 2011; Heekeren et al., 2005; Moll et al., 2002; Schaich Borg, Hynes, Van Horn, Grafton, & Sinnott-Armstrong, 2006) and confirm the importance of these structures in moral development (Decety, Michalska, & Kinzler, 2012). For reasons explained below, studies of moral judgment employing text-based narrative stimuli tend to implicate the entire DMN. Several studies highlight the importance of the insula in representing the aversiveness of moral transgressions (Baumgartner, Fischbacher, Feirabend, Luta, & Fehr, 2009; Decety, Michalska, & Kinzler, 2012; Greene, Nystrom, Engell, Darley, & Cohen, 2004; Schaich Borg, Lieberman, & Kiehl, 2008; Schaich Borg, Sinnott-Armstrong, Calhoun, & Kiehl, 2011). Others indicate that the representation of moral value, like other forms of value, depends on the brain's domain-general valuation mechanisms enabled by the frontostriatal pathway (Decety & Porges, 2011; Moll et al., 2006; Shenhav & Greene, 2010).

One of the most basic distinctions in moral evaluation is the one between intentional and accidental harm. (As Oliver Wendell Holmes Jr. famously observed, even a dog knows the difference between being tripped over and being kicked.) Young, Saxe, and colleagues have conducted a series of studies examining how the brain represents and applies this distinction in the context of moral judgment. Their work highlights the importance of the temporoparietal junction (TPJ) along with other DMN regions, which are widely implicated in ToM/mentalizing (Frith & Frith, 2006; Mitchell, 2009). The TPJ is especially sensitive to attempted harms (Koster-Hale, Saxe, Dungan, & Young, 2013; Young, Cushman, Hauser, & Saxe, 2007), which are wrong because of the agent's mental state, not the action's outcome. Disrupting TPJ activity results in a childlike (Piaget, 1965), "no harm no foul" pattern of judgment in which attempted harms are judged less harshly (Young, Camprodon, Hauser, Pascual-Leone, & Saxe, 2010). We see the same pattern in patients with VMPFC damage (Young, Bechara, et al., 2010) and split-brain patients (Miller et al., 2010), indicating that the use of mental state information in moral judgment depends, at least in part, on translating this information into an affective signal and on the integration of information across the cerebral hemispheres. Individuals with high-functioning autism exhibit a complementary pattern, "if harm, then foul," judging accidental harms unusually harshly (Moran et al., 2011). Accidental harms appear to set up a tension between outcome-based and intention-based harm. Consistent with this, such harms preferentially engage the frontoparietal control network (Miller & Cohen, 2001).

Puzzled Brains

We have considered the two most straightforward entry points into moral neuroscience: the unhealthy brains of people who act badly and the healthy brain's responses to prototypically bad acts. A third approach begins with moral dilemmas. Moral dilemmas are useful, not because they reflect everyday moral experience but because dilemmas, by their nature, pit competing processes against one another. They are high-contrast stimuli, analogous to the flashing checkerboards of vision scientists, and thus especially useful for revealing cognitive structure (Cushman & Greene, 2012).

The research described above emphasizes the role of emotion in moral judgment (Haidt, 2001), whereas traditional theories of moral development emphasize the role of controlled cognition (Kohlberg, 1969; Turiel, 2006). I and others have developed a dual-process (Chaiken & Trope, 1999; Kahneman, 2003) theory of moral judgment that synthesizes these perspectives (Greene et al., 2001; Greene, 2007, 2013). According to this theory both intuitive emotional responses and more controlled cognitive responses play crucial and, in some cases, competing roles. More specifically, this theory associates controlled cognition with utilitarian (or consequentialist) moral judgment aimed at promoting the "greater good" (Mill, 1861/1998) while associating automatic emotional responses with competing deontological judgments that are naturally justified in terms of rights or duties (Kant, 1785/1959).

We developed this theory in response to a long-standing philosophical puzzle known as the trolley problem (Foot, 1978; Thomson, 1985). In one version, which I call the *switch* case, one can save five people who are mortally threatened by a runaway trolley by hitting a switch. This will turn the trolley onto a side track, where it will run over and kill only one person instead. Here, most people approve of diverting the trolley (Petrinovich, O'Neill, & Jorgensen, 1993), a characteristically utilitarian judgment favoring the greater good. In the contrasting *footbridge* dilemma a runaway trolley once again threatens five people. The only way to save the five is to push a large person off a footbridge and into the trolley's path, stopping the trolley but killing the person pushed. (Yes, this will work, and, no, you cannot stop the trolley yourself.) Here, most people say that it is wrong to trade one life for five, consistent with the deontological perspective favoring the rights of the individual over the greater good. The question: Why do people typically say "yes" to hitting the switch, but "no" to pushing?

We hypothesized that this pattern of judgment reflects the outputs of distinct and (in some cases) competing neural systems (Greene et al., 2001). The more "personal"[2] harmful action in the *footbridge* case, pushing the man off the footbridge, triggers a relatively strong negative emotional response, whereas the relatively impersonal harmful action in the *switch* case does not. This predicts increased activity in emotion-related brain regions in response to "personal" dilemmas, such as the *footbridge* case, as compared to "impersonal" dilemmas, such as the *switch* case.

This emotional response can explain why people say "no" to pushing the man off the footbridge. But why do people say "yes" to hitting the switch? The answer seems obvious enough: hitting the switch saves more lives. We hypothesized that this utilitarian response depends on explicit cost-benefit reasoning enabled by the frontoparietal control network (Miller & Cohen, 2001), including the DLPFC. Thus, we predicted increased DLPFC activity in response to "impersonal" dilemmas, such as the *switch* case, in which this controlled response tends to dominate. Likewise, we predicted increased DLPFC activity when people override a negative emotional response in making a utilitarian judgment, as when people say "yes" to the *footbridge* dilemma.

We first tested this theory using functional MRI (fMRI) (Greene et al., 2001), contrasting a (rather heterogeneous) set of "personal" dilemmas with a set of (even more heterogeneous) "impersonal" dilemmas. (More recent studies have been better controlled, focusing on differing responses to "high-conflict" dilemmas such as the *footbridge* case.) We found that the "personal" dilemmas elicited increased activity in what is now known as the DMN (Buckner et al., 2008; Raichle et al., 2001), including large portions of medial prefrontal cortex, medial parietal cortex, and the TPJ, all of which had been previously associated with emotion (e.g., Maddock, 1999). In contrast, the "impersonal" dilemmas elicited relatively greater activity in the frontoparietal control network. Also as predicted our second fMRI experiment (Greene et al., 2004) found increased DLPFC activity for utilitarian judgment and increased amygdala activity for "personal" dilemmas. These results provided initial support for the dual-process theory, which has been both supported and refined by subsequent research using a broad range of methods.

In retrospect, the DMN's response to "personal" dilemmas is best interpreted as *related* to increased emotional engagement, but not as its proper neural substrate. The DMN is active when people are doing nothing in particular (hence "default") and is most reliably engaged by attention to non-present events, as in remembering the past, imaging the future, thinking about contents of other minds, and imaging hypothetical possibilities (Buckner et al., 2008; DeBrigard, Addis, Ford, Schacter, & Giovanello, 2013). Thus, if "personal" dilemmas preferentially engage the DMN, it is probably not because DMN activity reflects emotional engagement per se. Rather, it is because "personal" dilemmas make for especially gripping mental television, which may be both a cause and a consequence of their emotional salience. Consistent with this hypothesis, Amit and Greene (2012) found that individuals with more visual cognitive styles tend to make fewer utilitarian judgments in response to high-conflict personal dilemmas and that disrupting visual imagery while one is contemplating these dilemmas increases utilitarian judgment.

More direct evidence for the dual-process theory comes from studies of patients with emotion-related deficits. Mendez, Anderson, and Shapira (2005) found that patients with frontotemporal dementia, who are known for their "emotional blunting," were disproportionately likely to approve of the utilitarian action in the *footbridge* dilemma. Likewise, patients with VMPFC lesions make up to five times as many utilitarian judgments in response to standard high-conflict dilemmas (Ciaramelli, Muccioli, Ladavas, & di Pellegrino, 2007; Koenigs et al., 2007) and in response to dilemmas pitting familial duty against the greater good (e.g., your sister vs. five strangers) (Thomas, Croft, & Tranel, 2011). VMPFC patients also exhibit correspondingly weak physiological responses when making such judgments (Moretto, Làdavas, Mattioli, & Di Pellegrino, 2010), and healthy people who are more physiologically reactive are less utilitarian (Cushman, Murray, Gordon-McKeon, Wharton, & Greene, 2012). Low-anxiety psychopaths (Koenigs et al., 2012) and people with high levels of testosterone (Carney & Mason, 2010), which is associated with a higher tolerance for stress, tend to make more utilitarian judgments, as do people with alexithymia (Koven, 2011), a condition that reduces awareness of one's own emotional states. Here, the VMPFC seems to respond specifically to harmful behavior that is active and also intentional, rather than merely foreseen (Schaich Borg et al., 2006).

Other studies highlight the role of the amygdala. As noted above, individuals with psychopathic traits exhibit reduced amygdala responses to personal moral dilemmas (Glenn, Raine, & Schug, 2009). In healthy people amygdala activity tracks self-reported emotional responses to harmful transgressions and predicts deontological judgments in response to them

(Shenhav & Greene, 2013). Studies employing pharmacological interventions paint a consistent picture. Citalopram—a selective serotonin reuptake inhibitor (SSRI) that, in the short term, increases emotional reactivity through its influence on the amygdala and VMPFC, among other regions—increases deontological judgment (Crockett, Clark, Hauser, & Robbins, 2010). By contrast, lorazepam, an antianxiety drug, has the opposite effect (Perkins et al., 2012). Consistent with the effects of citalopram, variation in the serotonin transporter (*5-HTTLPR*) genotype (S alleles) predicts deontological judgment—but in response dilemmas in which the harm is a foreseen side effect (Marsh et al., 2011).

Most of the evidence linking controlled cognition to utilitarian judgment comes from behavioral studies beyond the scope of this chapter (e.g., Greene, Morelli, Lowenberg, Nystrom, & Cohen, 2008; Paxton, Ungar, & Greene, 2012.). However, a few neuroscientific studies, in addition to those described above (Greene et al., 2001, 2004), provide further evidence. Sarlo et al. (2012) examined the temporal dynamics of moral judgment using EEG and found a pattern consistent with the results of Greene et al. (2001, 2004). Here, *footbridge*-like dilemmas produced a stronger early neural response (P260) in regions consistent with VMPFC activity, whereas *switch*-like dilemmas elicited more utilitarian responses and a more pronounced later component consistent with the engagement of the frontoparietal control network. Also consistent with this, activity in the frontoparietal control network is associated with rejecting the deontological distinction between harmful acts and harmful omissions (Cushman et al., 2012). (See also Schaich Borg et al., 2006.) Likewise, VMPFC patients who tend to give more utilitarian responses are thought to do so because their capacity for explicit, cost–benefit reasoning remains intact (Koenigs et al., 2007).

A recent study (Shenhav & Greene, 2013) helps differentiate the functions of the amygdala and VMPFC in moral judgment. As noted above, the amygdala signal tracks with self-reports of negative emotional responses to harmful actions and predicts deontological condemnation of those actions. The VMPFC, however, does not. Instead, the VMPFC is most active when people have to make "all things considered" judgments, as compared to simply reporting on emotional reactions or utilitarian considerations. This suggests that the amygdala generates an initial negative response to personally harmful actions (consistent with Glenn, Raine, & Schug, 2009), whereas the VMPFC weighs that signal against a competing signal reflecting the utilitarian advantages of committing the harmful act. This is consistent with an evolving understanding of the VMPFC as a domain-general integrator of decision weights (Rangel & Hare, 2010). However, this leaves us with a puzzle: If the VMPFC is acting as a neutral broker among competing decision weights, then why does VMPFC damage so reliably increase utilitarian judgment? Our hypothesis is that the frontoparietal control network's explicit utilitarian reasoning can influence behavior independent of the VMPFC, whereas the amygdala's competing deontological signal requires the VMPFC's integration, at least when competing utilitarian considerations are in play. Thus, if this is correct, VMPFC damage favors utilitarian judgment, not by damaging a region with inherent deontological tendencies, but by damaging a pathway that is necessary for deontological judgment, but not utilitarian judgment, to prevail.

This integrative role for the VMPFC is consistent with its role in integrating other kinds of morally relevant information. Shenhav and Greene (2010) examined people's responses to dilemmas in which failing to save one person can allow one to save a group of others. We varied the size of the group and the probability of saving them. We found that neural sensitivity to the magnitude of the outcome (group size) in the ventral striatum predicts behavioral sensitivity to this variable, and we observed a parallel effect for outcome probability in the insula. The VMPFC, by contrast, responded to the interaction of these two variables, reflecting the probability-discounted magnitude of the moral consequences. In other words, the VMPFC represents "expected moral value," just as it represents expected value in self-interested economic decision making (Knutson et al., 2005). Thus, once again, we see a domain-general system—here, the frontostriatal pathway—operating in the context of moral judgment. This system evolved in mammals to value goods that tend to exhibit diminishing marginal returns. This may explain our puzzling (and highly consequential) tendency to regard the saving of human lives as exhibiting diminishing marginal returns, as if the hundredth life saved is somehow worth less than the first.

In an important theoretical development Cushman (2013) and Crockett (2013) have proposed that the dissociation between deontological and utilitarian/consequentialist judgment reflects a more general dissociation between model-free and model-based learning systems (Daw & Doya, 2006). Model-free learning mechanisms assign values to actions intrinsically based on past experience, whereas model-based learning mechanisms attach values to actions based on internal models of causal relations in the world. Thus, an action may seem intrinsically wrong because past experience has associated actions of that type (e.g., pushing people) with negative consequences (e.g., social disapproval), and yet the same action may seem right because it will, according to one's world model, produce optimal consequences (the saving of five lives instead of one). Thus, the fundamental tension in normative ethics, reflected in the competing philosophies of Kant and Mill, may find its origins in a competition between distinct, domain-general mechanisms for assigning values to actions.

Cooperative Brains

Research on altruism and cooperation does not always fall under the heading of "morality," but it could not be more central to our understanding of the moral brain. The most basic question about the cognitive neuroscience of altruism and cooperation is this: What neural processes enable and motivate people to be "nice"—that is, to pay costs to benefit others?

Consistent with our evolving story, the value of helping others, both in unidirectional altruism and bidirectional cooperation, is represented in the frontostriatal pathway. Activity in this pathway tracks the value of charitable contributions (Hare, Camerer, Knoepfle, O'Doherty, & Rangel, 2010; Moll et al., 2006), sharing resources with other individuals (Zaki & Mitchell, 2011), and cooperation (Rilling et al., 2007), maximizing benefits delivered by a distribution of resource (i.e., "efficiency"), and optimizing the subjective trade-off between efficiency and equality (Hsu, Anen, & Quartz, 2008). Likewise, this pathway tracks the value of punishing individuals who are insufficiently "nice" (de Quervain et al., 2004; Singer et al., 2006). As above, the DMN has a hand in altruism as well. TPJ volume (Morishima, Schunk, Bruhin, Ruff, & Fehr, 2012) and medial PFC activity (Rilling et al., 2007; Waytz, Zaki, & Mitchell, 2012) both predict altruistic behavior.

Thus, the brain uses its endogenous carrots—reward signals—to motivate cooperative behavior. It also uses its sticks—negative affective responses—to uncooperative behavior. Activity in the insula, known for its role in the representation of somatic states and the awareness of feelings (Craig, 2009), scales with the magnitude of the unfairness in unfair Ultimatum Game offers (Sanfey, Rilling, Aronson, Nystrom, & Cohen, 2003), predicts aversion to inequality in the distribution of resources (Hsu et al., 2008), and predicts egalitarian behavior and attitudes (Dawes et al., 2012). The insula and the amygdala both respond to the punishment of well-behaved people (Singer, Kiebel, Winston, Dolan, & Frith, 2004).

The dual-process tension between automatic and controlled processes is observed in a range of morally laden economic choices. Accepting unfair Ultimatum Game offers, despite their distastefulness, is associated with increased activity in the frontoparietal control network (Sanfey et al., 2003; Tabibnia, Satpute, & Lieberman, 2008). Perhaps surprisingly, VMPFC damage leads to increased *rejection* of unfair offers (Koenigs & Tranel, 2007). (Consistent with this, psychopaths do the same; see Koenigs, Kruepke, & Newman, 2010.) This may be because the VMPFC integrates signals responding both to unfairness and material gain (which compete in the Ultimatum Game) and because, in the absence of such signals, one applies a reciprocity rule. In a study of dishonesty, Greene and Paxton (2009) gave people repeated opportunities to gain money by lying about their accuracy in predicting the outcomes of coin-flips. Consistently honest subjects appeared to be "gracefully" honest, exhibiting no additional engagement of the frontoparietal control network in forgoing dishonest gains. By contrast, subjects who behaved dishonestly (as indicated by improbably high self-reported accuracy) exhibited increased control-related activity both when lying and when refraining from lying. A follow-up study (Abe & Greene, in press) traces these behavioral differences to response characteristics of the frontostriatal pathway. Baumgartner et al. (2009) describe a similar dual-process dynamic in which breaking promises involves increased engagement of the amygdala and the frontoparietal control network. (For a behavioral approach to dual-process cooperation, see also Rand, Greene, & Nowak, 2012.)

Cooperation depends on trust, which in turn requires evaluating individuals (Delgado, Frank, & Phelps, 2005; Singer et al., 2004) and groups (Phelps et al., 2000) as potential cooperation partners. Oxytocin, a neuropeptide known for its role in social attachment and affiliation in mammals (Insel & Young, 2001), appears to be important for both kinds of decisions. Intranasal administration of oxytocin increases investment in a "trust game" (Kosfeld, Heinrichs, Zak, Fischbacher, & Fehr, 2005) but also biases judgment and behavior toward ingroup members and against outgroup members (de Dreu et al., 2010: de Dreu, Greer, Van Kleef, Shalvi, & Handgraaf, 2011). Likewise, genetic variants associated with oxytocin are associated with increased prosocial behavior, particularly when the world is seen as threatening (Poulin, Holman, & Buffone, 2012).

From an evolutionary perspective the double-edged sword of human morality comes as no surprise. Morality evolved, not as device for universal cooperation, but as a competitive weapon, as a system for turning Me into Us, which in turn enables Us to outcompete Them. Morality's dark, tribalistic side is powerful, but there is no reason why it must prevail. The flexible thinking enabled by our enlarged prefrontal cortices may enable us to retain the best of our moral impulses while we transcend their inherent limitations (Greene, 2013; Pinker, 2011).

Looking Back and Ahead

How does the moral brain work? Answer: exactly the way we would expect it to work if we understand (1) which cognitive functions morality requires and (2) which cognitive functions are performed by the brain's core neural systems. On the one hand, this means that morality has no proprietary neural territory of its own. On the other hand, it means that the cognitive neuroscience of morality, beginning with the entry points described above, can teach important lessons about how the brain's core neural systems interact to solve complex problems.

From its inception cognitive neuroscience has focused on structure-function relationships. We have a general understanding of what various neural structures do, but when it comes to complex cognition, we are mostly blind to the specific information content shuttled about the brain. We know, for example, that the thought of pushing someone off a footbridge pushes our emotional buttons, but we know almost nothing about how we think such thoughts in the first place. However, with the advent of multivariate analysis methods (Kriegeskorte, Goebel, & Bandettini, 2006; Norman, Polyn, Detre, & Haxby, 2006), we may finally be ready to understand how the brain encodes and manipulates the *contents* of thoughts. When we finally do, we will learn a lot more about morality—and everything else as well.

Acknowledgments

Many thanks to Joshua Buckholtz, Joe Paxton, Adina Roskies, Walter Sinnott-Armstrong, and Liane Young for helpful comments.

Notes

1. The "weighty volume" referred to here is *The Cognitive Neurosciences*, fifth edition (M. Gazzaniga, Ed., Cambridge, MA: MIT Press, 2014), in which this chapter was originally published.

2. The personal/impersonal distinction (Greene et al., 2001) has been revised (Greene et al., 2009) since it was originally introduced. For present purposes, one can think of "personal" harms as ones in which the agent actively and intentionally harms the victim using the direct force of his or her muscles.

References

Abe, N., & Greene, J. D. (in press). Response to anticipated reward in the nucleus accumbens predicts behavior in an independent test of honesty. *Journal of Neuroscience*, in press.

Aharoni, E., Sinnott-Armstrong, W., & Kiehl, K. A. (2012). Can psychopathic offenders discern moral wrongs? A new look at the moral/conventional distinction. *Journal of Abnormal Psychology, 121*(2), 484.

Amit, E., & Greene, J. D. (2012). You see, the ends don't justify the means: Visual imagery and moral judgment. *Psychological Science, 23*(8), 861–868.

Anderson, S. W., Bechara, A., Damasio, H., Tranel, D., & Damasio, A. R. (1999). Impairment of social and moral behavior related to early damage in human prefrontal cortex. *Nature Neuroscience, 2*, 1032–1037.

Baumgartner, T., Fischbacher, U., Feierabend, A., Lutz, K., & Fehr, E. (2009). The neural circuitry of a broken promise. *Neuron, 64*(5), 756–770.

Bechara, A., Tranel, D., Damasio, H., & Damasio, A. R. (1996). Failure to respond autonomically to anticipated future outcomes following damage to prefrontal cortex. *Cerebral Cortex, 6*, 215–225.

Blair, R. J. (1995). A cognitive developmental approach to mortality: Investigating the psychopath. *Cognition, 57*, 1–29.

Blair, R. J. (2001). Neurocognitive models of aggression, the antisocial personality disorders, and psychopathy. *Journal of Neurology, Neurosurgery, and Psychiatry, 71*, 727–731.

Blair, R. J. (2007). The amygdala and ventromedial prefrontal cortex in morality and psychopathy. *Trends in Cognitive Sciences, 11*, 387–392.

Blair, R. J., Jones, L., Clark, F., & Smith, M. (1997). The psychopathic individual: A lack of responsiveness to distress cues? *Psychophysiology, 34*, 192–198.

Buckholtz, J. W., & Meyer-Lindenberg, A. (2012). Psychopathology and the human connectome: Toward a transdiagnostic model of risk for mental illness. *Neuron, 74*(6), 990–1004.

Buckholtz, J. W., Treadway, M. T., Cowan, R. L., Woodward, N. D., Benning, S. D., Li, R., et al. (2010). Mesolimbic dopamine reward system hypersensitivity in individuals with psychopathic traits. *Nature Neuroscience, 13*(4), 419–421.

Buckner, R. L., Andrews⊠Hanna, J. R., & Schacter, D. L. (2008). The brain's default network. *Annals of the New York Academy of Sciences, 1124*(1), 1–38.

Carney, D. R., & Mason, M. F. (2010). Decision making and testosterone: When the ends justify the means. *Journal of Experimental Social Psychology, 46*(4), 668–671.

Chaiken, S., & Trope, Y. (Eds.). (1999). *Dual-process theories in social psychology*. New York: Guilford Press.

Ciaramelli, E., Muccioli, M., Ladavas, E., & di Pellegrino, G. (2007). Selective deficit in personal moral judgment following damage to ventromedial prefrontal cortex. *Social Cognitive and Affective Neuroscience, 2*, 84–92.

Craig, A. D. (2009). How do you feel—now? The anterior insula and human awareness. *Nature Reviews Neuroscience, 10*(1), 59–70.

Crockett, M. J. (2013). Models of morality. *Trends in Cognitive Sciences, 17*(8), 363–366.

Crockett, M. J., Clark, L., Hauser, M. D., & Robbins, T. W. (2010). Serotonin selectively influences moral judgment and behavior through effects on harm aversion. *Proceedings of the National Academy of Sciences USA, 107*(40), 17433–17438.

Cushman, F. (2013). Action, outcome and value: A dual-system framework for morality. *Personality and Social Psychology Review, 17*(3), 273–292.

Cushman, F., & Greene, J. D. (2012). Finding faults: How moral dilemmas illuminate cognitive structure. *Social Neuroscience, 7*(3), 269–279.

Cushman, F., Murray, D., Gordon-McKeon, S., Wharton, S., & Greene, J. D. (2012). Judgment before principle: Engagement of the frontoparietal control network in condemning harms of omission. *Social Cognitive and Affective Neuroscience, 7*(8), 888–895.

Damasio, A. R. (1994). *Descartes' error: Emotion, reason, and the human brain*. New York: G. P. Putnam.

Darwin, C. (1871/2004). *The descent of man*. New York: Penguin.

Davis, M., & Whalen, P. J. (2001). The amygdala: Vigilance and emotion. *Molecular Psychiatry, 6*, 13–34.

Daw, N. D., & Doya, K. (2006). The computational neurobiology of learning and reward. *Current Opinion in Neurobiology, 16*(2), 199–204.

Dawes, C. T., Loewen, P. J., Schreiber, D., Simmons, A. N., Flagan, T., McElreath, R., et al. (2012). Neural basis of egalitarian behavior. *Proceedings of the National Academy of Sciences USA, 109*(17), 6479–6483.

De Brigard, F., Addis, D. R., Ford, J. H., Schacter, D. L., & Giovanello, K. S. (2013). Remembering what could have happened: Neural correlates of episodic counterfactual thinking. *Neuropsychologia, 51*(12), 2401–2414.

Decety, J., Michalska, K. J., & Kinzler, K. D. (2012). The contribution of emotion and cognition to moral sensitivity: A neurodevelopmental study. *Cerebral Cortex, 22*(1), 209–220.

Decety, J., & Porges, E. C. (2011). Imagining being the agent of actions that carry different moral consequences: An fMRI study. *Neuropsychologia, 49*(11), 2994–3001.

De Dreu, C. K., Greer, L. L., Handgraaf, M. J., Shalvi, S., Van Kleef, G. A., Baas, M., et al. (2010). The neuropeptide oxytocin regulates parochial altruism in intergroup conflict among humans. *Science, 328*(5984), 1408–1411.

De Dreu, C. K., Greer, L. L., Van Kleef, G. A., Shalvi, S., & Handgraaf, M. J. (2011). Oxytocin promotes human ethnocentrism. *Proceedings of the National Academy of Sciences USA, 108*(4), 1262–1266.

Delgado, M. R., Frank, R., & Phelps, E. A. (2005). Perceptions of moral character modulate the neural systems of reward during the trust game. *Nature Neuroscience, 8,* 1611–1618.

De Quervain, D. J., Fischbacher, U., Treyer, V., Schellhammer, M., Schnyder, U., Buck, A., et al. (2004). The neural basis of altruistic punishment. *Science, 305,* 1254–1258.

Foot, P. (1978). The problem of abortion and the doctrine of double effect. In *Virtues and Vices* (pp. 19–32). Oxford: Blackwell.

Frank, R. H. (1988). *Passions within reason: The strategic role of the emotions.* New York: W. W. Norton.

Frith, C. D., & Frith, U. (2006). The neural basis of mentalizing. *Neuron, 50*(4), 531–534.

Gintis, H. E., Bowles, S. E., Boyd, R. E., & Fehr, E. E. (2005). *Moral sentiments and material interests: The foundations of cooperation in economic life.* Cambridge, MA: MIT Press.

Glenn, A. L., Raine, A., Yaralian, P. S., & Yang, Y. (2010). Increased volume of the striatum in psychopathic individuals. *Biological Psychiatry, 67*(1), 52–58.

Glenn, A. L., Raine, A., & Schug, R. A. (2009). The neural correlates of moral decision-making in psychopathy. *Molecular Psychiatry, 14*(1), 5.

Glenn, A. L., Raine, A., Schug, R. A., Young, L., & Hauser, M. (2009). Increased DLPFC activity during moral decision-making in psychopathy. *Molecular Psychiatry, 14*(10), 909–911.

Grattan, L. M., & Eslinger, P. J. (1992). Long-term psychological consequences of childhood frontal lobe lesion in patient DT. *Brain and Cognition, 20*, 185–195.

Greene, J. D. (2007) The secret joke of Kant's soul. In W. Sinnott-Armstrong (Ed.), *Moral psychology*. Vol. 3: *The neuroscience of morality: Emotion, disease, and development* (pp. 35–79). Cambridge, MA: MIT Press.

Greene, J. (2011). Social neuroscience and the soul's last stand. In A. Todorov, S. Fiske, & D. Prentice (Eds.), *Social neuroscience: Toward understanding the underpinnings of the social mind*. New York: Oxford University Press.

Greene, J. (2013). *Moral tribes: Emotion, reason, and the gap between us and them*. New York: Penguin Press.

Greene, J. D., Cushman, F. A., Stewart, L. E., Lowenberg, K., Nystrom, L. E., & Cohen, J. D. (2009). Pushing moral buttons: The interaction between personal force and intention in moral judgment. *Cognition, 111*(3), 364–371.

Greene, J., & Haidt, J. (2002). How (and where) does moral judgment work? *Trends in Cognitive Sciences, 6*, 517–523.

Greene, J. D., Morelli, S., Lowenberg, K., Nystrom, L., & Cohen, J. D. (2008). Cognitive load selectively interferes with utilitarian moral judgment. *Cognition, 107*, 1144–1154.

Greene, J. D., Nystrom, L. E., Engell, A. D., Darley, J. M., & Cohen, J. D. (2004). The neural bases of cognitive conflict and control in moral judgment. *Neuron, 44*, 389–400.

Greene, J. D., & Paxton, J. M. (2009). Patterns of neural activity associated with honest and dishonest moral decisions. *Proceedings of the National Academy of Sciences USA, 106*(30), 12506–12511.

Greene, J. D., Sommerville, R. B., Nystrom, L. E., Darley, J. M., & Cohen, J. D. (2001). An fMRI investigation of emotional engagement in moral judgment. *Science, 293*, 2105–2108.

Haidt, J. (2001). The emotional dog and its rational tail: A social intuitionist approach to moral judgment. *Psychological Review, 108*, 814–834.

Haidt, J. (2012). *The righteous mind: Why good people are divided by politics and religion*. New York: Random House.

Hare, R. D. (1991). *The Hare psychopathy checklist-revised*. Toronto: Multi-Health Systems.

Hare, T. A., Camerer, C. F., Knoepfle, D. T., O'Doherty, J. P., & Rangel, A. (2010). Value computations in ventral medial prefrontal cortex during charitable decision making incorporate input from regions involved in social cognition. *Journal of Neuroscience, 30*(2), 583–590.

Harenski, C. L., Harenski, K. A., Shane, M. S., & Kiehl, K. A. (2010). Aberrant neural processing of moral violations in criminal psychopaths. *Journal of Abnormal Psychology, 119*(4), 863.

Harenski, C. L., Kim, S. H., & Hamann, S. (2009). Neuroticism and psychopathy predict brain activation during moral and nonmoral emotion regulation. *Cognitive, Affective & Behavioral Neuroscience, 9*(1), 1–15.

Hauser, M. (2006). The liver and the moral organ. *Social Cognitive and Affective Neuroscience, 1*, 214–220.

Heekeren, H. R., Wartenburger, I., Schmidt, H., Prehn, K., Schwintowski, H. P., & Villringer, A. (2005). Influence of bodily harm on neural correlates of semantic and moral decision-making. *NeuroImage, 24*, 887–897.

Hsu, M., Anen, C., & Quartz, S. R. (2008). The right and the good: Distributive justice and neural encoding of equity and efficiency. *Science, 320*, 1092–1095.

Insel, T. R., & Young, L. J. (2001). The neurobiology of attachment. *Nature Reviews. Neuroscience, 2*, 129–136.

Kahneman, D. (2003). A perspective on judgment and choice: Mapping bounded rationality. *American Psychologist, 58*, 697–720.

Kant, I. (1959). *Foundation of the metaphysics of morals*. Indianapolis: Bobbs-Merrill. (Original work published 1785.)

Kiehl, K. A. (2006). A cognitive neuroscience perspective on psychopathy: Evidence for paralimbic system dysfunction. *Psychiatry Research, 142*, 107–128.

Kiehl, K. A., Smith, A. M., Hare, R. D., Mendrek, A., Forster, B. B., Brink, J., et al. (2001). Limbic abnormalities in affective processing by criminal psychopaths as revealed by functional magnetic resonance imaging. *Biological Psychiatry, 50*, 677–684.

Koven, N. S. (2011). Specificity of meta-emotion effects on moral decision-making. *Emotion (Washington, D.C.), 11*(5), 1255.

Knutson, B., Taylor, J., Kaufman, M., Peterson, R., & Glover, G. (2005). Distributed neural representation of expected value. *Journal of Neuroscience, 25*(19), 4806–4812.

Koenigs, M., Kruepke, M., & Newman, J. P. (2010). Economic decision-making in psychopathy: A comparison with ventromedial prefrontal lesion patients. *Neuropsychologia, 48*(7), 2198–2204.

Koenigs, M., Kruepke, M., Zeier, J., & Newman, J. P. (2012). Utilitarian moral judgment in psychopathy. *Social Cognitive and Affective Neuroscience, 7*(6), 708–714.

Koenigs, M., & Tranel, D. (2007). Irrational economic decision-making after ventromedial prefrontal damage: Evidence from the Ultimatum Game. *Journal of Neuroscience, 27*, 951–956.

Koenigs, M., Young, L., Adolphs, R., Tranel, D., Cushman, F., Hauser, M., et al. (2007). Damage to the prefrontal cortex increases utilitarian moral judgements. *Nature, 446*, 908–911.

Kohlberg, L. (1969). Stage and sequence: The cognitive-developmental approach to socialization. In D. A. Goslin (Ed.), *Handbook of socialization theory and research* (pp. 347–480). Chicago: Rand McNally.

Kosfeld, M., Heinrichs, M., Zak, P. J., Fischbacher, U., & Fehr, E. (2005). Oxytocin increases trust in humans. *Nature, 435*, 673–676.

Koster-Hale, J., Saxe, R., Dungan, J., & Young, L. L. (2013). Decoding moral judgments from neural representations of intentions. *Proceedings of the National Academy of Sciences USA, 110*(14), 5648–5653.

Kriegeskorte, N., Goebel, R., & Bandettini, P. (2006). Information-based functional brain mapping. *Proceedings of the National Academy of Sciences USA, 103*(10), 3863–3868.

Maddock, R. J. (1999). The retrosplenial cortex and emotion: New insights from functional neuroimaging of the human brain. *Trends in Neurosciences, 22*, 310–316.

Marsh, A. A., Crowe, S. L., Henry, H. Y., Gorodetsky, E. K., Goldman, D., & Blair, R. J. R. (2011). Serotonin transporter genotype (5-HTTLPR) predicts utilitarian moral judgments. *PLoS ONE, 6*(10), e25148.

Marsh, A., Finger, E., Mitchell, D., Reid, M., Sims, C., Kosson, D., et al. (2008). Reduced amygdala response to fearful expressions in children and adolescents with callous-unemotional traits and disruptive behavior disorders. *American Journal of Psychiatry, 165*(6), 712–720.

Mendez, M. F., Anderson, E., & Shapira, J. S. (2005). An investigation of moral judgement in frontotemporal dementia. *Cognitive and Behavioral Neurology, 18*, 193–197.

Mill, J. S. (1861/1998). *Utilitarianism.* New York: Oxford University Press.

Miller, E. K., & Cohen, J. D. (2001). An integrative theory of prefrontal cortex function. *Annual Review of Neuroscience, 24*, 167–202.

Miller, M. B., Sinnott-Armstrong, W., Young, L., King, D., Paggi, A., Fabri, M., et al. (2010). Abnormal moral reasoning in complete and partial callosotomy patients. *Neuropsychologia*, *48*(7), 2215–2220.

Mitchell, J. P. (2009). Inferences about mental states. *Philosophical Transactions of the Royal Society of London. B, Biological Sciences*, *364*(1521), 1309–1316.

Moll, J., de Oliveira-Souza, R., Eslinger, P. J., Bramati, I. E., Mourão-Miranda, J., Andreiuolo, P. A., et al. (2002). The neural correlates of moral sensitivity: A functional magnetic resonance imaging investigation of basic and moral emotions. *Journal of Neuroscience*, *22*(7), 2730–2736.

Moll, J., Eslinger, P. J., & de Oliveira-Souza, R. (2001). Frontopolar and anterior temporal cortex activation in a moral judgment task: Preliminary functional MRI results in normal subjects. *Arquivos de Neuro-Psiquiatria*, *59*, 657–664.

Moll, J., Krueger, F., Zahn, R., Pardini, M., de Oliveira-Souza, R., & Grafman, J. (2006). Human fronto-mesolimbic networks guide decisions about charitable donation. *Proceedings of the National Academy of Sciences USA*, *103*, 15623–15628.

Moran, J. M., Young, L. L., Saxe, R., Lee, S. M., O'Young, D., Mavros, P. L., et al. (2011). Impaired theory of mind for moral judgment in high-functioning autism. *Proceedings of the National Academy of Sciences USA*, *108*(7), 2688–2692.

Moretto, G., Làdavas, E., Mattioli, F., & Di Pellegrino, G. (2010). A psychophysiological investigation of moral judgment after ventromedial prefrontal damage. *Journal of Cognitive Neuroscience*, *22*(8), 1888–1899.

Morishima, Y., Schunk, D., Bruhin, A., Ruff, C. C., & Fehr, E. (2012). Linking brain structure and activation in temporoparietal junction to explain the neurobiology of human altruism. *Neuron*, *75*(1), 73–79.

Norman, K. A., Polyn, S. M., Detre, G. J., & Haxby, J. V. (2006). Beyond mind-reading: Multi-voxel pattern analysis of fMRI data. *Trends in Cognitive Sciences*, *10*(9), 424–430.

Oxford, M., Cavell, T. A., & Hughes, J. N. (2003). Callous/unemotional traits moderate the relation between ineffective parenting and child externalizing problems: A partial replication and extension. *Journal of Clinical Child and Adolescent Psychology*, *32*, 577–585.

Parkinson, C., Sinnott-Armstrong, W., Koralus, P. E., Mendelovici, A., McGeer, V., & Wheatley, T. (2011). Is morality unified? Evidence that distinct neural systems underlie moral judgments of harm, dishonesty, and disgust. *Journal of Cognitive Neuroscience*, *23*(10), 3162–3180.

Paxton, J. M., Ungar, L., & Greene, J. D. (2012). Reflection and reasoning in moral judgment. *Cognitive Science*, *36*(1), 163–177.

Perkins, A. M., Leonard, A. M., Weaver, K., Dalton, J. A., Mehta, M. A., Kumari, V., & Ettinger, U. (2012). A dose of ruthlessness: Interpersonal moral judgment is hardened by the anti-anxiety drug lorazepam. *Journal of Experimental Psychology. General*, *142*(3), 612–620.

Pessoa, L. (2010). Emotion and cognition and the amygdala: From "what is it?" to "what's to be done?" *Neuropsychologia*, *48*(12), 3416–3429.

Petrinovich, L., O'Neill, P., & Jorgensen, M. (1993). An empirical study of moral intuitions: Toward an evolutionary ethics. *Journal of Personality and Social Psychology*, *64*, 467–478.

Phelps, E. A., O'Connor, K. J., Cunningham, W. A., Funayama, E. S., Gatenby, J. C., Gore, J. C., et al. (2000). Performance on indirect measures of race evaluation predicts amygdala activation. *Journal of Cognitive Neuroscience*, *12*(5), 729–738.

Piaget, J. (1965). *The moral judgment of the child*. New York: Free Press.

Pinker, S. (2011). *The better angels of our nature: Why violence has declined*. New York: Viking.

Poulin, M. J., Holman, E. A., & Buffone, A. (2012). The neurogenetics of nice receptor genes for oxytocin and vasopressin interact with threat to predict prosocial behavior. *Psychological Science*, *23*(5), 446–452.

Pujol, J., Batalla, I., Contreras-Rodríguez, O., Harrison, B. J., Pera, V., Hernández-Ribas, R., et al. (2012). Breakdown in the brain network subserving moral judgment in criminal psychopathy. *Social Cognitive and Affective Neuroscience*, *7*(8), 917–923.

Raichle, M. E., MacLeod, A. M., Snyder, A. Z., Powers, W. J., Gusnard, D. A., & Shulman, G. L. (2001). A default mode of brain function. *Proceedings of the National Academy of Sciences USA*, *98*(2), 676–682.

Raine, A., & Yang, Y. (2006). Neural foundations to moral reasoning and antisocial behavior. *Social Cognitive and Affective Neuroscience*, *1*, 203–213.

Rand, D. G., Greene, J. D., & Nowak, M. A. (2012). Spontaneous giving and calculated greed. *Nature*, *489*(7416), 427–430.

Rangel, A., Camerer, C., & Montague, P. R. (2008). A framework for studying the neurobiology of value-based decision making. *Nature Reviews. Neuroscience*, *9*(7), 545–556.

Rangel, A., & Hare, T. (2010). Neural computations associated with goal-directed choice. *Current Opinion in Neurobiology*, *20*(2), 262–270.

Rilling, J., Glenn, A., Jairam, M., Pagnoni, G., Goldsmith, D., Elfenbein, H., et al. (2007). Neural correlates of social cooperation and non-cooperation as a function of psychopathy. *Biological Psychiatry*, *61*, 1260–1271.

Sanfey, A. G., Rilling, J. K., Aronson, J. A., Nystrom, L. E., & Cohen, J. D. (2003). The neural basis of economic decision-making in the Ultimatum Game. *Science, 300,* 1755–1758.

Sarlo, M., Lotto, L., Manfrinati, A., Rumiati, R., Gallicchio, G., & Palomba, D. (2012). Temporal dynamics of cognitive–emotional interplay in moral decision-making. *Journal of Cognitive Neuroscience, 24*(4), 1018–1029.

Saver, J., & Damasio, A. (1991). Preserved access and processing of social knowledge in a patient with acquired sociopathy due to ventromedial frontal damage. *Neuropsychologia, 29,* 1241–1249.

Schaich Borg, J., Hynes, C., Van Horn, J., Grafton, S., & Sinnott-Armstrong, W. (2006). Consequences, action, and intention as factors in moral judgments: An fMRI investigation. *Journal of Cognitive Neuroscience, 18,* 803–817.

Schaich Borg, J., Lieberman, D., & Kiehl, K. A. (2008). Infection, incest, and iniquity: Investigating the neural correlates of disgust and morality. *Journal of Cognitive Neuroscience, 20,* 1–19.

Schaich Borg, J., Sinnott-Armstrong, W. (2013). Do psychopaths make moral judgments? In K. Kiehl & W. Sinnott-Armstrong (Eds.), *Handbook on psychopathy and law* (pp. 107–128). New York: Oxford University Press.

Schaich Borg, J., Sinnott-Armstrong, W., Calhoun, V. D., & Kiehl, K. A. (2011). Neural basis of moral verdict and moral deliberation. *Social Neuroscience, 6*(4), 398–413.

Schultz, W., Dayan, P., & Montague, P. R. (1997). A neural substrate of prediction and reward. *Science, 275,* 1593–1599.

Shenhav, A., & Greene, J. D. (2010). Moral judgments recruit domain-general valuation mechanisms to integrate representations of probability and magnitude. *Neuron, 67*(4), 667–677.

Shenhav, A., & Greene, J. D. (2013). Integrative moral judgment: Dissociating the roles of the amygdala and ventromedial prefrontal cortex. *Journal of Neuroscience, 34*(13), 4741–4749.

Singer, T., Kiebel, S., Winston, J., Dolan, R., & Frith, C. (2004). Brain response to the acquired moral status of faces. *Neuron, 41,* 653–662.

Singer, T., Seymour, B., O'Doherty, J. P., Stephan, K. E., Dolan, R. J., & Frith, C. D. (2006). Empathic neural responses are modulated by the perceived fairness of others. *Nature, 439,* 466–469.

Tabibnia, G., Satpute, A. B., & Lieberman, M. D. (2008). The sunny side of fairness: Preference for fairness activates reward circuitry (and disregarding unfairness activates self-control circuitry). *Psychological Science, 19,* 339–347.

Thomas, B. C., Croft, K. E., & Tranel, D. (2011). Harming kin to save strangers: Further evidence for abnormally utilitarian moral judgments after ventromedial prefrontal damage. *Journal of Cognitive Neuroscience, 23*(9), 2186–2196.

Thomson, J. (1985). The trolley problem. *Yale Law Journal, 94*, 1395–1415.

Turiel, E. (2006). Thought, emotions and social interactional processes in moral development. In M. Killen & J. Smetana (Eds.), *Handbook of moral development* (pp. 1–30). Mahwah, NJ: Lawrence Erlbaum Associates.

Viding, E., Blair, R. J., Moffitt, T. E., & Plomin, R. (2005). Evidence for substantial genetic risk for psychopathy in 7-year-olds. *Journal of Child Psychology and Psychiatry, and Allied Disciplines, 46*, 592–597.

Waytz, A., Zaki, J., & Mitchell, J. P. (2012). Response of dorsomedial prefrontal cortex predicts altruistic behavior. *Journal of Neuroscience, 32*(22), 7646–7650.

Young, L., Bechara, A., Tranel, D., Damasio, H., Hauser, M., & Damasio, A. (2010). Damage to ventromedial prefrontal cortex impairs judgment of harmful intent. *Neuron, 65*(6), 845–851.

Young, L., Camprodon, J. A., Hauser, M., Pascual-Leone, A., & Saxe, R. (2010). Disruption of the right temporoparietal junction with transcranial magnetic stimulation reduces the role of beliefs in moral judgments. *Proceedings of the National Academy of Sciences USA, 107*(15), 6753–6758.

Young, L., Cushman, F., Hauser, M., & Saxe, R. (2007). The neural basis of the interaction between theory of mind and moral judgment. *Proceedings of the National Academy of Sciences USA, 107*(15), 6753–6758.

Young, L., & Dungan, J. (2012). Where in the brain is morality? Everywhere and maybe nowhere. *Social Neuroscience, 7*(1), 1–10.

Zaki, J., & Mitchell, J. P. (2011). Equitable decision making is associated with neural markers of intrinsic value. *Proceedings of the National Academy of Sciences USA, 108*(49), 19761–19766.

13   Neuromodulators and the (In)stability of Moral Cognition

Molly J. Crockett and Regina A. Rini

Ancient moral codes may be written in stone, but human moral cognition is malleable. Moral judgments and decisions are susceptible to influence from a variety of factors, many of which are nonnormative in the sense that they are not relevant to the moral dilemma under consideration. For example, people make harsher moral judgments in the presence of disgusting smells (Schnall, Haidt, Clore, & Jordan, 2008) and are more likely to cheat when the lights are low (Zhong, Bohns, & Gino, 2010). Many of these incidental influences on morality have been reviewed elsewhere (Greene, 2011, 2014; Huebner, Dwyer, & Hauser, 2009). This chapter explores how and why physiological changes in the brain and body can influence human moral cognition.

One especially striking example of nonnormative influences on moral judgments comes from a study conducted with Israeli judges. Danziger and colleagues (2011) were interested in testing empirically the common trope that "justice is what the judge ate for breakfast" (Kozinski, 1992). To investigate this question the researchers examined over a thousand sequential parole decisions made by experienced judges and tested whether the judges' two daily food breaks influenced their decisions. Strikingly, they found that judges were substantially more likely to grant parole when the decision took place directly following a food break compared with when the decision took place long after a food break. The effect of food breaks remained significant even when controlling for legally relevant factors such as the prisoner's history of repeat offenses and whether he was enrolled in a rehabilitation program (Danziger, Levav, & Avnaim-Pesso, 2011).

The fact that something as trivial as a snack break can influence hugely consequential judicial decisions is deeply problematic if we intend for our legal systems to operate on the basis of normative principles. It is therefore essential that we investigate how such extraneous factors influence moral judgment and decision making. Studying the brain can advance this agenda by illuminating the mechanisms through which the environment can affect moral cognition.

One potential mechanism through which extraneous factors influence moral cognition is the context-sensitive modulation of neuronal activity by *neuromodulator* systems. Neuromodulators are chemicals that modify neuronal dynamics, excitability, and synaptic function. These include neurotransmitters (e.g., serotonin, norepinephrine, acetylcholine, and dopamine) as well as hormones (e.g., testosterone, oxytocin, vasopressin). These chemical systems may serve the function of preparing organisms to interact optimally with the environment, adaptively shaping behavior to fit the current context. Activation of one or more neuromodulator systems is an efficient way to globally alter the computational properties of neural networks (Robbins & Arnsten, 2009). Recent work in neuroscience has demonstrated that manipulating the function of neuromodulators in the laboratory can influence moral cognition in humans (Crockett & Fehr, 2013, 2014). Here, we review this evidence and explore its normative implications.

How Neuromodulators Shape Moral Cognition

Moral Judgment

How do people *judge* whether an action is morally permissible, and how is moral judgment shaped by neuromodulators? Perhaps the most widely used tool for probing human moral judgment is the (in)famous *trolley problem*. In one variant ("push"), a trolley is heading down a track toward five workers, who will be killed if no action is taken. An unwitting protagonist (call him Joe) is standing on a footbridge overlooking the tracks, along with another person wearing a large, heavy backpack. If Joe pushes the backpacker off the footbridge and onto the tracks, the backpacker's body will stop the trolley before it hits the five workers, killing him but saving the five workers. Participants reading the scenario are asked to judge whether it is morally acceptable for Joe to push the backpacker off the footbridge.

Moral dilemmas like these generate conflicting responses from two major traditions in normative ethics. *Consequentialism* judges the moral acceptability of actions based on their outcomes, so according to this tradition, it is morally acceptable, indeed even required, for Joe to kill one person in order to save five others. *Deontology*, on the other hand, judges the acceptability of actions based on factors other than outcomes, such as whether the action involves treating a person as a means to an end. Certain actions, such as killing an innocent bystander as a means of saving others, are strictly prohibited, no matter the potential benefits; according to this tradition, it is unacceptable for Joe to kill one to save five in this way.

Much ink has been spilled in the quest to understand when, why, and how people adopt consequentialist versus deontological perspectives in moral judgment (Crockett, 2013; Cushman, 2013; Greene, 2014; Mikhail, 2007). More recently, researchers have begun to investigate how manipulating neuromodulator systems influences judgment in moral dilemmas (see table 13.1). In a typical design participants are randomly assigned to receive a drug or a placebo, and subsequently make a series of moral judgments in a set of moral dilemmas like the trolley problem. In some studies these dilemmas are divided into "personal" dilemmas that involve emotionally salient harms (such as the "push" scenario above) and "impersonal" dilemmas

Table 13.1

Neuromodulation of moral judgment

Treatment	Effect on neuromodulator	Personal dilemmas	Impersonal dilemmas	References
Citalopram, 30 mg (SSRI)	↑ serotonin	↑ deontological	—	Crockett et al., 2010
Propranolol, 40 mg (β-blocker) Atomoxetine, 60 mg (SNRI)	↓ norepinephrine ↑ norepinephrine	↑ deontological —	— —	Terbeck et al., 2013 Crockett et al., 2010
Trier Social Stress Task Public speech anticipation	↑ stress ↑ stress	↑ deontological ↑ deontological	— ↑ deontological	Youssef et al., 2012 Starcke et al., 2012
Intranasal oxytocin, 24 IU	↑ oxytocin	n/t	↑ deontological (ingroup only)	De Dreu et al., 2011
Sublingual testosterone, 0.5 mg	↑ testosterone	↓ deontological (inevitable harms, covariate 2D:4D)	—	Montoya et al., 2013

Abbreviations: SSRI, selective serotonin reuptake inhibitor; SNRI, selective norepinephrine reuptake inhibitor; IU, intranasal units; 2D:4D, second- to fourth-digit ratio, a measure of prenatal testosterone exposure; —, no effect; n/t, not tested.

where harms are less salient (e.g., the "switch" scenario, in which Joe has the option to flip a switch to divert the trolley onto a different set of tracks where there is one worker instead of five).

In the first of such studies Crockett, Clark, Hauser, and Robbins (2010) investigated how the neuromodulator serotonin influences moral judgment. Serotonin has long been implicated in social behavior (Insel & Winslow, 1998); in general, impaired serotonin function has been associated with aggression and antisocial behavior, whereas intact or enhanced serotonin function has been associated with prosocial behavior. To test how serotonin influences moral judgment, Crockett and colleagues examined the effects of citalopram (a selective serotonin reuptake inhibitor, which enhances serotonin function by prolonging its actions in the synapse) on judgments in personal and impersonal scenarios. Citalopram promoted deontological responding, thereby reducing subjects' willingness to endorse harming one to save many others (Crockett et al., 2010). This effect was selective to emotionally salient personal scenarios and was stronger in individuals who scored higher on an independent measure of empathy. These results are consistent with other work suggesting that serotonin facilitates aversive processing (Crockett, Clark, & Robbins, 2009; Dayan & Huys, 2009), which could result in a serotoninergic enhancement of harm aversion in social settings (Siegel & Crockett, 2013).

Other studies have examined the role of norepinephrine in moral judgment. Norepinephrine is a neuromodulator thought to play a role in sharpening attention in response to arousal and stress (Robbins & Arnsten, 2009). Terbeck et al. (2013) tested the effects of blocking the β-adrenergic receptor on judgments in personal and impersonal dilemmas. Noradrenergic blockade increased deontological responding in personal but not impersonal scenarios, suggesting that norepinephrine may normally facilitate consequentialist responding (Terbeck et al., 2013). However, enhancing noradrenergic function with the norepinephrine reuptake inhibitor atomoxetine had no effect on moral judgment (Crockett et al., 2010).

Both serotonergic and noradrenergic neurons are stimulated by acute stress (Robbins & Arnsten, 2009). A few recent studies have examined how stress influences moral judgments. Youssef and colleagues (2012) showed that acute stress induced via the Trier Social Stress Test increased deontological responding in personal, but not impersonal, dilemmas. Starcke and colleagues (2012), using a different method of acute stress induction (anticipated public speech), also found that stress increased deontological responding, but this effect was present in both personal and impersonal dilemmas (Starcke, Ludwig, & Brand, 2012).

Another neuromodulator that is released in response to stress is the hormone oxytocin. Commonly referred to by catchy monikers such as the "love hormone" or "moral molecule," oxytocin plays many important roles in social behavior, although its function is far more complex than popular accounts imply (Bartz, Zaki, Bolger, & Ochsner, 2011). One account, proposed by Taylor (2006), is that during stress, oxytocin promotes a "tend and befriend" social affiliation response. Consistent with this account De Dreu and colleagues report that oxytocin administration increases deontological responding in moral dilemmas—reducing subjects' approval of harming one to save many—but only when the target of harm is an ingroup member (De Dreu, Greer, Kleef, Shalvi, & Handgraaf, 2011).

Testosterone is a hormone thought to be involved in the pursuit of social dominance (Eisenegger, Haushofer, & Fehr, 2011). The effects of testosterone on moral judgment appear to run in the opposite direction of oxytocin; high baseline testosterone levels are associated with decreased deontological responding in moral dilemmas (Carney & Mason, 2010), and testosterone administration decreased deontological responding, although the effects were specific to individuals who likely experienced higher prenatal testosterone exposure, as indicated by the second- to fourth-digit ratio (Montoya et al., 2013).

Considering these findings together, a few broad conclusions emerge. First, the majority of studies have reported neuromodulatory effects on moral judgment that are selective to personal dilemmas. This suggests that neuromodulators influence moral judgment through emotional channels. Second, the bulk of the evidence is consistent with the idea that stress shifts moral judgment toward a deontological style by stimulating the release of monoamines and hormones that may have synergistic effects. Finally, it is worth noting that these results should be interpreted with caution. Most of these studies have been carried out in relatively small samples; the reported effect sizes are small; few effects have been replicated; and most have used the same small set of moral dilemmas, which themselves have important limitations (Christensen & Gomila, 2012). Further work is needed to replicate these basic findings and extend them to a broader range of moral dilemmas.

Moral Decision Making

How do neuromodulators shape moral *decisions* about whether to behave ethically? Although it is challenging to capture ethical decision-making processes in the laboratory, behavioral paradigms have been used to measure ethically relevant behaviors such as aggression, generosity, and cooperation (table 13.2). Early work in this area investigated the effects of neuromodulators on laboratory measures of aggression. These tasks measure subjects' responses to provocation by an opponent, typically in the form of imposed monetary losses, electric shocks, or loud noises. Aggression is operationalized as the level of stimulus intensity or monetary loss delivered in return

Table 13.2
Neuromodulation of moral decision making

Treatment	Effect on neuromodulator	Aggression	Generosity	Cooperation	References
ATD	↓ serotonin	↑			Bjork et al., 1999
		↑			Bjork et al., 2000
		↑			Dougherty et al., 1999
		↑			Marsh et al., 2002
		↑			Moeller et al., 1996
				↓	Wood et al., 2006
		↑			Crockett et al., 2008
		↑			Crockett et al., 2013
Paroxetine, 40mg (SSRI)	↑ serotonin	↓			Berman et al., 2009
Citalopram, 10 mg (SSRI)				↑	Tse & Bond, 2002
Citalopram, 30 mg (SSRI)		↓			Crockett et al., 2010
MDMA, 125 mg			↑		Hysek et al., 2013

l-DOPA, 300 mg	↑ dopamine	↓		Pedroni et al., 2014
Trier Social Stress Test	↑ stress	↑		von Dawans et al., 2012
		↓		Vinkers et al., 2013
Intranasal oxytocin, 24 IU (40 IU)	↑ oxytocin		↑ (avoidant attachment)	De Dreu, 2012
			↑	Rilling et al., 2012
			↑ (familiar)	Declerck et al., 2010
			↑ (ingroup)	De Dreu et al., 2010
			↑	Israel et al., 2012
		↑		Chang et al., 2012
		↑		Barraza et al., 2011
Sublingual testosterone, 0.5mg	↑ testosterone	↑		Eisenegger et al., 2010
			↑ (2D:4D)	van Honk et al., 2012

Abbreviations: ATD, acute tryptophan depletion; SSRI, selective serotonin reuptake inhibitor; l-DOPA, levodopa; IU, intranasal units; 2D:4D, second- to fourth-digit ratio, an index of prenatal testosterone exposure.

to the opponent. Several studies have shown that manipulating serotonin influences behavior in these paradigms. Depleting the chemical precursor of serotonin, tryptophan, increases aggression (Bjork, Dougherty, Moeller, & Swann, 2000; Cleare & Bond, 1995; Crockett et al., 2013; Crockett, Clark, Tabibnia, Lieberman, & Robbins, 2008; Dougherty, Bjork, Marsh, & Moeller, 1999; Marsh, Dougherty, Moeller, Swann, & Spiga, 2002; Moeller et al., 1996; Pihl et al., 1995). Aggressive responding decreases following fenfluramine (Cherek & Lane, 1999), tryptophan augmentation (Marsh et al., 2002), and SSRI administration (Berman, McCloskey, Fanning, Schumacher, & Coccaro, 2009; Crockett et al., 2010), all of which enhance serotonin function.

Generosity is typically measured with the dictator game, in which subjects are given a sum of money and have the opportunity to donate any amount to another person or a charity. Generosity is enhanced by the serotonin-releasing agent MDMA (Hysek et al., 2013). Testosterone administration increases generosity only when the recipient has the option to punish, suggesting that testosterone influences strategic social concerns (Eisenegger, Naef, Snozzi, Heinrichs, & Fehr, 2010). Meanwhile, enhancing dopaminergic neurotransmission with L-DOPA decreases generosity (Pedroni, Eisenegger, Hartmann, Fischbacher, & Knoch, 2014). Oxytocin has more complex effects on generosity. Barraza and colleagues report no effect of oxytocin on decisions to donate to charity; however, among subjects who chose to donate, those who received oxytocin donated more money than those who received placebo (Barraza, McCullough, Ahmadi, & Zak, 2011). A study in rhesus monkeys found that oxytocin increased generous choices to reward another when the alternative was to reward no one, but oxytocin also increased selfish choices to reward oneself when the alternative was to reward another monkey (Chang, Barter, Ebitz, Watson, & Platt, 2012). Evidence for acute stress effects on generosity is also mixed; one study reported increased generosity following acute stress (von Dawans, Fischbacher, Kirschbaum, Fehr, & Heinrichs, 2012), whereas another reported reduced generosity following stress (Vinkers et al., 2013). One possible explanation for these conflicting findings is that von Dawans et al. studied generosity toward other participants in the lab (who could have been construed as ingroup members), whereas Vinkers et al. studied donations to UNICEF (which could have been construed as providing for outgroup members). If stress promotes generosity toward ingroup members but reduces generosity toward outgroup members, this could account for the observed pattern of results (Vinkers et al., 2013).

Preferences for social cooperation have been studied using the prisoner's dilemma and the public goods game. In these games subjects have the option to cooperate, which results in higher average payoffs for the group as a whole, or to defect, which results in higher average payoffs for oneself. Enhancing serotonin function with the SSRI citalopram increases cooperation (Tse & Bond, 2002), whereas depleting the serotonin precursor tryptophan decreases cooperation (Wood, Rilling, Sanfey, Bhagwagar, & Rogers, 2006). Testosterone administration increases cooperation—but only among individuals with lower prenatal testosterone exposure (van Honk, Montoya, Bos, van Vugt, & Terburg, 2012). Meanwhile, oxytocin generally increases cooperation (Rilling et al., 2012), but in some studies the effects of oxytocin interact with individual difference variables such as attachment avoidance (De Dreu, 2012). In addition, oxytocin appears to selectively enhance cooperation with familiar others or ingroup members (Declerck, Boone, & Emonds, 2013; De Dreu et al., 2010; but see also Israel, Weisel, Ebstein, & Bornstein, 2012).

From these findings it becomes apparent that serotonin and oxytocin both generally promote prosocial decision making; these two systems may even have synergistic effects in light of evidence that they interact in the brain (Dölen, Darvishzadeh, Huang, & Malenka, 2013). However, there are not yet enough data on the effects of testosterone, stress, and dopamine manipulations to draw any firm conclusions. Moreover, it is clear that the effects of neuromodulators on moral decision making are highly sensitive to changes in context and individual personality traits. Further work is required to flesh out the nature of these interactions.

Discussion: Normative Implications

Neuromodulators have been shown to affect moral judgment and deci-sion making in a number of different ways. Some scholars have recently proposed that we could use this knowledge to improve ourselves and our societies (Douglas, 2008; Persson & Savulescu, 2012). Aside from the fact that we are not even close to understanding the precise ways in which neu-romodulators shape moral cognition, applying this knowledge is far from straightforward. Think again about Danziger and colleagues' (2011) study of judges. We can all agree that one's chance at parole should not come down to a judge's snack schedule. Suppose we came to understand the neu-romodulators involved in this effect, and we could prescribe "stabilizer" drugs for judges that ensured they would make the same judgments no matter when they had last eaten. This could seem like an improvement in justice. But notice that to do this we would have to decide which type of judge is the *better* one. Do we want judges who judge as if they were hungry and so are tougher on potential parolees? Or do we want judges who judge as if they were sated and so are less tough? We can all agree that judges should judge consistently—but consistently in which direction? And how would we decide?

These are normative questions about what *should* be done. If neuromod-ulators are ever going to be of any practical use, we shall have to address these normative questions. We do not claim to be able to resolve any of these questions here. Instead, we can only try to make clear two important sets of considerations for future work on the normative implications of neuromodulation.

First: *who decides* which neuromodulators to use? It might seem obvious that individuals should be allowed to choose for themselves which substances they ingest, but this point runs into problems with moral neuromodulation. Suppose that you are convinced that deontological moral judgments are gravely morally wrong. Would you agree that individuals should be able to choose for themselves to take substances that shift them toward *more* deontological judgments? In fact, you might think that individuals should be *compelled* to take neuromodulators promoting consequentialist judgments.

Obviously such ideas trigger important worries about liberty and authenticity (Harris, 2011); we have strong background assumptions that people should be in control of their minds and their own decision making. But moral neuromodulation is not like, say, drinking lattes to get a caffeine buzz. Because moral neuromodulation is specifically about *morality*, it is directly concerned with *interpersonal* decisions with great significance to people beyond the individual. We already recognize limitations on individual discretion when it comes to ingesting certain substances with morally problematic consequences (such as alcohol before driving, or strong hallucinogens at any time). If moral neuromodulators cut directly to affecting moral decisions, is it so obvious that their use should be wholly up to individual choice?

A second question is: what are the implications of neuromodulation for our understanding of *morality itself*? The evidence reviewed here shows that our moral judgments and decisions are malleable and contingent. That is, if you had different levels of serotonin, oxytocin, or testosterone, then you might disapprove of some things that now seem right to you and accept some of what now seems wrong. Because neurochemical levels fluctuate naturally, in response to everyday events such as changes in diet or stress, even without artificially manipulating your brain chemistry, your judgments might shift. Most of us want to believe that our moral judgments and decisions are far less contingent than this. Some may feel strongly that what we consider *right* and *wrong* should not depend on the happenstance of neurochemistry. The very concept of moral neuromodulation may therefore bring with it an unnerving sense of uncertainty in some of our most important choices.

A natural suggestion, then, is that we should seek to think about morality from some morally "neutral" neurochemical state. But what would that mean? All of our thinking—every brain event—is facilitated by some combination of neurochemicals. How do we decide which state is the "neutral" one? This is the problem of the snacking judges again, applied more generally: we know that we do not want our moral commitments to fluctuate arbitrarily, but do we want to be like the sated judges or like the hungry judges? Even if it seems to you that, say, being a tough judge is a good thing, *this* opinion is the result of some combination of neurochemicals. How do you know *that* combination is the right one?

All of this means that we will have difficulty answering questions about *how* to decide which (if any) neuromodulators to use. Absurdly, one can imagine rival moral philosophers prescribing rival drugs to one another: "if only you'd take this drug, then you'd agree with me about consequentialism (and that you ought to take this drug!)." But this problem seems inescapable—moral neuromodulation is centrally about altering our tendencies toward certain types of judgments and decisions, and we have no basis for choosing such alterations other than those very same judgments and decisions. This is obviously a knotty problem, which in some ways will seem familiar to moral philosophers. If moral neuromodulation becomes a practical possibility, it is a problem that will have to be addressed by everyone: scientists, policy makers, and all morally conscientious people.

References

Barraza, J. A., McCullough, M. E., Ahmadi, S., & Zak, P. J. (2011). Oxytocin infusion increases charitable donations regardless of monetary resources. *Hormones and Behavior, 60*(2), 148–151.

Bartz, J. A., Zaki, J., Bolger, N., & Ochsner, K. N. (2011). Social effects of oxytocin in humans: Context and person matter. *Trends in Cognitive Sciences, 15*(7), 301–309.

Berman, M. E., McCloskey, M. S., Fanning, J. R., Schumacher, J. A., & Coccaro, E. F. (2009). Serotonin augmentation reduces response to attack in aggressive individuals. *Psychological Science, 20*(6), 714–720.

Bjork, J. M., Dougherty, D. M., Moeller, F. G., Cherek, D. R., & Swann, A. C. (1999). The effects of tryptophan depletion and loading on laboratory aggression in men: time course and a food-restricted control. *Psychopharmacology, 142*(1), 24–30.

Bjork, J. M., Dougherty, D. M., Moeller, F. G., & Swann, A. C. (2000). Differential behavioral effects of plasma tryptophan depletion and loading in aggressive and nonaggressive men. *Neuropsychopharmacology, 22*(4), 357–369.

Carney, D. R., & Mason, M. F. (2010). Decision making and testosterone: When the ends justify the means. *Journal of Experimental Social Psychology, 46*(4), 668–671.

Chang, S. W. C., Barter, J. W., Ebitz, R. B., Watson, K. K., & Platt, M. L. (2012). Inhaled oxytocin amplifies both vicarious reinforcement and self reinforcement in rhesus macaques (*Macaca mulatta*). *Proceedings of the National Academy of Sciences USA, 109*(3), 959–964.

Cherek, D. R., & Lane, S. D. (1999). Effects of *d,l*-fenfluramine on aggressive and impulsive responding in adult males with a history of conduct disorder. *Psychopharmacology, 146*(4), 473–481.

Christensen, J. F., & Gomila, A. (2012). Moral dilemmas in cognitive neuroscience of moral decision-making: A principled review. *Neuroscience and Biobehavioral Reviews, 36*(4), 1249–1264.

Cleare, A. J., & Bond, A. J. (1995). The effect of tryptophan depletion and enhancement on subjective and behavioural aggression in normal male subjects. *Psychopharmacology, 118*(1), 72–81.

Crockett, M. J. (2013). Models of morality. *Trends in Cognitive Sciences, 17*(8), 363–366.

Crockett, M. J., Apergis-Schoute, A., Herrmann, B., Lieberman, M. D., Muller, U., Robbins, T. W., et al. (2013). Serotonin modulates striatal responses to fairness and retaliation in humans. *Journal of Neuroscience, 33*(8), 3505–3513.

Crockett, M. J., Clark, L., Hauser, M. D., & Robbins, T. W. (2010). From the cover: Serotonin selectively influences moral judgment and behavior through effects on harm aversion. *Proceedings of the National Academy of Sciences USA, 107*(40), 17433–17438.

Crockett, M. J., Clark, L., & Robbins, T. W. (2009). Reconciling the role of serotonin in behavioral inhibition and aversion: Acute tryptophan depletion abolishes punishment-induced inhibition in humans. *Journal of Neuroscience, 29*(38), 11993–11999.

Crockett, M. J., Clark, L., Tabibnia, G., Lieberman, M. D., & Robbins, T. W. (2008). Serotonin modulates behavioral reactions to unfairness. *Science, 320*(5884), 1739.

Crockett, M. J., & Fehr, E. (2013). Pharmacology of economic and social decision making. In P. Glimcher & E. Fehr (Eds.), *Neuroeconomics: Decision making and the brain* (pp. 255–275). New York: Academic Press.

Crockett, M. J., & Fehr, E. (2014). Social brains on drugs: Tools for neuromodulation in social neuroscience. *Social Cognitive and Affective Neuroscience, 9*(2), 250–254.

Cushman, F. (2013). Action, outcome, and value: A dual-system framework for morality. *Personality and Social Psychology Review, 17*(3), 273–292.

Danziger, S., Levav, J., & Avnaim-Pesso, L. (2011). Extraneous factors in judicial decisions. *Proceedings of the National Academy of Sciences USA, 108*(17), 6889–6892.

Dayan, P., & Huys, Q. J. M. (2009). Serotonin in affective control. *Annual Review of Neuroscience, 32*(1), 95–126.

Declerck, C. H., Boone, C., & Emonds, G. (2013). When do people cooperate? The neuroeconomics of prosocial decision making. *Brain and Cognition, 81*(1), 95–117.

De Dreu, C. K. W. (2012). Oxytocin modulates the link between adult attachment and cooperation through reduced betrayal aversion. *Psychoneuroendocrinology, 37*(7), 871–880.

De Dreu, C. K. W. D., Greer, L. L., Handgraaf, M. J. J., Shalvi, S., Kleef, G. A. V., Baas, M., et al. (2010). The neuropeptide oxytocin regulates parochial altruism in intergroup conflict among humans. *Science, 328*(5984), 1408–1411.

De Dreu, C. K. W. D., Greer, L. L., Kleef, G. A. V., Shalvi, S., & Handgraaf, M. J. J. (2011). Oxytocin promotes human ethnocentrism. *Proceedings of the National Academy of Sciences USA, 108*(4), 1262–1266.

Dölen, G., Darvishzadeh, A., Huang, K. W., & Malenka, R. C. (2013). Social reward requires coordinated activity of nucleus accumbens oxytocin and serotonin. *Nature, 501*(7466), 179–184.

Dougherty, D. M., Bjork, J. M., Marsh, D. M., & Moeller, F. G. (1999). Influence of trait hostility on tryptophan depletion-induced laboratory aggression. *Psychiatry Research, 88*(3), 227–232.

Douglas, T. (2008). Moral enhancement. *Journal of Applied Philosophy, 25*(3), 228–245.

Eisenegger, C., Haushofer, J., & Fehr, E. (2011). The role of testosterone in social interaction. *Trends in Cognitive Sciences, 15*(6), 263–271.

Eisenegger, C., Naef, M., Snozzi, R., Heinrichs, M., & Fehr, E. (2010). Prejudice and truth about the effect of testosterone on human bargaining behaviour. *Nature, 463*(7279), 356–359.

Greene, J. D. (2011). Emotion and morality: A tasting menu. *Emotion Review, 3*(3), 227–229.

Greene, J. (2014). *Moral tribes: Emotion, reason and the gap between us and them.* London: Atlantic Books.

Harris, J. (2011). Moral enhancement and freedom. *Bioethics, 25*(2), 102–111.

Huebner, B., Dwyer, S., & Hauser, M. (2009). The role of emotion in moral psychology. *Trends in Cognitive Sciences, 13*(1), 1–6.

Hysek, C. M., Schmid, Y., Simmler, L. D., Domes, G., Heinrichs, M., Eisenegger, C., et al. (2013). MDMA enhances emotional empathy and prosocial behavior. *Social Cognitive and Affective Neuroscience*, nst161. doi:10.1093/scan/nst161.

Insel, T. R., & Winslow, J. T. (1998). Serotonin and neuropeptides in affiliative behaviors. *Biological Psychiatry, 44*(3), 207–219.

Israel, S., Weisel, O., Ebstein, R. P., & Bornstein, G. (2012). Oxytocin, but not vasopressin, increases both parochial and universal altruism. *Psychoneuroendocrinology, 37*(8), 1341–1344.

Kozinski, A. (1992). What I ate for breakfast and other mysteries of judicial decision making. *Loyola of Los Angeles Law Review, 26*, 993.

Marsh, D. M., Dougherty, D. M., Moeller, F. G., Swann, A. C., & Spiga, R. (2002). Laboratory-measured aggressive behavior of women: Acute tryptophan depletion and augmentation. *Neuropsychopharmacology, 26*(5), 660–671.

Mikhail, J. (2007). Universal moral grammar: Theory, evidence and the future. *Trends in Cognitive Sciences, 11*(4), 143–152.

Moeller, F. G., Dougherty, D. M., Swann, A. C., Collins, D., Davis, C. M., & Cherek, D. R. (1996). Tryptophan depletion and aggressive responding in healthy males. *Psychopharmacology, 126*(2), 97–103.

Montoya, E. R., Terburg, D., Bos, P. A., Will, G.-J., Buskens, V., Raub, W., et al. (2013). Testosterone administration modulates moral judgments depending on second-to-fourth digit ratio. *Psychoneuroendocrinology, 38*(8), 1362–1369.

Pedroni, A., Eisenegger, C., Hartmann, M. N., Fischbacher, U., & Knoch, D. (2014). Dopaminergic stimulation increases selfish behavior in the absence of punishment threat. *Psychopharmacology, 231*(1), 135–141.

Persson, I., & Savulescu, J. (2012). *Unfit for the future: The need for moral enhancement.* Oxford: Oxford University Press.

Pihl, R. O., Young, S. N., Harden, P., Plotnick, S., Chamberlain, B., & Ervin, F. R. (1995). Acute effect of altered tryptophan levels and alcohol on aggression in normal human males. *Psychopharmacology, 119*(4), 353–360.

Rilling, J. K., DeMarco, A. C., Hackett, P. D., Thompson, R., Ditzen, B., Patel, R., et al. (2012). Effects of intranasal oxytocin and vasopressin on cooperative behavior and associated brain activity in men. *Psychoneuroendocrinology, 37*(4), 447–461.

Robbins, T. W., & Arnsten, A. F. T. (2009). The neuropsychopharmacology of fronto-executive function: Monoaminergic modulation. *Annual Review of Neuroscience, 32*(1), 267–287.

Schnall, S., Haidt, J., Clore, G. L., & Jordan, A. H. (2008). Disgust as embodied moral judgment. *Personality and Social Psychology Bulletin, 34*(8), 1096–1109.

Siegel, J. Z., & Crockett, M. J. (2013). How serotonin shapes moral judgment and behavior. *Annals of the New York Academy of Sciences, 1299*(1), 42–51.

Starcke, K., Ludwig, A.-C., & Brand, M. (2012). Anticipatory stress interferes with utilitarian moral judgment. *Judgment and Decision Making, 7*(1). Retrieved from http://journal.sjdm.org/11/11729/jdm11729.html.

Taylor, S. E. (2006). Tend and befriend biobehavioral bases of affiliation under stress. *Current Directions in Psychological Science, 15*(6), 273–277.

Terbeck, S., Kahane, G., McTavish, S., Savulescu, J., Levy, N., Hewstone, M., et al. (2013). Beta adrenergic blockade reduces utilitarian judgement. *Biological Psychology, 92*(2), 323–328.

Tse, W., & Bond, A. (2002). Serotonergic intervention affects both social dominance and affiliative behaviour. *Psychopharmacology, 161*(3), 324–330.

Van Honk, J., Montoya, E. R., Bos, P. A., van Vugt, M., & Terburg, D. (2012). New evidence on testosterone and cooperation. *Nature, 485*(7399), E4–E5.

Vinkers, C. H., Zorn, J. V., Cornelisse, S., Koot, S., Houtepen, L. C., Olivier, B., et al. (2013). Time-dependent changes in altruistic punishment following stress. *Psychoneuroendocrinology, 38*(9), 1467–1475.

Von Dawans, B., Fischbacher, U., Kirschbaum, C., Fehr, E., & Heinrichs, M. (2012). The social dimension of stress reactivity: Acute stress increases prosocial behavior in humans. *Psychological Science, 23*(6), 651–660.

Wood, R. M., Rilling, J. K., Sanfey, A. G., Bhagwagar, Z., & Rogers, R. D. (2006). Effects of tryptophan depletion on the performance of an iterated Prisoner's Dilemma game in healthy adults. *Neuropsychopharmacology, 31*(5), 1075–1084.

Youssef, F. F., Dookeeram, K., Basdeo, V., Francis, E., Doman, M., Mamed, D., et al. (2012). Stress alters personal moral decision making. *Psychoneuroendocrinology, 37*(4), 491–498.

Zhong, C.-B., Bohns, V. K., & Gino, F. (2010). Good lamps are the best police. Darkness increases dishonesty and self-interested behavior. *Psychological Science, 21*(3), 311–314.

V Psychopathic Immorality

14   Immorality in the Adult Brain

Rheanna J. Remmel and Andrea L. Glenn

For most adults, in everyday life, behaving morally is something that requires little thought. Most people do not steal or take advantage of others or utilize shady business practices to get what they need to survive. However, a certain subset of adults has little trouble doing many of these things. These individuals lack the moral conscience that most adults take for granted, and evidence suggests that their brains may reflect this difference. In this chapter we explore the different ways that morality has been studied and measured as well as the ways that it has been found to be deficient in psychopathy and antisocial personality disorder. We begin with a short overview of these constructs.

Psychopathy and Antisocial Personality Disorder

According to the DSM-5 (American Psychiatric Association, 2013), antisocial personality disorder (ASPD) is a personality disorder "characterized by a pervasive pattern of disregard for, and violation of, the rights of others." ASPD is the closest DSM diagnosis to psychopathy, which is not in the DSM, but is described by researchers as a personality disorder characterized by interpersonal, affective, and behavioral features that include manipulation, grandiosity, shallow affect, deception, lack of empathy, impulsivity, and irresponsibility (Hare, 2003). Whereas most psychopaths would qualify for a diagnosis of ASPD, not all individuals with ASPD can be considered psychopaths. A somewhat unique characteristic of psychopathy is that it tends to involve increased instrumental aggression (Blair, 2007a). Instrumental aggression consists of controlled and purposeful aggression with intent to achieve a goal. It is not preceded by strong emotion, as is reactive aggression. Instrumental aggression tends to be premeditated, whereas reactive aggression is impulsive and is usually in response to feeling threatened or provoked. Psychopaths can engage in reactive aggression too, but they tend to engage in higher rates of predatory violence (which involves instrumental aggression) than nonpsychopaths (Williamson, Hare, & Wong, 1987), especially for more serious crimes such as murder. This distinction is relevant to the study of morality in psychopaths because it demonstrates a pronounced lack of concern for the well-being of others rather than simply a failure to regulate emotions of anger. This could be a result of deficits in brain regions important for emotion and morality, leaving psychopaths undeterred from performing harmful actions for their own advantage (Glenn & Raine, 2009). However, not all psychopaths commit violent crimes. Antisocial behavior also includes acts that are nonviolent and nonillegal, such as manipulation and lying. These types of antisocial behaviors are more common than violent or illegal acts.

Conceptually, psychopathy can generally be divided into four facets: interpersonal, affective, lifestyle, and antisocial behavior. Some researchers group these into two factors, so that Factor 1 psychopathy refers to the interpersonal/affective traits, and Factor 2 refers to the antisocial/lifestyle traits. The most commonly used instrument to measure psychopathy, especially in forensic settings, is the Psychopathy Checklist-Revised (PCL-R; Hare, 2003). The PCL-R is a semistructured interview and comprehensive records review, resulting in a score that can range from 0 to 40. Self-report measures have also been developed primarily for use in nonincarcerated populations and include the Levenson Self-Report Psychopathy Scale (LSRP; Levenson, Kiehl, & Fitzpatrick, 1995) and the Psychopathic Personality Inventory (PPI; Lilienfeld & Andrews, 1996).

Although many of the interpersonal and affective aspects of psychopathy are not included in the current criteria for ASPD, the conceptualization of ASPD is moving closer to psychopathy, as evidenced by the DSM-5 introducing the "alternative DSM-5 model for personality disorders." This model includes more emphasis on the interpersonal and affective elements typically associated more with psychopathy than with ASPD (such as empathy, callousness, deceitfulness, and hostility). It also includes a specifier to clarify if the individual presents "with psychopathic features." In future editions this conceptualization may be included in the diagnostic criteria, which would close the gap slightly between the two disorders.

Morality and Moral Decision Making

Morality can be measured and tested in several different ways, including the prisoner's dilemma (PD) and classic trolley problems (Greene, Sommerville, Nystrom, Darley, & Cohen, 2001). The PD describes a situation in which two men have been accused of robbing a bank and murdering a bank employee. They are being held separately for questioning and cannot communicate. At least one of them has to confess in order to be convicted. If neither of them confesses, they will both be charged with a lesser crime and will both serve five years in prison. If one testifies while the other refuses, the man who testifies gets the charges dropped, and the other will serve 30 years in prison. If both testify, each will get 10 years in prison. The best overall outcome is for both to remain silent, which is a cooperation strategy. However, the best outcome for each of them individually is if they testify, and the other person does not, which is a cheating strategy. This dilemma has been adapted for laboratory settings by adding monetary values instead of prison sentences, as well. It represents a one-time interaction with another individual.

The trolley problems differentiate between personal and impersonal moral dilemmas. In the impersonal dilemma a trolley is coming down a track and is headed toward five people, who will all die if it hits them. You can pull a lever that will switch it onto a different track, killing just one person and saving the five. In the personal dilemma you are standing next to a large stranger on a footbridge overlooking the tracks between the trolley and the five people. You can push the stranger off the bridge and onto the tracks, and his body will stop the trolley. Again, this will kill one person and save five. Although the outcome of each scenario is the same, most people say that you ought to pull the lever to move the trolley but that you ought not to push the man off the bridge to stop it (Greene, Sommerville, Nystrom, Darley, & Cohen, 2001). The difference seems to be that people feel more personal responsibility for pushing the man off the bridge than for pulling the lever and killing the person. In an fMRI study of these dilemmas in a "normal" population, Greene and colleagues (2001) found that the medial frontal gyrus, posterior cingulate gyrus, and bilateral angular gyrus (also referred to as the superior temporal gyrus) were all significantly more active in the personal than the impersonal conditions. These three areas may make up a network that responds when an individual makes a decision that has some sort of personal moral significance.

In other early fMRI research on the pathways activated while a person is viewing moral stimuli compared to unpleasant nonmoral stimuli, Moll and colleagues (2002) found that the right medial orbitofrontal cortex (OFC), the medial frontal gyrus, and the cortex surrounding the right angular gyrus were all activated. The researchers also found a specific and significant increase in functional connectivity between the OFC and the angular gyrus, the precuneus, and the medial frontal gyrus. These changes were seen only when contrasting moral stimuli over unpleasant stimuli, and not vice versa. This shows that although we might find moral violations to be unpleasant to look at, we still process them differently than stimuli that are unpleasant but not moral in nature. Moral stimuli are processed uniquely in the brain.

The ventromedial prefrontal cortex (vmPFC) also seems to play a part in morality, but only when it involves a social-emotional component. In a study of patients with vmPFC lesions (Koenigs et al., 2007), the patients showed normal patterns of responding when moral dilemmas did not feature competing considerations between aggregate welfare and harm to others. On moral dilemmas that did feature competition between these features, vmPFC patients showed significantly higher rates of utilitarian judgments (choosing to save the many over the few, even when this entailed doing something very emotionally aversive such as killing a child). These results show the importance of the vmPFC in the processes of emotional responding and moral decision making.

Aside from the ventromedial portion the greater medial prefrontal/frontopolar cortex also seems to play a role in several different moral situations. It has been implicated in viewing moral versus nonmoral pictures (Moll et al., 2002) and in trying to decrease the emotional response to moral violation pictures (Harenski & Hamann, 2006). It has also been shown to be active while a person is reading (Moll et al., 2005) and listening to morally disgusting statements compared to nonmorally disgusting statements (de Oliveira-Souza & Moll, 2000), in moral versus semantic decision making (Heekeren, Wartenburger, Schmidt, Schwintowski, & Villringer, 2003), and in moral versus nonmoral judgment (Borg, Hynes, Van Horn, Grafton, & Sinnott-Armstrong, 2006). It seems that these areas of the brain may be important to morality because they help process the emotional and social components of moral stimuli.

In addition to frontal and limbic structures, some temporal lobe structures have been implicated in morality. The superior temporal sulcus (adjacent to the angular gyrus/superior temporal gyrus) seems to play a role in distinguishing moral stimuli (Moll et al., 2002) from other unpleasant stimuli, as well as distinguishing personal from impersonal moral dilemmas (Greene et al., 2004). When comparing utilitarian to nonutilitarian responses, Greene and colleagues (2004) also found that the right angular gyrus, right middle temporal gyrus, and left inferior temporal gyrus were more active when an individual made utilitarian judgments.

These areas, including the various regions of the PFC, the angular gyrus, and the posterior cingulate, have been consistently associated with processing morally relevant information using various methods and across multiple laboratories. Together these regions have been referred to as the "moral neural network," although other areas of the brain may play important roles as well (Raine & Yang, 2006). To varying extents these areas have been found to be deficient in antisocial populations such as psychopaths and individuals with ASPD. Specifically, functional deficits in the right OFC, left dlPFC, and right anterior cingulate cortex (ACC) were found to be highly associated with antisocial behavior in a meta-analysis of forty-three studies on antisocial populations (Yang & Raine, 2009).

Morality and Moral Deficits in Psychopathy and ASPD

Immoral behavior such as lying and criminality are central behavioral traits of psychopathy and ASPD. Brain-imaging studies from forensic and community samples have shown that individuals with psychopathy and ASPD share a host of brain deficits and that some of them overlap with areas that have been implicated in morality or moral judgment. If deficits in these areas are present, that may mean that these individuals lack the social emotions such as guilt and shame that normally arise if a moral violation is committed. Lacking these negative emotional responses, the individual would then not become conditioned against performing such actions in the future and may persist with antisocial behavior.

Another possible explanation is that psychopaths have an impaired ability to recognize cues that others are in distress. Blair and colleagues (2004) showed psychopathic and control-group inmates facial expressions of varying intensities and found that psychopaths made more incorrect emotion identifications than did the control group. This was especially true for fearful faces, which psychopaths were less likely to correctly identify even at

100% intensity (the most fearful faces presented). If psychopaths cannot reliably detect when a person is fearful, then they may be less likely to refrain from harming that person. These two explanations are not mutually exclusive; psychopaths may be deficient in both distress recognition and emotion processing to some extent. Individual differences in these areas may help explain why people with varying levels of psychopathic traits behave in various immoral ways.

In other behavioral research Glenn and colleagues (Glenn, Koleva, Iyer, Graham, & Ditto, 2010) found that people high on psychopathy are less likely than those low on psychopathy to base their self-concepts on moral traits. This suggests that reduced moral identity might contribute to more immoral behavior, possibly in part because having a strong sense of moral identity produces motivation for one to behave morally. Behaviorally, though, some research suggests that psychopaths are able to distinguish between an action that is morally "right" and one that is morally "wrong" as well as nonpsychopaths do (Cima, Tonnaer, & Hauser, 2010). Aharoni, Sinnot-Armstrong, and Kiehl (2012) also found that psychopathic offenders performed as well as nonpsychopathic offenders in determining a moral from a conventional violation. Cima and colleagues (2010) contend that such results show that deficient moral reasoning or behavior in psychopathy is not a result of faulty logic or lack of knowledge but, rather, the result of abnormal emotional processing. Because psychopaths seem to commit immoral acts that harm others without feeling remorse, it seems that even though they might cognitively know that these acts are considered immoral, they do not feel the expected emotional reaction that would stop most people from committing such acts. This may suggest abnormal amygdala functioning, which has been supported by other findings on psychopathy (Motzkin, Newman, Kiehl, & Koenigs, 2011; Carré, Hyde, Neumann, Viding, & Hariri, 2013).

The amygdala is crucial in stimulus-reinforcement learning and emotion processing, especially fear. Carré et al. (2013) found that in psychopaths, those scoring higher on the interpersonal facet of psychopathy demonstrated reduced amygdala responsivity to fearful faces. This supports the idea that psychopathic traits are associated with lower responding to the distress of others (Blair, 2007a), but it may also suggest that psychopaths do not initially recognize the fearful expressions of others. Either way, failure to correctly process the fearful faces of others may mean that psychopaths are less deterred from committing immoral acts. Amygdala dysfunction may also result in an inability to effectively learn that immoral acts result in unpleasant consequences, and this impaired learning could explain why psychopaths commit immoral acts regardless of the negative consequences.

Blair (2007b) argues that dysfunction within the amygdala and ventromedial prefrontal cortex (vmPFC) can explain why reinforcement-based decision making, including moral decision making, is impaired in psychopaths. The amygdala, as discussed above, is important for stimulus-reinforcement learning, and especially fear-based learning. Stimuli that are feared tend to be avoided, and many types of moral violations (injuring another, theft, murder) result in consequences, such as jail time, that most people fear. Through deficits in the amygdala, or in the connections between the amygdala and the OFC and PFC, psychopaths may lack this characteristic fear-based learning which teaches them that things that harm others are bad and should be avoided. If amygdala dysfunction is to blame, then fearful stimuli and emotions may be processed incorrectly or fear-based learning be impaired. Even if amygdala function is intact, if connections between the amygdala and the OFC/PFC are deficient, then the information that the amygdala processes ("this action will result in negative consequences") will not be transmitted properly to the OFC and PFC, which use the information to make a decision or plan of action (e.g., "I should not perform the action, and that way I will avoid the negative consequences"). In this way poor connectivity between areas of the brain can be just as important as dysfunction in the regions themselves.

Further evidence for amygdala dysfunction in psychopathy comes from comparing emotional and nonemotional moral decisions. Using the trolley problems described above, Glenn and colleagues (Glenn, Raine, & Schug, 2009) examined brain functioning during emotional, moral-personal dilemmas (such as pushing the man off the bridge to stop the trolley), nonemotional, moral-impersonal dilemmas (such as pulling the lever to switch the trolley to another track), and nonmoral dilemmas. Individuals with higher psychopathy scores showed decreased amygdala activity during emotional moral decision making compared to those with low psychopathy scores. This lack of amygdala reactivity may mean that they do not find the thought of directly harming another individual to be as emotionally aversive, which might help explain why they are more willing to harm others.

Areas of the frontal cortex are also important for moral judgment, and evidence shows that some of these areas are deficient in antisocial populations. Laakso and colleagues (2002) found that violent offenders diagnosed with ASPD showed reduced volumes in the left orbitofrontal cortex and the left dorsolateral prefrontal cortex (dlPFC).

In a study examining aversive conditioning psychopaths showed the least activation (compared to a typical control group and a group with social phobia) in the orbitofrontal cortex, as well as the bilateral insula, anterior cingulate, right amygdala, and left dlPFC (Veit et al., 2002). And in a community sample of individuals high on psychopathy, de Oliveira-Souza and colleagues (2008) found reductions in gray matter concentration (GMC) in the frontopolar cortex (FPC), which is associated with moral judgments (Borg, Hynes, VanHorn, Grafton, & Sinnott-Armstrong, 2006; Moll et al., 2001), along with reductions in several other regions of the brain. The authors also found that reductions in GMC were significantly and inversely related to Total and Factor 1 scores on the PCL:SV, a shortened version of the PLC-R used for screening purposes. This suggests that a region of the brain that has been implicated in morality may contribute to the interpersonal and affective traits of psychopathy but not necessarily with the lifestyle and antisocial traits.

Additional moral reasoning research on antisocial populations supports these findings. In a study using the PD, Rilling and colleagues (2007) found that when choosing to cooperate (contrasted with choosing to defect and take advantage of another person), psychopathy was positively correlated with activity in the dlPFC and negatively correlated with activity in the OFC. Individuals with higher psychopathy scores showed less activation in the OFC when cooperating than when defecting and showed more activation in the dlPFC when cooperating. In another study asking psychopaths to respond (to endorse an action or not) to moral dilemmas, they showed decreased activation in the medial frontal cortex when compared to controls (Pujol et al., 2012). These results seem to suggest that the medial and lateral prefrontal cortex and the orbitofrontal cortex, which have been implicated in morality and moral decision making, appear to function differently in psychopaths than in nonpsychopaths.

Researchers have also found correlations between psychopathy and the difference in activation in the dlPFC and medial PFC during moral compared to nonmoral decisions (Reniers et al., 2012), although this was significant only at the trend level. Rather than previous studies, many of which examined psychopathy as a categorical variable, these findings, although they fall short of being statistically significant, show that these variables may be more dimensional in nature. This would mean that a person who shows a high level of psychopathic traits would show very impaired brain functioning during moral situations, whereas a person with a low level of psychopathic traits would show less impairment. More research in this area is needed to claim that this type of relationship exists, but it is an interesting new direction for research on morality in psychopathy.

Further evidence for frontal cortex dysfunction comes from a study asking psychopathic and nonpsychopathic offenders to rate the severity of moral or nonmoral violations presented in photographs. Compared to psychopaths, nonpsychopaths showed greater activity in the ventromedial prefrontal cortex (vmPFC) as well as the anterior temporal cortex during moral relative to nonmoral and neutral picture viewing (Harenski, Harenski, Shane, & Kiehl, 2010). Additionally, nonpsychopaths exhibited a positive association between severity ratings of moral violations and amygdala activity, which psychopaths did not, and psychopaths showed a negative association between severity ratings and posterior temporal activity, which nonpsychopaths did not. These results generally show that psychopaths use different areas of the brain to process moral decisions than do nonpsychopaths, as well as showing that the prefrontal cortex specifically may be underactive. Previous research has shown that the vmPFC is crucial in encoding the emotional value of a stimulus (Koenigs et al., 2007; Rolls, 2000) and is implicated in different types of moral decision making from simple judgments to complex dilemmas (Prehn et al., 2008). If the prefrontal cortex in psychopaths is unable to correctly encode when a stimulus is emotional in nature, then the amygdala may not be recruited in processing the stimulus. In some cases this may explain why amygdala reactivity to emotional stimuli is low in individuals with psychopathy.

In addition to the amygdala and the frontal cortex, Pujol and colleagues (2012) found that psychopaths showed reduced activation compared to control subjects in the posterior cingulate cortex. Participants were shown a narrative moral dilemma and an accompanying image and then asked to provide a "yes" or "no" answer to whether the character in the dilemma should commit an act. The authors also report that psychopaths showed reduced activity in the bilateral hippocampi/amygdala junction compared to nonpsychopaths, which they suggest relates to Blair's (2010) theory that deficient processing of fear-related stimuli is largely responsible for aggression in psychopathy. Most individuals who contemplate performing a violent act would feel aversive emotions at the thought, or fear of the possible consequences, and this fear could be enough for them to decide not to commit the act. But if a psychopath does not experience this fear, it could help explain why psychopaths commit immoral acts that most other people do not.

The angular gyrus is another area important to morality that has been implicated in antisocial behavior. Using positron emission tomography (PET), Raine, Buchsbaum, and LaCasse (1997) found that a sample of murderers showed lower glucose metabolism than controls in the left angular gyrus. Kiehl and colleagues (2004) compared a group of psychopathic criminals to a group of nonpsychopathic, noncriminal control subjects and found that the psychopathic group showed lower activation during an abstract language-processing task. Additionally, researchers performing visual analysis of single photon emission computerized tomography (SPECT) and MRI data of violent criminals have found reduced blood flow in the temporal lobe, including the angular gyrus (Soderstrom, Tullberg, Wikkelsö, Ekholm, & Forsman, 2000), which they hypothesize might contribute to the violence and aggression these individuals have shown.

Conclusion

The evidence described above shows that some of the areas of the brain that are involved in processing moral stimuli are abnormal or deficient in psychopathy and ASPD, which might go a long way in explaining why psychopaths and individuals with ASPD seem to more frequently commit moral violations. Deficits in brain regions such as the amygdala, the prefrontal cortex (including the orbitofrontal cortex), and the angular gyrus are found in multiple studies using different methodologies with antisocial populations. These areas are all implicated in different parts of the moral reasoning process, and individual differences in these areas might help explain why psychopaths and individuals with ASPD exhibit differences in the type of antisocial behavior in which they engage, its frequency, and their feelings in response to it.

References

Aharoni, E., Sinnot-Armstrong, W., & Kiehl, K. A. (2012). Can psychopathic offenders discern moral wrongs? A new look at the moral/conventional distinction. *Journal of Abnormal Psychology, 121*(2), 484–497.

American Psychiatric Association. (2013). *Diagnostic and statistical manual of mental disorders* (5th ed.). Arlington, VA: American Psychiatric Publishing.

Blair, R. J. R. (2007a). The amygdala and ventromedial prefrontal cortex in morality and psychopathy. *Trends in Cognitive Sciences, 111*(9), 387–392.

Blair, R. J. R. (2007b). Dysfunctions of medial and lateral orbitofrontal cortex in psychopathy. *Annals of the New York Academy of Sciences, 1121*, 461–479.

Blair, R. J. R. (2010). Psychopathy, frustration, and reactive aggression: The role of ventromedial prefrontal cortex. *British Journal of Psychology, 101*, 383–399.

Blair, R. J. R., Mitchell, D. G. V., Peschardt, K. S., Colledge, E., Leonard, R. A., Shine, J. H., et al. (2004). Reduced sensitivity to others' fearful expressions in psychopathic individuals. *Personality and Individual Differences, 37*, 1111–1122.

Borg, J. S., Hynes, C., Van Horn, J., Grafton, S., & Sinnott-Armstrong, W. (2006). Consequences, action, and intention as factors in moral judgments: An fMRI investigation. *Journal of Cognitive Neuroscience, 18*(5), 803–817.

Carré, J. M., Hyde, L. W., Neumann, C. S., Viding, E., & Hariri, A. R. (2013). The neural signatures of distinct psychopathic traits. *Social Neuroscience, 8*, 122–135.

Cima, M., Tonnaer, F., & Hauser, M. D. (2010). Psychopaths know right from wrong but don't care. *Social Cognitive and Affective Neuroscience, 5*, 59–67.

de Oliveira-Souza, R., Hare, R. D., Bramati, I. E., Garrido, G. J., Ignácio, F. A., Tovar-Moll, F., et al. (2008). Psychopathy as a disorder of the moral brain: Fronto-temporo-limbic grey matter reductions demonstrated by voxel-based morphometry. *NeuroImage, 40*, 1202–1213.

de Oliveira-Souza, R., & Moll, J. (2000). The moral brain: Functional MRI correlates of moral judgment in normal adults. *Neurology, 54*, 252.

Glenn, A. L., Koleva, S., Iyer, R., Graham, J., & Ditto, P. H. (2010). Moral identity in psychopathy. *Judgment and Decision Making, 5*(7), 497–505.

Glenn, A. L., & Raine, A. (2009). Psychopathy and instrumental aggression: Evolutionary, neurobiological, and legal perspectives. *International Journal of Law and Psychiatry, 32*, 253–258.

Glenn, A. L., Raine, A., & Schug, R. A. (2009). The neural correlates of moral decision-making in psychopathy. *Molecular Psychiatry, 14*, 5–9.

Greene, J. D., Nystrom, L. E., Engell, A. D., Darley, J. M., & Cohen, J. D. (2004). The neural bases of cognitive conflict and control in moral judgment. *Neuron, 44,* 389–400.

Greene, J. D., Sommerville, R. B., Nystrom, L. E., Darley, J. M., & Cohen, J. D. (2001). An fMRI investigation of emotional engagement in moral judgment. *Science, 293*(5537), 2105–2108.

Hare, R. D. (2003). *The Hare Psychopathy Checklist-Revised* (2nd ed.). Toronto: Multi-Health Systems.

Harenski, C. L., & Hamann, S. (2006). Neural correlates of regulation negative emotions related to moral violations. *NeuroImage, 30*(1), 313–324.

Harenski, C. L., Harenski, K. A., Shane, M. S., & Kiehl, K. A. (2010). Aberrant neural processing of moral violations in criminal psychopaths. *Journal of Abnormal Psychology, 119*(4), 863–874.

Heekeren, H. R., Wartenburger, I., Schmidt, H., Schwintowski, H., & Villringer, A. (2003). An fMRI study of simple ethical decision-making. *Neuroreport, 14*(9), 1215–1219.

Kiehl, K. A., Smith, A. M., Mendrek, A., Forster, B. B., Hare, R. D., & Liddle, P. F. (2004). Temporal lobe abnormalities in semantic processing by criminal psychopaths as revealed by functional magnetic resonance imaging. *Psychiatry Research: Neuroimaging, 130,* 27–42.

Koenigs, M., Young, L., Adolphs, R., Tranel, D., Cushman, F., Hauser, M., et al. (2007). Damage to the prefrontal cortex increases utilitarian moral judgements. *Nature, 446*(7138), 908–911.

Laakso, M. P., Gunning-Dixon, F., Vaurio, O., Repo-Tiihonen, E., Soininen, H., & Tiihonen, J. (2002). Prefrontal volumes in habitually violent subjects with antisocial personality disorder and type 2 alcoholism. *Psychiatry Research: Neuroimaging, 114,* 95–102.

Levenson, M., Kiehl, K., & Fitzpatrick, C. (1995). Assessing psychopathic attributes in a noninstitutionalized population. *Journal of Personality and Social Psychology, 68,* 151–158.

Lilienfeld, S. O., & Andrews, B. P. (1996). Development and preliminary validation of a self-report measure of psychopathic personality traits in noncriminal populations. *Journal of Personality Assessment, 66,* 488–524.

Moll, J., de Oliveira-Souza, R., Eslinger, P. J., Bramati, I. E., Mourão-Miranda, J., Andreiuolo, P. A., et al. (2002). The neural correlates of moral sensitivity: A functional magnetic resonance imaging investigation of basic and moral emotions. *Journal of Neuroscience, 22*(7), 2730–2736.

Moll, J., de Oliveira-Souza, R., Moll, F. T., Ignácio, F. A., Bramati, I. E., Caparelli-Dáquer, E. M., et al. (2005). The moral affiliations of disgust: A functional MRI study. *Cognitive and Behavioral Neurology, 18*(1), 68–78.

Moll, J., Eslinger, P. J., & de Oliveira-Souza, R. (2001). Frontopolar and anterior temporal cortex activation in a moral judgment task: Preliminary functional MRI results in normal subjects. *Arquiva Neuropsychiatria, 59*(3-B), 657–664.

Motzkin, J. C., Newman, J. P., Kiehl, K. A., & Koenigs, M. (2011). Reduced prefrontal connectivity in psychopathy. *Journal of Neuroscience, 31,* 17348–17357.

Prehn, K., Wartenburger, I., Mériau, K., Scheibe, C., Goodenough, O.R., Villringer, A., et al. (2008). Individual differences in moral judgment competence influence neural correlates of socio-normative judgments. *Social Cognitive and Affective Neuroscience, 3,* 33–46.

Pujol, J., Batalla, I., Contreras-Rodríguez, O., Harrison, B.J., Pera, V., Hernández-Ribs, R., et al. (2012). Breakdown in the brain network subserving moral judgment in criminal psychopathy. *Social, Cognitive and Affective Neuroscience, 7,* 917–923.

Raine, A., Buchsbaum, M., & LaCasse, L. (1997). Brain abnormalities in murderers indicated by positron emission tomography. *Biological Psychiatry, 42,* 495–508.

Raine, A., & Yang, Y. (2006). Neural foundations to moral reasoning and antisocial behavior. *Social Cognitive and Affective Neuroscience, 1*(3), 203–213.

Reniers, R. L. E. P., Corcoran, R., Völlm, B. A., Mashru, A., Howard, R., & Liddle, P. F. (2012). Moral decision-making, ToM, empathy and the default mode network. *Biological Psychology, 90,* 202–210.

Rilling, J. K., Glenn, A. L., Jairam, M. R., Pagnoni, G., Goldsmith, D. R., Elfenbein, H. A., et al. (2007). Neural correlates of social cooperation and non-cooperation as a function of psychopathy. *Biological Psychiatry, 61,* 1260–1271.

Rolls, E. T. (2000). The orbitofrontal cortex and reward. *Cerebral Cortex, 10,* 284–294.

Soderstrom, H., Tullberg, M., Wikkelsö, C., Ekholm, S., & Forsman, A. (2000). Reduced regional cerebral blood flow in non-psychotic violent offenders. *Psychiatry Research: Neuroimaging, 98,* 29–41.

Veit, R., Flor, H., Erb, M., Hermann, C., Lotze, M., Grodd, W., et al. (2002). Brain circuits involved in emotional learning in antisocial behavior and social phobia in humans. *Neuroscience Letters, 328,* 233–236.

Williamson, S., Hare, R. D., & Wong, S. (1987). Violence: Criminal psychopaths and their victims. *Canadian Journal of Behavioural Science, 19,* 454–462.

Yang, Y., & Raine, A. (2009). Prefrontal structural and functional brain imaging findings in antisocial, violent, and psychopathic individuals: A meta-analysis. *Psychiatry Research: Neuroimaging, 174,* 81–88.

15   The Moral Brain: Psychopathology

Caroline Moul, David Hawes, and Mark Dadds

This chapter considers two systems that are fundamental to human behavior: learning and the allocation of attention. We review the evidence to suggest that there may be deficits in these systems in a subset of children with antisocial behavior problems—those with high levels of callous-unemotional traits—and explore how altered function of these systems might contribute to the development of immoral behavior.

Viewing immoral behavior through a developmental lens helps to demystify adult perpetrators and provides a mechanism by which small and subtle differences in cognitive function can, over time, interact with environmental factors to interfere with the development of moral behavior.

Morals can be violated by anyone of any age. Children can often be purposefully immoral and intentionally cruel. Indeed, behaviors demonstrated by toddlers are frequently immoral; toddlers are generally self-focused and unconcerned with the effects of their behavior on others. Very young children are, essentially, driven by their own desires and their own needs, having not yet developed the cognitive capabilities, or theory of mind, to understand that their experience of the world is not shared by everyone else. As the understanding that other people have different experiences, thoughts, and feelings develops, so too does a child's ability and inclination to consider other people and to adapt his or her behavior toward social and moral standards. It is the failure of this development that appears in a large part to characterize the forms of psychopathology that we associate with moral dysfunction.

So, the question underlying the etiology of immoral behavior really asks why normal development fails. Why do some people learn to stop breaking moral rules in childhood and others continue to do so well into adulthood? This chapter focuses on psychopathologies of immoral behavior from a developmental perspective: which psychopathologies of childhood are associated with a lack of moral development, and what are the cognitive and emotional processes that drive these disorders?

Psychopathologies Associated with Immoral Behavior

The psychopathologies of childhood most strongly associated with immoral behavior are oppositional defiant disorder (ODD) and conduct disorder (CD). These disorders are relatively common; around one in every twenty children (at least one child per classroom) will meet diagnostic criteria for ODD or CD. ODD describes a pattern of defiant behaviors that can include anger and hostility, disobedience, temper tantrums, and spitefulness and that occur more frequently than would be typical for the child's age and cause serious and significant impairment in the child's life and for that child's parents, teachers, siblings, and peers. The criteria for CD includes behaviors that demonstrate more serious antisocial behaviors and violations of others' rights, such as violence and aggression, physical cruelty, destruction of property, deceitfulness, and theft.

Impulsivity and Emotion Dysregulation

Children diagnosed with ODD or CD often display difficulties with impulsivity and emotion regulation. These children may be described as "hot-blooded"; they are quick to get angry or upset in situations in which they then act without thinking of the consequences. These children express remorse and regret when they have lost their temper and have done something wrong and may be frustrated and upset by their own lack of ability to control their behavior. Although impulsivity and emotion dysregulation are the most common features of antisocial behavior in childhood, and indeed in adulthood, they are not addressed in this chapter. The reason for their exclusion from a discussion of psychopathologies of morality is that they are not, in and of themselves, specific to immoral behavior—impulsive children are as likely to be impulsively prosocial as they are antisocial. There are, however, some children (and adults) who appear to engage in antisocial acts not because they fail to control their impulses but because they simply desire to do so. It is these people, and the processes underlying their behavior, that are the focus of this chapter.

Callous-Unemotional Traits

Children with antisocial behavior problems such as ODD or CD may be characterized by the types of personality traits we associate with psychopathy. Psychopathy is generally considered as comprising two main factors. The first factor concerns features of personality and interpersonal style including diminished guilt and remorse, the manipulation of others, a shallow affect, superficial charm, and a lack of empathy. The second factor concerns antisocial behaviors and lifestyle factors such as irresponsibility and juvenile delinquency. Factor 1 traits—aspects of personality—vary normally within the general population and may not always be accompanied by factor 2 behaviors.

In children and adolescents these factor 1 features of personality are referred to as callous-unemotional (CU) traits, and they have been introduced into DSM-5 as a specifier to the diagnosis of conduct disorder. It is too early to say whether CU traits in childhood are the same traits with the same etiology as psychopathic personality traits in adulthood. However, it has been demonstrated that the occurrence of high levels of CU traits is a marker for a particularly severe and chronic course of antisocial behavior. Furthermore, there is growing evidence that CU traits are associated with poor response to current interventions for antisocial behavior problems. Frick, Ray, Thornton, and Kahn, 2014 (2014) identified twenty studies comparing outcomes of treatment for youth with and without CU traits, eighteen of which reported poorer outcomes for youths with CU traits. This finding has been replicated in samples ranging from early childhood through adolescence and is evident when controlling for baseline severity of behavior problems. Thus, it appears that CU traits interact with the mechanisms of change through which these interventions operate.

Evidence also shows that there is a strong genetic component to both CU traits (Viding, Jones, Paul, Moffitt, & Plomin, 2008) and psychopathic personality traits (Beaver, Barnes, May, & Schwartz, 2011). Research has implicated a number of neurochemical systems in the etiology of psychopathy including, but not limited to, the serotonin system and the oxytocin system; and abnormalities in the function of specific brain regions such as the amygdala and ventromedial prefrontal cortex have been linked to psychopathy and CU traits (see Blair, 2010, for a review). As research suggests that CU traits and psychopathic traits have similar genetic and neurological correlates, it is understandable that CU traits are often considered the developmental analogue of psychopathy.

The Development of Callous-Unemotional Traits

So, why do some children develop CU traits? What causes them to become vindictive, manipulative, and seemingly emotionally cold? It is a difficult question, and its complexity allows it to be addressed from a variety of angles. Philosophers might question whether people are born "evil"; theologians may query the role of God; criminologists may look to the family environment; and sociologists may point to the social context. So, what if we were to simplify the question? For example, why do some people put their hands in front of their mouth when they sneeze and others don't? It is not an instinctive reaction; babies do not do it. Why then is it so commonplace? From a psychological perspective, the answer is simple: we are taught to do so. During early childhood a parent or a teacher will have repeatedly reminded the child to cover his or her mouth when they sneeze. The child may have been chastised for not covering his or her mouth. An inquisitive child will have asked why it is important and will have been told about germs and how stopping the passing on of those germs is a considerate thing to do. Perhaps that child also saw the expression of disgust on someone's face when he or she was sneezed on. Maybe the child him- or herself disliked the sensation of receiving a full-force sneeze in the face. It is likely that, together and over time, all these scenarios have developed the child's response to cover his or her mouth when a sneeze was imminent.

What if we were to learn more complex moral behaviors in the same way? What processes are required? If we look back at our example of sneezing, then two fundamental processes stand out. First, the child must be able to learn, that is, to form an association between one thing and another. The child must have been able to associate being told off with not putting her hand in front of her mouth and to associate being praised with putting her hand in front of her mouth. The child must also have been able to associate the action of sneezing—without a covering hand—with other people's facial or verbal expressions of disgust or anger. Without these very simple associations, appropriate new behaviors cannot be learned—children would just experience the world as one in which they sneeze and that, independently of this, sometimes they are praised, sometimes they are told off, and sometimes people get angry or disgusted with them for no apparent reason. It is the association between the child's behavior and the reaction of others that allows consistent behavior change to occur. Presuming that the child enjoys being praised, dislikes being chastised, and feels bad when other people get angry or upset with him, the child will adapt his behavior so as to maximize praise and minimize punishment. In other words he will learn to cover his mouth when he sneezes.

The second process that is fundamental to behavior change is the allocation of attention. What if, when the child sneezed in the face of her parent, the child failed to notice the look of anger and disgust displayed on the parent's face? What if the child failed to see her parent wipe the spittle from her face? What if the child only paid attention to the verbal reprimand? And what if that reprimand was not specific such as: "Eurgh, that was disgusting. Don't do that!"? Then, how does the child know that she is being chastised for not covering her mouth rather than just for sneezing in general? Without the appropriate allocation of attention, social cues are missed, and the specific features of association formation can be lost.

It is easy to see how a child with deficits in learning and in the allocation of attention may fail to learn how to appropriately control a sneeze. But, surely a child with such profound deficits would stand out? Wouldn't he struggle in school and have severe developmental delay? Children with antisocial behavior problems and high levels of CU traits are not typically characterized by developmental delay (Allen, Briskman, Humayun, Dadds, & Scott, 2013). Similarly, adult psychopaths are not known for having learning difficulties or below-average IQ. Despite this, both adults with high levels of psychopathic personality traits and children with high levels of CU traits have been shown to have subtle yet distinct deficits in both associative learning and attentional processes that point to a neural basis for the development of psychopathy (see Moul, Killcross, & Dadds, 2012, for a review). The remainder of this chapter considers how deficits in these two processes manifest and how they might contribute to the development of psychopathic personality traits.

Deficits in Associative Learning

Individuals with high levels of psychopathic personality traits show no deficiencies in the ability to form an association; that is, they are able to learn that a response to a certain stimulus results in a particular outcome. A psychopath's deficit in associative learning is, however, very subtle. Imagine a scenario in which a response to a given stimulus has, every time you have encountered that stimulus, resulted in a certain outcome. Pressing the spacebar on your computer keyboard, for example, has always resulted in a space appearing in the text you are typing. An association has been formed between the spacebar (stimulus) and a space appearing in your text (outcome) each time the spacebar is pressed (response). Now, imagine a scenario where, unbeknownst to you, a virus has infected your computer. Now, when you press the spacebar, the last letter you typed is deleted—the outcome associated with the stimulus has been changed. Most people may press the spacebar one or two more times before ceasing to do so—as they fear that continuing to press the spacebar will delete more of their work. In contrast, someone with high levels of psychopathic personality traits is more likely to continue pressing the spacebar for longer even though the outcome associated with the action is deleterious. Experimentally, it has been reliably shown in a variety of tasks that psychopaths are poorer at updating their responses as a result of changing outcomes associated with a stimulus. Tasks of passive avoidance, for example, have demonstrated that people with high levels of psychopathic traits will continue to select a stimulus (rather than not select it) when that stimulus was originally paired with a positive outcome but has since become paired with a punishing outcome (e.g., Newman, Patterson, Howland, & Nichols, 1990). Psychopaths show an analogous deficit when two stimuli (e.g., a red button and a yellow button) are presented simultaneously. In these "response-reversal" paradigms one button is associated with a reward (winning points) while the other button is associated with a loss or punishment (losing points). Both psychopaths and nonpsychopaths quickly learn to press the rewarding button and to not press the punishing button at the start of the experiment. In a typical response-reversal paradigm, at a point during the task, the outcomes associated with each button swap over so the rewarding button becomes punishing and the punishing button becomes rewarding. Nonpsychopaths quickly learn to alter their responses accordingly—they will press the button that is now rewarding but that was previously punishing and avoid pressing the button that used to be rewarding but is now punishing. Just as in tasks of passive avoidance, as compared to nonpsychopaths, psychopaths take significantly longer to appropriately alter their responses once the switch in outcomes has occurred (e.g., Budhani, Richell, & Blair, 2006).

To take a real-world example, imagine a boy with high levels of CU traits who is making his friends laugh by doing a comedic impersonation of a classmate. Perhaps when the boy first does the impression, the classmate he is mimicking finds it funny and laughs along with everyone else. After a while, however, the classmate starts to feel embarrassed and gets upset at the continuing impersonation. Another child might quickly notice this and stop the impersonation to save the classmate's feelings, but the boy with high levels of CU traits is insensitive to the change in the outcome (the classmate's response changing from laughing to sadness) and so continues to embarrass his classmate in front of his friends. He becomes, albeit inadvertently, a bully.

Deficits in the Allocation of Attention

The allocation of attention also plays an important role in learning. Attentional deficits have been demonstrated both in adults with high levels of psychopathic traits and in children with high levels of CU traits. As with deficiencies in associative learning, the problems with the allocation of attention demonstrated by psychopaths are subtle. For example, one of the most reliable findings in psychopathy research is that of a reduced conditioned-fear response.

Fear-Potentiated Startle

Normally, if a neutral stimulus (e.g., a tone) is consistently paired with an aversive stimulus (e.g., a loud noise) then the neutral stimulus becomes predictive of the aversive stimulus. Thus, when the neutral stimulus is presented, the animal expects the aversive cue to follow and is made to feel fearfully expectant. If the aversive stimulus does then occur (the loud noise plays), the animal displays a fear-response that is more exaggerated (conditioned-fear response) than if the aversive stimulus (the loud noise) had occurred in the absence of the cue (unconditioned fear-response). It has been reliably demonstrated, however, that psychopaths fail to show this fear-potentiated startle response (Newman, Curtin, Bertsch, & Baskin-Sommers, 2010), although they do demonstrate normal unconditioned fear responses (Birbaumer et al., 2005).

Interestingly, it has recently been shown that this deficit in the conditioned-fear response in psychopaths is moderated by attention. Newman et al. (2010) found that under conditions in which participants were instructed to focus on nonpredictive aspects of the stimuli, adult psychopaths demonstrated a reduced conditioned-fear response that replicated previous findings. However, when the participants were told specifically to attend to the fear-relevant features of the stimuli, the psychopaths' deficit in the conditioned-fear response normalized. Importantly, the psychopaths were able, before the instructions to attend to the relevant features of the stimuli they were given, to identify which feature of the stimulus was predictive. Consequently, the manipulation served only to direct their attention and did not alter their knowledge of the relationships between stimuli.

Emotion Recognition

The allocation of attention has also been demonstrated to be important in the accurate recognition of emotions. Both adult psychopaths and children with high levels of CU traits show deficits in emotion recognition, most reliably in the recognition of fear (see Marsh & Blair, 2008, for a meta-analysis). Compared to people without high levels of psychopathic traits, those *with* high levels of psychopathic traits are more likely to incorrectly label a face expressing fear as expressing a different type of emotion. Fear is the only main emotion in which the meaning of the expression is expressed by the eyes only. Fearfully widened eyes are the key feature of a fearful face. Indeed, there is an evolutionary argument for why the eyes are the key to a fearful expression. The expression of fear is more than just a portrayal of an emotion; it is also a very efficient and important method of communication. If someone is expressing fear it is likely that there is a considerable threat in the near vicinity. In order to determine where this threat is coming from and so enable you to run in the opposite direction, all you have to do is look to see where the gaze of the fearful person is directed. In this way the expression of fear communicates information that would have been critical to our survival in the early evolution of our species.

It has been shown that psychopaths and children with high levels of CU traits exhibit a deficit in the ability to correctly identify fearful faces unless their attention is directed toward the eye region of the fearful face. When boys with antisocial behavior problems and high levels of CU traits were simply told to attend to the eye region of emotional faces, their previously evident fear-recognition deficit disappeared (Dadds et al., 2006). It is likely that this specific fear-recognition deficit is a marker of a more profound deficit in the allocation of attention to socially relevant cues. Indeed, boys with high levels of CU traits and antisocial behavior problems have been shown to display less eye contact with their mothers than children with antisocial behavior problems without concurrent high levels of CU traits (Dadds et al., 2012). A lack of attention to socially relevant cues would, of course, impact on a child's ability to form associations between his behavior and how it affects other people and would make it harder to learn when it was necessary to modify antisocial behaviors.

From Minor Deficits to Psychopathy

In all, the development of behaviors characteristic of those with high levels of CU traits may be simply a product of a lack of development, a lack of learning alternative behaviors, a lack of awareness early in childhood of the impact of their behavior on others. They may manifest a lack of understanding, perhaps, of a world in which they seem to function normally when they are, in fact, disabled by a cluster of subtle and complex cognitive deficits that over time snowball into robust and deleterious personality, and sometimes behavioral, traits.

These deficits, of course, do not exist in a vacuum but, rather, are constantly interacting with and influencing the social environment. For example, a child who makes less eye contact with his parent than a sibling does may be considered by the parent to be less engaging and less affectionate, leading to a deterioration in parental warmth—eroding the very dimension of parenting that is of most proximal importance to these children. In all, the development of psychopathy may be a real-life example of the childhood game of "telephone." Cognitive differences may cause small, seemingly insignificant errors in the child's understanding of behavior and social interaction that serve to mold his environment and compound his difficulties to result in behaviors and personality traits that, by the time adulthood is reached, may be so immoral as to appear unfathomable.

The lack of empathy that characterizes psychopathy and children with high levels of CU traits is one feature that has received considerable interest from both researchers and clinicians and points to complex interactions between cognitive and emotional deficits and the social environment. As with the aforementioned deficits in attention and learning, the difficulties with empathy demonstrated by children with high levels of CU traits are not straightforward. Empathy can be thought of as comprising two components: cognitive empathy and affective empathy. Cognitive empathy refers to the ability to understand how another person might feel in a given situation; for example, knowing someone might feel sad if her dog had just died or understanding that a classmate might feel angry if she were told off by the teacher for something she had not done. Affective empathy describes the sharing of an emotion with another person, for example, feeling like you want to cry when watching a sad film or feeling anxious for a friend who is about to take an examination and is displaying nervousness. Research has demonstrated that children with high levels of CU traits have no difficulties with cognitive empathy but display deficits in affective empathy (Jones, Happé, Gilbert, Burnett, & Viding, 2010). It is possible that this deficit is driven by a lack of awareness or of attention to the emotional states of others and may be reversed under conditions in which the child is directed to pay attention to the emotional cues of others such as facial expression, body posture, and tone of voice. Interestingly, fMRI research has found that these deficits in affective empathy are mirrored by neural function. Youths with psychopathic traits showed less amygdala activation than healthy youth in response to seeing images of pain inflicted on other people, but there were no differences in amygdala activation between the groups when the participant was asked to imagine that the pain was being inflicted on himself (Marsh et al., 2013). This finding suggests that the normal neurological underpinnings of the development of empathy may be altered or, at the very least, developmentally delayed in these adolescents. Research such as this helps us understand why a child with high levels of CU traits may be at a greater risk of developing, and maintaining, antisocial and aggressive behavior problems—such a child may be less sensitive to the impact, both emotional and physical, that his behavior has on others. In this regard empathy can be thought of as a stepping-stone in the pathway from basic neural and cognitive functions to the manifestation of complex personality and behavioral traits.

Concluding Remarks

The deficits in simple cognitive functions described in this chapter provide the basis of a pathway along which immoral behavior may develop.

Basic functions such as learning and attention are the tools psychologists use to solve problems of behavior and to inform the development of treatments. This chapter presents a possible explanation for psychopathologies of immorality—for the development of callous-unemotional traits. Sociology, theology, and other disciplines concerned with immoral behavior have their own sets of tools to approach the problem and provide different possible explanations. Undoubtedly, both the combination of knowledge and the interactions among differing approaches allow for the best solutions. After all, a child does not develop in a void; his or her behavior is not solely a function of the subtle nuances of his or her cognitive processes. A child's behavior is also influenced by that child's physical environment, by his or her social environment, by that child's beliefs and prior experiences, by his or her family dynamics, and by his or her successes and failures.

When we think of psychopathologies associated with immoral behavior, we tend to imagine the worst case scenarios. We tend to view criminal psychopaths as "monsters" who are somehow alien to the rest of the human race. By going to the extremes of immoral behaviors we make it difficult for ourselves to imagine the developmental course of such behaviors. When we take a step back, however, and consider basic psychological processes that lie at the root of these disorders, where these behaviors may have begun and how they may manifest in childhood, we can envisage a more innocent foundation on which improved interventions and treatments may be built.

References

Allen, J. L., Briskman, J., Humayun, S., Dadds, M. R., & Scott, S. (2013). Heartless and cunning? Intelligence in adolescents with antisocial behavior and psychopathic traits. *Psychiatry Research, 210*(3), 1147–1153.

Beaver, K. M., Barnes, J. C., May, J. S., & Schwartz, J. A. (2011). Psychopathic personality traits, genetic risk, and gene-environment correlations. *Criminal Justice and Behavior, 38*(9), 896–912.

Birbaumer, N., Veit, R., Lotze, M., Erb, M., Hermann, C., Grodd, W., et al. (2005). Deficient fear conditioning in psychopathy: A functional magnetic resonance imaging study. *Archives of General Psychiatry, 62*(7), 799–805.

Blair, R. J. R. (2010). Neuroimaging of psychopathy and antisocial behavior: A targeted review. *Current Psychiatry Reports, 12*(1), 76–82.

Budhani, S., Richell, R. A., & Blair, R. J. R. (2006). Impaired reversal but intact acquisition: Probabilistic response reversal deficits in adult individuals with psychopathy. *Journal of Abnormal Psychology, 115*(3), 552–558.

Dadds, M. R., Allen, J. L., Oliver, B. R., Faulkner, N., Legge, K., Moul, C., et al. (2012). Love, eye contact and the developmental origins of empathy v. psychopathy. *British Journal of Psychiatry, 200*(3), 191–196.

Dadds, M. R., Perry, Y., Hawes, D. J., Merz, S., Riddell, A. C., Haines, D. J., et al. (2006). Attention to the eyes and fear-recognition deficits in child psychopathy. *British Journal of Psychiatry, 189*(Sep), 280–281.

Frick, P. J., Ray, J. V., Thornton, L. C., & Kahn, R. E. (2014). Can callous-unemotional traits enhance the understanding, diagnosis, and treatment of serious conduct problems in children and adolescents? A comprehensive review. *Psychological Bulletin, 140*(1), 1–57.

Jones, A. P., Happé, F. G. E., Gilbert, F., Burnett, S., & Viding, E. (2010). Feeling, caring, knowing: Different types of empathy deficit in boys with psychopathic tendencies and autism spectrum disorder. *Journal of Child Psychology and Psychiatry, and Allied Disciplines, 51*(11), 1188–1197. doi:10.1111/j.1469-7610.2010.02280.x.

Marsh, A. A., & Blair, R. J. R. (2008). Deficits in facial affect recognition among antisocial populations: A meta-analysis. *Neuroscience and Biobehavioral Reviews, 32*(3), 454–465.

Marsh, A. A., Finger, E. C., Fowler, K. A., Adalio, C. J., Jurkowitz, I. T., Schechter, J. C., et al. (2013). Empathic responsiveness in amygdala and anterior cingulate cortex in youths with psychopathic traits. *Journal of Child Psychology and Psychiatry, and Allied Disciplines, 54*(8), 900–910. doi:10.1111/jcpp.12063.

Moul, C., Killcross, S., & Dadds, M. R. (2012). A model of differential amygdala activation in psychopathy. *Psychological Review, 119*(4A), 789–806.

Newman, J. P., Curtin, J. J., Bertsch, J. D., & Baskin-Sommers, A. R. (2010). Attention moderates the fearlessness of psychopathic offenders. *Biological Psychiatry, 67*(1), 66–70.

Newman, J. P., Patterson, C. M., Howland, E. W., & Nichols, S. L. (1990). Passive avoidance in psychopaths: The effects of reward. *Personality and Individual Differences, 11*(11), 1101–1114.

Viding, E., Jones, A. P., Paul, J. F., Moffitt, T. E., & Plomin, R. (2008). Heritability of antisocial behaviour at 9: Do callous-unemotional traits matter? *Developmental Science, 11*(1), 17–22.

VI Considerations and Implications for Justice and Law

16   Neuroscience versus Phenomenology and the Implications for Justice

Thalia Wheatley

The more we learn about the brain, the further we seem to get from the picture painted by our own subjective experience. Visual perception appears richly detailed yet relies on sparse and impoverished images. Consciousness appears as a continuous stream but is in fact punctuated by episodes spent in an inattentional twilight. Our feelings of agency suggest that we are freer to act than physical laws allow. Increasingly, neuroscience reveals a disconnection between what it *feels like* to be human and the machinery otherwise responsible for our thoughts and actions. What does this disconnection mean for moral responsibility—a concept historically rooted in intuitions about character and agency? Here I suggest that our conceptions of moral responsibility must increasingly align with science over phenomenology. This realignment should promote greater humility and compassion while leaving much of our judicial system intact.

Reading about the brain is an odd experience. It means reading about the very neural mechanisms responsible for choosing to read about those neural mechanisms in the first place. Considered deeply, this leads to the inevitable question: "But if my neurons chose to read this book, then who am I? *Am I just a collection of neurons*?" The answer, according to most neuroscientists, is *"Yes."*

Those who answer in the negative do so only because there are other biological mechanisms that should be included, such as glial cells. Either way, neuroscientists are confident that there is nothing metaphysical *up there* doing the work.

Why does it not feel like this? If the human brain is a "lump of warm wet matter" (Donald, 1990), why does that lump make it feel as if there is something else, namely a "me," doing the work? The purpose of subjective experience is one of the deepest mysteries of science and promises to remain a mystery for a good while longer. However, an equally important question is far more tractable: If we agree that warm wet matter causes thought and behavior (even if we do not understand why), then how should this change concepts such as free will and moral responsibility? This chapter delves into this question in three parts. First, I review some facts about the brain. Second, I explore what it feels like to be human (phenomenology). Third, I suggest that concepts of free will and morality should increasingly align with our understanding about the brain. Finally, I conclude that such a conceptual evolution would change little about the legal system but should impel firm but humane treatment for those who have committed crimes.

How the Brain Works

The Neuron

Consider a simple question: *Can a neuron choose when or whether to fire?* The answer, equally simple, is "No." Just as a red blood cell cannot choose whether or not to release oxygen, a neuron cannot choose whether to fire; it fires given a sufficient change in its membrane voltage. This is true whether the neuron exists in the brain of an astrophysicist or a sea slug (Kandel, 1976).

Many Neurons

The major difference between the sea slug and the human brain lies in scale. A sea slug has around 18,000 neurons, whereas a human has around 85 billion. Likewise, the simple question can be scaled up: Can 85 billion hierarchically and complexly organized neurons choose whether to fire? That is, even if neuron A cannot choose to fire by itself, perhaps it gains choice from being in an interconnected system in which its membrane voltage is affected by the firings of other neurons. Does adding spatial and temporal complexity to a nervous system give neuron A (or any other neuron in that system) choice over whether to fire? The answer still appears to be "No." Neurons together have no more power to choose whether to fire than neurons individually.

This is important because if the basic unit of the nervous system cannot choose whether or not to fire, and our brains are a collection of these basic units, then the lay view of free will becomes very difficult to reconcile. We return to this point later.

The Pull of Human Exceptionalism

The inordinate complexity of the human brain includes emergent properties and, with them, new capacities for learning, art, insight, and planning. Human beings have created a world that can oppose and manage the twists of nature. This observation often leads to an assertion of *human exceptionalism*: the view that humans are categorically different from other animals. And yet, there is no bright line in evolution. Our brains are wet matter with the same general architecture as those of our nearest evolutionary relatives. Uniquely human capacities are not unique all the way down but appear to harness earlier mammalian capacities.

In his book *The Accidental Mind* David Linden (2007) offers a useful analogy for the evolution of the human brain. He describes a racecar designer given the challenge of creating the fastest, most efficient, and most beautiful racecar in the world. As the designer is about to agree to the challenge, a few odd stipulations emerge. First, the designer must construct the racecar on the chassis of a Model T. Second, everything that has been put on a car since the Model T must be included in the design, and it must be on all the time (e.g., the eight-track cassette player). It becomes clear that the resulting design can be nothing more than a kluge. The beautiful racecar must also be a living fossil of automotive history.

Like the racecar built on a Model T, our brains are built on frog, rat, and monkey parts. The difference is that these abilities appear to be scaled up, combined in new ways, and repurposed for novel tasks (Anderson, 2010; Dehaene & Cohen, 2007; Gallese & Lakoff, 2005; Parkinson, Liu, & Wheatley, 2014; Parkinson & Wheatley, 2013).

Take the ability to think twenty years into the future. Considering the remote future appears to be a uniquely human capacity. Indeed Professor Dan Gilbert at Harvard suggests that it is the best ending for the sentence that all psychologists hope to answer by the end of their careers—the sentence that begins, *"The human being is the only animal that..."* He writes:

it is for good reason that most psychologists put off completing The Sentence for as long as they can. ... I have never before written The Sentence, but I'd like to do so now, with you as my witness. *The human being is the only animal that thinks about the future.* (Gilbert, 2005; italics original)

As far as science has ascertained, thinking of the remote future is a uniquely human capacity. However, that is not to say that it appeared out of the blue in the course of evolution. We recently published a paper that suggests that thinking about time utilizes the same computations as thinking about spatial distances, a primitive faculty shared with many other animals (Parkinson, Liu, & Wheatley, 2014). Viewing a close object evoked a similar brain pattern as viewing words referring to soon time (e.g., "a few seconds from now"); viewing a far object evokes a similar brain pattern as viewing words referring to later time (e.g., "months from now"). This suggests that even something as putatively human as prospection appears to be built on evolutionarily older neural computations—in this case distance perception. The same abilities that appear to separate us from other animals also connect us to them through shared ancient mechanisms.

What It Feels Like To Be Human

It does not feel as though we have scaled up primate brains. It does not feel as if our neurons are firing due to voltage changes, without any choice in the matter. Instead, phenomenology is inherently dualist. It feels as though there is "someone" in our heads deciding what we think, feel, and do. And it feels as though that someone can change its mind at any moment. Science has shown repeatedly, however, that things are not necessarily what they seem.

We Do Not See What We Think We See

If you put down this book and look around, it would seem as though you see a richly detailed scene. In reality only a small fraction of the visual field is viewed in full color and high resolution at any one time, and this small window of acuity moves around as one's eyes saccade about three times per second (Mack & Rock, 1998). This is a very different picture from our visual phenomenology. As Andy Clark put it: "My conscious experience, clearly, is not of a small central pool of colour surrounded by a vague and out of focus expanse of halftones. Things look coloured all the way out" (Clark, 2002).

The inconsistency between what we *actually see* and our impressions of visual richness has been coined "The Grand Illusion" (Blackmore, Brelstaff, Nelson, & Troscianko, 1995; Dennett, 1991, 1992, 1998; O'Regan, 1992; Rensink, O'Regan, & Clark, 1997) and was illustrated dramatically in a study by O'Regan (1990). In this study participants were seated in front of a computer screen that appeared to present a solid page of text. Participants' eye movements were tracked, and they were asked to read the text. Participants proceeded to do so, noticing nothing out of the ordinary. However, in actuality, the text to the left and right of where each participant was looking was constantly changing to gibberish. Participants never noticed this bizarre fact because the text where they happened to be looking was normal: the eye tracking made certain that wherever they looked appeared in high resolution. Clearly, the moment-by-moment internal representations of the visual system are far more sparse and impoverished than it would seem based on our subjective experience.

Apparent motion illusions similarly reveal how our subjective experience is underpinned by unconscious mental gymnastics. First described by Max Wertheimer in 1912, the *phi phenomenon* refers to the illusory motion perceived when still images or lights are presented in rapid succession. If a light is flashed briefly followed by a second light a short distance away, the perception will be of a single light having moved from the first location to the second. When the lights are of different colors, the "moving" light appears to change color in the middle of its trajectory (Kohlers & vonGrunau, 1976). What is fascinating about this illusion is the simple fact that the light *appears to start moving right away*. This could not have happened, of course, because the "moving" light could not know the direction in which to move until the second light flashed. This simple illusion reveals an equally simple but deep truth: conscious experience is delayed relative to unconscious sensation (Madl, Baars, & Franklin, 2011). This temporal buffer allows the brain a window of time to construct the best story of what it thinks probably happened a moment earlier, and *that* construction—not veridical sensation—comprises the content of conscious experience. Work by Shiffrar and Freyd (1990 suggests that this unconscious perceptual construction even incorporates high-level knowledge about the world. In their study participants saw sequentially presented slides of human poses. In one example the first slide showed a person standing with her legs together, whereas the second slide showed the same person standing with her legs crossed. Showing these pairs in rapid succession produced the illusion of leg movement—our old friend, the phi phenomenon. Interestingly, however, the kind of illusory movement that a participant reported perceiving depended on the temporal interval between slides. A very short interval between slides (<175 ms) led to reports of seeing the shortest path of motion: one leg moving through the other leg to get to the other side.

When the interval was longer (>175 ms), subjects reported seeing the longer, *anatomically correct* path of motion: one leg moving around the other to get to the other side. Either way the motion perceived was illusory, but with a little extra time, the brain created a *better* illusion.

We Are Not Aware That We Are Not Aware

Not only do we feel that we are visually aware of our surroundings, we assume that our conscious awareness is temporally continuous. This again appears to be at least somewhat illusory. Jonathan Smallwood and Jonathan Schooler (2006) demonstrated that our minds wander—defined as "engaging in thoughts unrelated to the task at hand"—an astonishing 30 percent of our waking hours. Even more disconcerting is that we appear to be unaware of the fact that we are mind wandering about a third of the time we are doing so. Experimenters sporadically interrupted reading participants to ask whether they were mind wandering at that very moment, while they read a long text. About 10 percent of the time, participants replied that they were indeed mind wandering *but had not yet caught themselves*. As Schooler put it: "It is daunting to think that we're slipping in and out so frequently and we never notice that we were gone" (*New York Times* interview, 2010).

We Are Not as Free as We Think

Although college students are surprised by visual illusions and the mind-wandering studies of Smallwood and Schooler, they are not unwilling to accept them. In contrast the lay concept of free will is far more difficult to unseat. I take the lay view to be simply this: "Our conscious selves can decide, in the moment, to do otherwise." This feels like an undeniable truth in the same way that it feels that there is a self, "up there," irreducible to a large collection of cells.

Given all that we know about visual illusions, perhaps we should be able to consider that our phenomenological experience *of being a free agent in the moment* might not be correct. After all, we can understand that the lines of a Muller-Lyer illusion are the same length despite our compelling subjective experience otherwise. However, the feeling of conscious will is integral to the feeling of acting on the world. It is onerous to contemplate the possibility that purposeful action may be initiated independently from our subjective experience.

Over a hundred years ago William James suggested this very dissociation:

The willing terminates with the prevalence of the idea; and whether the act then follows or not is a matter quite immaterial, so far as the willing itself goes. I will to write and the act follows. I will to sneeze, and it does not. I will that the distant table slide over the floor towards me; it also does not. My willing representation can no more instigate my sneezing center than it can instigate the table to activity. But in both cases it is as true and good willing as it was when I willed to write. (James, 1892, p. 415)

A growing body of evidence suggests that William James was correct: the feeling of doing can be dissociated and manipulated independently from the actual doing (Bick & Kinsbourne, 1987; Biran & Chatterjee, 2004; Bode et al., 2011; Nisbett & Wilson, 1977; Ramachandran & Blakeslee, 1998; Soon, Brass, Heinze, & Haynes, 2008; Wegner & Wheatley, 1999; Wegner, 2002). As would be expected from this dissociation, several disorders are marked by unwarranted, misattributed, or disconnected feelings of agency for action (e.g., schizophrenia, alien hand syndrome, phantom limb syndrome, Parkinson disease).

Neuroscience versus Phenomenology (and What It Means for Justice)

Neuroscientific research makes an uncontroversial, unequivocal, and long-held conclusion: thought and action are physically realized in neural activity. Another uncontroversial fact is that neurons cannot choose, either individually or collectively, whether or not to send electrical signals. Although these facts may seem banal, they quickly lead to a logical conclusion that many find unsettling: if neurons cause thought and behavior but cannot themselves choose whether to fire, then how do we gain free will? It *feels* as if our conscious selves can decide, in the moment, to do otherwise. But if a neuron cannot decide whether or not to fire—and the brain is comprised of neurons—then *how can we, our conscious selves, be the ones deciding*? In the absence of a metaphysical or other as yet unaccounted for scientific explanation, the common view of free will appears to violate the laws of physics.

The immediate concern many people have when neuroscience weighs in on free will is that criminals would go free under the rationale that they are not responsible for their behavior: "Their neurons made them do it." What are the actual legal implications of considering how the brain works?

The, perhaps surprising, answer is that much of the legal system would stay exactly the same.

Imprisonment and punishment, for example, would still make sense. The person still committed the crime and society must be kept safe; therefore, some form of separation from society must be an option. Also, the brain learns and prioritizes the avoidance of pain, which makes punishment as a deterrent—both for the individual and in witnesses to the punishment—sensible. The same applies to guilt and blame. Although blame may not make logical sense (can we blame neurons for firing?), as aversive social feedback it serves as another deterrent that can shape behavior going forward. So what would change?

The concept of retributive justice—the notion of giving people their "just deserts" for crimes committed—is a kind of legalized vengeance based on the belief that the criminal mind could have chosen otherwise. It is thus the legal notion most at odds with our current understanding of the brain. And yet any punishment, regardless of flawed retributive logic, may have its utility. In the words of Supreme Court Justice Oliver Wendell Holmes Jr.:

> If I were having a philosophical talk with a man I was going to have hanged (or electrocuted) I should say "I don't doubt that your act was inevitable for you but to make it more avoidable by others we propose to sacrifice you to the common good. You may regard yourself as a soldier dying for your country if you like. But the law must keep its promises." (Holmes, 1997, p. 216)

Neuroscience data are unlikely to cause big changes to the machinery of our legal system. The real promise of neuroscience is that it will afford a deeper understanding of our biology that will promote both humility and more humane treatment of others.

Toward Humility and Humanity

In 2003 Jeffrey Burns and Russell Swerdlow published a case study that became an instant classic in debates of moral responsibility. They reported on a 40-year-old public school teacher who had a clean record until 2000, when he began to amass an expanding collection of pornographic magazines that eventually extended to child pornography. He hid this collection from his wife but was eventually removed from his home after he began making sexual advances toward his stepdaughter. He was diagnosed as having pedophilia and found guilty of child molestation. After being sentenced to inpatient rehabilitation, he was expelled due to having solicited sexual favors from staff and other patients. The evening before he was to be sentenced to prison, he went to the ER at a local hospital complaining of a headache. He was admitted to a psychiatric service after expressing suicidal ideation and his fear that he would rape his landlady. Understandably, given the proximity of his court date, the doctors were suspicious of duplicity. However, the next day he was seen by a neurologist who noticed that he walked with his feet spaced widely apart, his balance was off, and he had problems identifying things in his left visual field—signs of real neurological impairment. The neurologist scheduled a magnetic resonance imaging (MRI) scan.

The tumor was hard to miss. The size of an egg, it was pushing deep into his prefrontal cortex. Surgical resection was scheduled. If there were any doubts that the tumor was connected to his poor behavior, those doubts were dispelled after surgery. Post-surgery he stopped downloading pornography and completed the rehabilitation program from which he had been expelled weeks earlier. He was no longer a threat and returned to his home and his wife. About a year later, however, he developed a persistent headache and started collecting pornography again. This time, he returned to the hospital. The MRI showed tumor regrowth which was again resected.

Until very recently this man's future would have been grim. The tumor would have festered, leading to a rapid decline in his mental and physical health. Indeed, this is very nearly what happened. Instead, neuroimaging intervened and revealed what everyone had been missing. The tumor had caused the bad behavior and was treatable. Instead of a lifetime of misery at the expense of the state, he lives a relatively normal life with intermittent neurological monitoring.

The tumor case is dramatic, but its logic can be broadly applied. By definition, a person who commits a crime has the neurological machinery that made the crime not only possible but, at that moment, inevitable. Better understanding of this machinery will lead to better estimates of recidivism and thus to a better understanding of how to keep people safe. This understanding will not make crime any less horrible, but it may provide a dose of humility that comes from understanding the frailty of the human brain and that neurons cannot choose whether or not to fire. There, but for the grace of God, go I.

In their paper titled "For the law, neuroscience changes nothing and everything," Joshua Greene and Jonathan Cohen (2004), wrote:

Free will as we ordinarily understand it is an illusion generated by our cognitive architecture. Retributivist notions of criminal responsibility ultimately depend on this illusion, and, if we are lucky, they will give way to consequentialist ones, thus radically transforming our approach to criminal justice. At this time, the law deals firmly but mercifully with individuals whose behaviour is obviously the product of forces that are ultimately beyond their control. Some day, the law may treat all convicted criminals this way. That is, humanely.

Conclusion

The Dalai Lama (Tenzen, 2005) once said: "…understanding the nature of reality is pursued by means of critical investigation: if scientific analysis were conclusively to demonstrate certain claims in Buddhism to be false, then we must accept the findings of science and abandon those claims." Can we too abandon our strongest-held intuitions if neuroscience demonstrates that they are false? We can now look inside the head and begin to observe the machinery at work. It is an observation that is simultaneously illuminating and disquieting. It teaches us that the feeling of being a conscious self—a self who can choose *right now* to stop reading this sentence—has little real efficacy. Yet, as unsettling as this may seem, it does not call for the gutting of our entire judicial system. What it does call for is an evolved approach to justice that is grounded in scientific understanding, an approach that is firm, humble, and humane.

References

Anderson, M. L. (2010). Neural reuse: A fundamental organizational principle of the brain. *Behavioral and Brain Sciences, 33*, 245–266.

Bick, P. A., & Kinsbourne, M. (1987). Auditory hallucinations and subvocal speech in schizophrenic patients. *American Journal of Psychiatry, 144*, 222–225.

Biran, I., & Chatterjee, A. (2004). Alien hand syndrome. *Archives of Neurology, 64*, 292–294.

Blackmore, S. J., Brelstaff, G., Nelson, K., & Troscianko, T. (1995). Is the richness of our visual world an illusion? Transsaccadic memory for complex scenes. *Perception, 24*, 1075–1081.

Bode, S., He, A. H., Soon, C. S., Trampel, R., Turner, R., & Haynes, J. D. (2011). Tracking the unconscious generation of free decisions using ultra-high field fMRI. *PLoS ONE, 6*(6), e21612.

Burns, J. M., & Swerdlow, R. H. (2003). Right orbitofrontal tumor with pedophilia symptom and constructional apraxia sign. *Archives of Neurology, 60*, 437–440.

Clark, A. (2002). Is seeing all it seems? Action, reason and the grand illusion. *Journal of Consciousness Studies, 9*, 181–202.

Dehaene, S., & Cohen, L. (2007). Cultural recycling of cortical maps. *Neuron, 56*, 384–398.

Dennett, D. C. (1991). *Consciousness explained*. Boston: Little, Brown.

Dennett, D. C. (1992). "Filling in" versus finding out: A ubiquitous confusion in cognitive science. In H. L. Pick, Jr., P. van den Broek, & D. C. Knill (Eds.), *Cognition: Conceptual and methodological issues* (pp. 33–49). Washington, DC: American Psychological Association.

Dennett, D. C. (1998). No bridge over the stream of consciousness. *Behavioral and Brain Sciences, 21*(6), 753–754.

Donald, M. J. (1990). Quantum theory and the brain. *Proceedings of the Royal Society of London. A, 427*, 43–93.

Gallese, V., & Lakoff, G. (2005). The brain's concepts: The role of the sensory-motor system in conceptual knowledge. *Cognitive Neuropsychology, 22*, 455–479.

Gilbert, D. T. (2005). *Stumbling on happiness*. New York: Random House.

Greene, J., & Cohen, J. (2004). For the law, neuroscience changes nothing and everything. *Philosophical Transactions of the Royal Society of London. B, Biological Sciences, 359*, 1775–1785.

Holmes, O. W. J. (1997). In R. A. Posner (Ed.), *The essential Holmes: Selections from the letters, speeches, judicial opinions, and other writings of Oliver Wendell Holmes, Jr.* Chicago. University of Chicago Press.

James, W. (1892). *Psychology—Briefer course*. New York: Henry Holt. (New edition, 1961, New York: Harper & Row.)

Kandel, E. R. (1976). *Cellular basis of behavior*. San Francisco: W. H. Freeman.

Kohlers, P. A., & von Grunau, M. (1976). Shape and color in apparent motion. *Vision Research, 16*, 329–335.

Linden, D. J. (2007). *The accidental mind: How brain evolution has given us love, memory, dreams, and God*. Cambridge, MA: Harvard University Press.

Mack, A., & Rock, I. (1998). *Inattentional blindness*. Cambridge, MA: MIT Press.

Madl, T., Baars, B. J., & Franklin, S. (2011). The timing of the cognitive cycle. *PLoS ONE, 6*, e14803. doi:10.1371/journal.pone.0014803.

Nisbett, R., & Wilson, T. D. (1977). Telling more than we can know: Verbal reports on mental processes. *Psychological Review, 84*, 231–259.

O'Regan, J. K. (1990). Eye movements and reading. In E. Kowler (Ed.), *Eye movements and their role in visual and cognitive processes*. Amsterdam: Elsevier.

O'Regan, J. K. (1992). Solving the "real" mysteries of visual perception: The world as an outside memory. *Canadian Journal of Psychology, 46*, 461–488.

Parkinson, C., Liu, S., & Wheatley, T. (2014). A common cortical metric for spatial, temporal and social distance. *Journal of Neuroscience, 34*, 1979–1987.

Parkinson, C., & Wheatley, T. (2013). Old cortex, new contexts: Re-purposing spatial perception for social cognition. *Frontiers in Human Neuroscience, 7*, 645.

Ramachandran, V. S., & Blakeslee, S. (1998). *Phantoms in the brain*. New York: William Morrow.

Rensink, R. A., O'Regan, J. K., & Clark, J. J. (1997). To see or not to see: The need for attention to perceive changes in scenes. *Psychological Science, 8*(5), 368–373.

Shiffrar, M., & Freyd, J. J. (1990). Apparent motion of the human body. *Psychological Science, 1*, 257–267.

Smallwood, J., & Schooler, J. W. (2006). The restless mind. *Psychological Bulletin, 132*(6), 946–958.

Soon, C. S., Brass, M., Heinze, H. J., & Haynes, J. D. (2008). Unconscious determinants of free decisions in the human brain. *Nature Neuroscience, 11*, 543–545.

Tenzen, G. (Dalai Lama). (2005). The universe in a single atom: The convergence of science and spirituality. New York: Morgan Road Books.

Wegner, D. M. (2002). *The illusion of conscious will*. Cambridge, MA: MIT Press.

Wegner, D. M., & Wheatley, T. (1999). Apparent mental causation: Sources of the experience of will. *American Psychologist, 54*, 480–492.

17 The Equivocal Relationship between Morality and Empathy

Jean Decety and Jason M. Cowell

There is broad consensus that empathy is a fundamental component of our socioemotional and interpersonal lives. Empathy plays a vital role in social interaction from bonding between mother and child to appreciating conspecific subjective psychological states. Empathy-related processes are thought to motivate prosocial behavior, inhibit aggression, and provide the foundation for care-based morality. The lack of empathy is a hallmark characteristic of psychopathy and, in these individuals, is associated with callous disregard for the well-being of others coupled with an inability to experience remorse and guilt. Moreover, research with both healthy people and patients with neurological damage indicates that some forms of moral decisions (utilitarian judgments[1]) can be the consequence of a lack of empathic concern.

Empathy, however, is not always a direct avenue to moral behavior. Indeed, at times it can interfere with morality by introducing partiality, for instance by favoring ingroup members. But empathy can provide the emotional fire and a push toward seeing a victim's suffering end, irrespective of group membership and culturally determined dominance hierarchies, preventing rationalization of injustice and derogation. For instance, studies in social psychology have shown that empathy and justice are two independent motives, each with its own unique goal, and that in resource-allocation situations in which these two motives conflict, empathy can become a source of immoral behavior (Batson, Klein, Highberger, & Shaw, 1995).

The goal of this chapter is to illuminate the complex relation between morality and empathy, drawing on theories and empirical research in evolutionary biology, developmental psychology, and social neuroscience. We first specify as clearly and systematically as possible what the concepts of morality and empathy encompass. In clarifying these notions we highlight the ultimate and proximate causes of empathy and morality, showing that the former has older evolutionary roots in parental care, affective communication, and social attachment than morality, which is more recent and relies on both affective and cognitive processes. Evolution has tailored the mammalian brain to be sensitive and responsive to the emotional states of others, especially from one's offspring and members of one's social group, and thus, empathy has some unfortunate features that can directly conflict with moral behavior such as implicit group preferences. From these characterizations of morality and empathy, using data drawn from behavioral, developmental, and functional neuroimaging studies, we next consider how empathy can result in immoral judgment and behavior. We then argue that perspective taking can be successfully used to reduce group partiality, expand the circle of empathic concern from the tribe to all humanity, and facilitate the general upholding of principles of justice. Finally we conclude that it is better in the future not to use the slippery concept of empathy.

Morality

The umbrella of morality includes concepts such as justice, fairness, and rights and consists of prescriptive norms regarding how people should treat one another (Killen & Rutland, 2011). Alternatively, Haidt and Kesebir (2010) contend that morality encompasses the full array of psychological mechanisms that are active in the moral lives of people across cultures. Rather than stating the content of moral issues (e.g., justice and welfare), this definition specifies the function of moral systems as interlocking sets of values, virtues, norms, practices, and identities that work together to suppress or to regulate selfishness and make cooperative social life possible. Regardless of the definition, it is clear that a central focus of morality is the judgment of the rightness or wrongness of acts or behaviors that knowingly cause harm to people.

Across history many scholars from Darwin (1871) to Wilson (2002) have claimed that morality is an evolved aspect of human nature. Such a claim is well supported when it comes to the role of emotion in moral cognition. It is indeed highly plausible that moral emotions (e.g., guilt, shame) have an evolutionary history because they contribute to fitness in shaping decisions and actions when living in complex social groups. Trivers (1985) proposed that certain emotional responses may have led our ancestors to adopt a tit-for-tat strategy (reciprocal altruism) in which liking motivates the initiation of altruistic partnerships, anger and moral indignation motivate the withdraw of help from free rides, and guilt dissuades from taking more than what one gives. Reinforcement of moral behaviors minimizes criminal behavior and social conflict (Joyce, 2006), and moral norms provide safeguards against possible safety or health infringements (Begossi, Hanazaki, & Ramos, 2004).

Moral philosophy has been dominated by two general perspectives that are often opposed to one another. Utilitarianism views right actions as those that maximize the greater good (of the group over the individual) and maximize aggregate happiness. Deontological views contend that action should be evaluated against conformity to abstract rules such as the injunction against treating others as means to an end. In translating these theories to the psychological and neural domains, attempts have been made to characterize moral behavior using insights from both theories. Primarily, proponents of dual systems have forwarded a notion of morality wherein the cognitive system and the affective systems interact in moral judgment, with the former system also conceptualized as cool, controlled, and analytic, whereas the latter are automatic, intuitive, and emotion driven. In addition to this dual-system theory prevalent in contemporary social neuroscience, approaches from a developmental perspective have argued for a more complex view of morality. In particular, the social domain theory suggests that the development of morality stems from social relationships having to do with social problems, conflicts, and struggles (e.g., concern with avoiding harm, benefiting others, issues of physical violence, emotional hurt, unequal treatment, unfairness, social injustices, upholding rights, and violating rights) (Turiel, 2014).

Findings in moral psychology indicate that moral cognition is related to affective and emotional processing (Greene & Haidt, 2002). In addition, many moral judgments are surprisingly robust to demographic differences. People are sensitive to some of the same moral principles independent of gender, age, ethnicity, and religious views (Young & Saxe, 2011). It is important to note that although this objectivist view seems to be the prevalent contemporary theory, some scholars favor moral relativism. In particular, Prinz (2008) argues that moral values are based on emotional responses, which are inculcated by culture and not hard wired through natural selection.

A growing body of developmental research suggests that the capacity to evaluate others based on their prosocial and antisocial actions operates within the first year of life and is sensitive to many of the same factors that constrain adults' social and moral judgments, including the role of mental states and context in distinguishing good and bad behavior (Hamlin, 2014). This developmental work provides strong empirical support for claims that human capacities for moral evaluation are rooted in basic systems that evolved in the context of cooperation necessary for communal living.

Investigations into the neuroscience of morality have begun to shed light on the neural mechanisms underpinning moral cognition (Young & Dungan, 2012). Functional neuroimaging and lesion studies converge to indicate that moral evaluations arise from the integration of cognitive and affective systems and involve the posterior superior temporal sulcus (pSTS/TPJ), amygdala, insula, ventromedial prefrontal cortex (vmPFC), dorsolateral prefrontal cortex (dlPFC), and medial prefrontal cortex (mPFC) (Buckholtz & Marois, 2012; Decety, Michalska, & Kinzler, 2012a; Fumagalli & Priori, 2012). Importantly, the timing of the neural processing underpinning implicit moral computations associated with the perception of harm is extremely fast, as demonstrated by studies employing high-density event-related potentials (Decety & Cacioppo, 2012; Yoder & Decety, 2014). Current source density maxima in the right pSTS/TPJ, as fast as 62 ms post-stimulus, first distinguished intentional versus accidental harm. Later responses in the amygdala (122 ms) and ventromedial prefrontal cortex (182 ms), respectively, were evoked by the perception of intentional (but not accidental) harmful actions, indicative of fast information processing associated with these early stages of moral sensitivity.

What has become clear from neuroscience research is that these systems are not domain specific; rather, they support more domain-general processing such as affective arousal, attention, intention understanding, and decision making (Decety & Howard, 2013). Thus, there is no unique moral center in the brain (see figure 17.1). For instance, a functional MRI study showed that moral judgment of harm, dishonesty, and sexual disgust are instantiated in dissociable neural systems that are engaged differentially depending on the type of transgression being evaluated (Parkinson et al., 2011). The only overlapping activation across all morally laden scenarios in that study was the dmPFC, a region not specifically involved in the decision

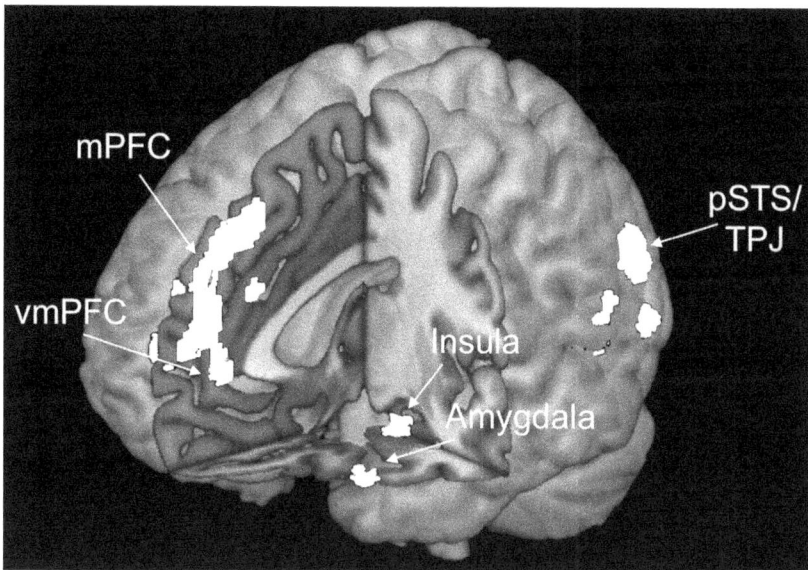

Figure 17.1
The moral brain comprises a large functional network of interconnected regions, including the medial prefrontal cortex (mPFC), dorsolateral prefrontal cortex, posterior superior temporal sulcus (pSTS) at the junction with the temporoparietal cortex (TPJ), ventromedial prefrontal cortex (vmPFC), insula, and amygdala. None of these regions is unique to morality. Rather, the functions implemented in these regions control different behavioral processes. Results from meta-analysis of 70 functional MRI studies of morality.

of wrongness, rather robustly associated with self-referential processing, thinking about other people (i.e., theory of mind), and processing ambiguous information. Importantly, some of the regions involved in moral evaluations overlap with a salience network associated with orienting toward and facilitating the processing of personally and motivationally salient social information (Harsay, Spaan, Wijnen, & Ridderinkhof, 2012), and during moral reasoning this information is used to direct activity within the executive control network.

Taken together, investigations of the neural, behavioral, and developmental bases of morality paint a strong picture of a constructivist view, an interaction of domain general systems, including executive control/ attentional, perspective taking, decision making, and emotional processing networks.

The Experience of Empathy

Similar to morality, empathy has been defined in multiple ways using various criteria (Batson, 2009). The number of competing conceptualizations circulating the literature has created a serious problem with the study of empathy by making it difficult to keep track of which process or mental state the term is being used to refer to in any given discussion (Coplan, 2011). These different conceptualizations refer to distinct psychological processes that vary, sometimes widely, in their function, phenomenology, biological mechanisms, and effects, including the relationships between empathy and moral behavior.

Empathy is complex and multifaceted, and some of these facets (emotional sharing and empathic concern) are not specific to humans (Decety, 2011a, 2011b). The ability to model the emotions of others and to react appropriately when interacting within a social group confers a number of evolutionarily positive advantages (e.g., increased ability to communicate and detect distress in group members and then to use this information to protect them). However, it is unlikely that nonhuman animals are aware of their feelings and emotions; most of the neural mechanisms that underlie empathy and caring are present in mammals and, as in humans, are affected by various interpersonal and social factors.

The Three Pillars of Empathy

In the past decade studies in affective and social neuroscience have contributed to informing theories of empathy, which can be conceptualized as a multidimensional construct composed of dissociable neurocognitive facets that interact and operate in parallel fashion, including affective, motivational, and cognitive components (Decety & Sveltova, 2012). The affective component of empathy reflects the capacity to share or become affectively aroused by others' emotions (at least in valence, tone, and relative intensity), commonly referred to as *emotional contagion* or *affective resonance*. The motivational component of empathy (empathic concern) corresponds to the urge to care for another's welfare. Finally, cognitive empathy is similar to the construct of perspective taking, the ability to consciously put oneself into the mind of another individual and imagine what that person is thinking or feeling. Each of these emotional, motivational, and cognitive facets of empathy has a different relationship with moral cognition.

Emotional Sharing

One of the most rudimentary mechanisms involved in empathy is the seemingly automatic, bottom-up process that allows the contagion of a conspecific affective state. Emotional sharing has old evolutionary roots in birds and mammals, is hardwired in the brain, and is thought to be a proxy for prosocial behavior (Decety & Sveltova, 2012). It plays a fundamental role in generating the motivation to care and help another individual in distress. It is also relatively independent of mind-reading and perspective-taking capacities. In naturalistic studies young children with high empathic disposition are more readily aroused vicariously by another's sadness, pain, or distress, but at the same time they possess greater capacities for emotion regulation such that their own negative arousal motivates rather than over-whelms their desire to alleviate the other's distress (Nichols, Sveltova, & Brownell, 2009). Developmental research has found that young (preverbal) children are not only moved by others' emotional states, but they attri-bute distress and pain in conjunction with their comforting behavior and recognize what the target is distressed about (Roth-Hanania, Davidov, & Zahn-Waxler, 2011).

Numerous neuroimaging studies have demonstrated that when indi-viduals are exposed to facial expressions of pain, sadness, or emotional dis-tress, or even when they imagine others in pain, brain regions involved in the first-hand experience of pain are activated (see Lamm, Decety, & Singer, 2011, for a meta-analysis). These regions include the ACC, the ante-rior midcingulate cortex (aMCC), anterior insula, supplementary motor area, amygdala, somatosensory cortex, and periaqueductal gray area (PAG). Importantly, the magnitude of multimodal (neurons) response in aMCC, insula, and somatosensory cortex correlates with perceived saliency of stimuli, regardless of their sensory modality (Legrain, Iannetti, Plaghki, & Mouraux, 2011). This multimodal response is related to bottom-up pro-cesses involved in saliency detection, arousal, and attentional capture (see figure 17.2) and plays a critical role in perceiving and responding to signs of distress in others.

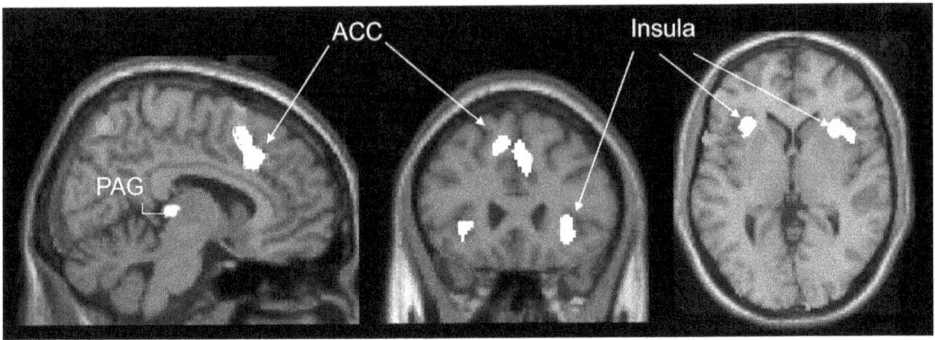

Figure 17.2
Neural regions that are associated with increased neurohemodynamic activation when individuals view another person in physical pain or emotional distress (i.e., emotional sharing). These regions include the anterior cingulate cortex (ACC), anterior insula, and periaqueductal gray (PAG) in the brainstem. Some studies also report activation of the somatosensory cortex and amygdala (not shown). Empathic concern involves the hypothalamus, especially the medial preoptic area, a region that is heavily interconnected with the ACC and the orbitofrontal and ventromedial prefrontal cortex, as well as the amygdala and brainstem.

Neuroimaging studies have reported that children and adolescents with disruptive psychopathic traits, who lack empathy, show reduced activity to the pain of others within the neural structures (ACC, insula, and amygdala) typically implicated in affective responses to others' pain and distress (Cheng, Hung, & Decety, 2012; Lockwood et al., 2013; Marsh et al., 2013). This blunted response likely contributes to callous disregard for the rights and feelings of others.

Empathic Concern
Another aspect often referred to as empathy is, in reality, *empathic concern*. The motivation to care for another evolved in a great many species because it promoted genetic fitness, which in mammalian species depends on the ability of conspecifics to communicate with each other, sharing information about their emotions and intentions and appropriately responding to their offspring's or relatives' needs (Bell, 2001). All social mammals depend on other conspecifics for survival and reproduction, particularly parental care, which is a necessity to infant survival and development.

Depending on each species the level of care varies, but the underlying neural circuitry for responding to infants (especially signals of vulnerability and need) is universally present and highly conserved. Converging evidence from animal research (Insel & Young, 2001), neuroimaging studies in healthy individuals, and lesion studies in neurological patients (Shamay-Tsoory, 2009) demonstrates that caring for others employs a large array of systems neural mechanisms extending beyond the cortex, including the brainstem, amygdala, hypothalamus, basal ganglia, anterior cingulate cortex (ACC), and vmFPC cortex. It also involves the autonomic nervous system, hypothalamic-pituitary-adrenal axis, and endocrine and hormonal systems (particularly oxytocin and vasopressin) that regulate bodily states, emotion, and social attachment. These systems underlying attachment appear to exploit the strong, established physical pain and reward systems, borrowing aversive signals associated with pain to indicate when relationships are threatened (Eisenberger, 2011).

It is worth noting that the vmPFC is not necessary for affective responses per se (it is not activated by the mere exposure to the pain of others and not by witnessing accidental harm), but it is critical when affective responses are shaped by conceptual information about specific outcomes (Roy, Shohamy, & Wager, 2012). Furthermore, individual differences in empathic concern have been shown to predict the magnitude of response in vmPFC in some moral contexts but not in others. Specifically, higher empathic concern was related to greater activity in vmPFC in moral evaluations where guilt was induced but not in moral evaluations where compassion was induced (Zahn, de Oliveira-Souza, Bramati, Garrido, & Moll, 2009). In addition, anatomical lesions and functional dysfunctions of the vmPFC and its reciprocal connections with the amygdala lead to a lack of empathic concern, inappropriate social behavior, a diminished sense of guilt, and immoral behavior (Sobhani & Bechara, 2011).

Importantly, this motivation to care for others is both deeply rooted in our biology and yet flexible. People can experience empathic concern for a wide range of "others" when cues of vulnerability and need are highly salient, including nonhumans, and in our Western culture, particularly pet animals such as puppies (Batson, 2012). Indeed, neuroscience research has shown that regions involved in perceiving the distress of other humans are similarly recruited when witnessing the distress of domesticated animals (Franklin et al., 2013).

Affective Perspective Taking

The final component of empathy is affective perspective taking. *Perspective taking* has been linked to social competence and social reasoning. It is a vital ability to navigate social interactions because it enables us to understand and predict the behaviors of other conspecifics and is, arguably, a defining difference between humans and other species. Perspective taking requires going outside of one's usual self-centeredness to view the world from another's subjective vantage point. The neural circuits supporting perspective taking overlap with those underpinning theory of mind and executive functions and include the medial prefrontal cortex and the posterior superior temporal sulcus (pSTS) at the temporoparietal junction (TPJ). There is general consensus among scholars that the ability to adopt and entertain the psychological perspective of others has a number of important consequences, including empathic concern (Batson, 2009). However, adopting the perspective of another person, in particular someone from another social group, is cognitively demanding and hence requires additional attentional resources and working memory, thus taxing executive function. Neuroscience research demonstrates that, when individuals adopt the perspective of another, neural circuits common to the ones underlying first-person experiences are activated to a large extent as well (Jackson, Brunet, Meltzoff, & Decety, 2006; Lamm, Meltzoff, & Decety, 2010; Ruby & Decety, 2004). However, taking the perspective of another produces increased activation in regions of the frontal cortex that are implicated in executive function, in particular to hold separate representations for self and other. In a neuroimaging study conducted by Lamm, Batson, and Decety (2007), participants watched video clips featuring patients undergoing painful medical treatment and were asked to either put themselves explicitly in the shoes of the patient (imagine self) or to focus on the patient's feelings and affective expressions (imagine other). Explicitly projecting oneself into an aversive situation led to higher personal distress, which was associated with enhanced activation in the amygdala and ACC, whereas focusing on the emotional and behavioral reactions of another in distress was accompanied by higher empathic concern, lower personal distress, increased activity in the executive attention network, vmPFC, and reduced amygdala response. Interestingly, individuals with high levels of psychopathy, when requested to imagine how another person would feel while viewing pictures depicting physical pain, fail to activate the vmPFC. Instead, they exhibit an atypical pattern of brain activation and effective connectivity between the anterior insula and amygdala with the ventromedial prefrontal cortex (Decety, Skelly, & Kiehl, 2013; Decety, Chen, Harenski, & Kiehl, 2013).

Altogether, developmental research, behavioral and functional neuroimaging studies, as well as lesion studies provide converging evidence that empathy is comprised of dissociable components that, at least partially, rely on distinct neural networks and that have unique evolutionary ultimate causes (Decety, Norman, Berntson, & Cacioppo, 2012b).

Empathy Is a Limited Resource

The building blocks of human psychology—cognition, emotion, motivation—have been shaped by the demands of social interdependence. Due to this obligatory interdependence, humans, like other social species, have evolved characteristics specifically adapted for group living. As empathy has evolved in the context of parental care and group living, it has produced some unfortunate features.

Empirical studies with animals and humans demonstrate kin and ingroup preferences in the detection and reaction to signs of distress. For instance rodents do not react indiscriminately to other conspecifics in distress. Female mice had higher fear responses (freezing behavior) when exposed to the pain of a close relative than when exposed to the pain of a more distant relative (Jeon et al., 2010). Another investigation found that female mice approaching a dyad member in physical pain led to less writhing from the mouse in pain. These beneficial effects of social approach were seen only when the mouse was a cagemate of the mouse in pain rather than a stranger (Langford et al., 2010). These results replicate previous findings reporting reduced pain sensitivity in mice when interacting with siblings but no such analgesic effect when mice interact with a stranger (D'Amato & Pavone, 1993). To test if genetic relatedness alone can motivate helping, a new study fostered rats from birth with another strain and found that, as adults, fostered rats helped strangers of the fostering strain but not rats of their own strain (Ben-Ami Bartal, Rodgers, Bernardes Sarria, Decety, & Mason, 2014). Thus, strain familiarity, even to one's own strain, is required for the expression of prosocial behavior.

Although empathy is one of the earliest social emotional competencies to develop (Davidov, Zahn-Waxler, Roth-Hanania, & Knafo, 2013), children do not display empathy and concern toward all people equally. Instead they show bias toward individuals and members of groups with which they identify. For instance young children two years of age display more concern-related behaviors toward their mother than toward unfamiliar individuals. In line with the ingroup hypothesis, 8-year-old children were more likely to be emotionally reactive toward their ingroup members compared with members of the outgroup, and dispositional empathy (as well as social anxiety) was positively correlated with group identification (Masten, Gillen-O'Neel, & Spears Brown, 2010). Moreover, children (aged three to nine years) view social categories as marking patterns of intrinsic interpersonal obligations; that is, they view people as intrinsically obligated only to their own group members and consider within-group harm as wrong regardless of explicit rules, but they view the wrongness of between-group harm as contingent on the presence of such rules (Rhodes & Chalik, 2013). These results regarding the nonobligatory nature of between-group harm contradict the prevalent notion from social domain theory that moral transgression about harm is unalterable (and contextually independent) from as young as preschool age (Smetana, 1981).

The mere assignment of individuals to arbitrary groups elicits evaluative preferences for ingroup relative to outgroup members that impacts empathy. In one behavioral study participants were assigned to artificial groups and required to perform pain intensity judgments of stimuli depicting bodily injuries from self, ingroup, and outgroup perspectives. Participants rated the stimuli as more painful when they had to adopt the perspective of an ingroup member as compared to their own perspective, whereas the outgroup perspective did not induce different responses to the painful stimuli as compared to the self perspective. Moreover, the ratings differences between the painful and nonpainful pictures were more important in the ingroup perspective than in the outgroup perspective (Montalan, Lelard, Godefroy, & Mouras, 2012).

Take, for instance, a recent study by Morton and Postmes (2011) in which British Caucasian participants were read a summary of the atrocities committed by British Caucasians against African slaves and asked about their guilt toward these actions and their categorization of the relationship between British and African nations. Opposing a commonsense view that conceptualizing nations as a single, shared humanity would predict greater remorse toward these actions, the individuals who viewed the British and

African nations as two separate races felt greater guilt over historic transgressions and had lesser expectations of forgiveness. Moreover, in another study, people's relative levels of economic well-being were found to shape their beliefs about what is right or wrong. In that study upper-class individuals were more likely to make calculated, dispassionate moral judgments in dilemmas in which utilitarian choices were at odds with visceral moral intuitions (Cote, Piff, & Willer, 2013). In this way the lower empathy of upper-class individuals ironically led them to make moral decisions that were more likely to maximize the greatest good for the greatest number. In short, straightforward predictions between empathy and morality are not possible and appear to be governed by contextual influences.

Further evidence from studies with adults suggests that although empathy does not necessarily change notions of fairness (e.g., what is the just action in a certain situation), it does change the decision an individual will make. In one such study (Batson et al., 1995), college students required to assign a good and a bad task to two individuals overwhelmingly endorsed random assignment (i.e., a coin flip) as the fairest means for deciding who would be assigned with the bad task. However, when asked to consider the feelings of a worker who had recently suffered hardship, students readily offered the good task to that worker rather than using random assignment.

Neuroimaging studies reveal that the neural network implicated in empathy for the pain of others is either strengthened or weakened by interpersonal variables, implicit attitudes, and group preferences. Activity in the pain neural network is significantly enhanced when individuals view their loved ones in pain compared to strangers (Cheng, Chen, Lin, Chou, & Decety, 2010). Affective sharing is moderated by a priori implicit attitudes toward conspecifics. For example, study participants were significantly more sensitive to the pain of individuals who had contracted AIDS as the result of a blood transfusion as compared to individuals who had contracted AIDS as the result of their illicit drug addiction (sharing needles), as evidenced by higher pain sensitivity ratings and greater hemodynamic activity in the ACC, insula, and PAG, although the intensity of pain on the facial expressions was strictly the same across all videos (Decety, Echols, & Correll, 2009). Another study found evidence for a modulation of empathic neural responses by racial group membership (Xu, Zuo, Wang, & Han, 2009). Notably, the response in the ACC to perception of others in pain decreased remarkably when participants viewed faces of racial outgroup members relative to racial ingroup members. This effect was comparable in Caucasian and Chinese subjects and suggests that modulations of empathic neural responses by racial group membership are similar in different ethnic groups. Another study demonstrated that the failures of an ingroup member are painful, whereas those of a rival outgroup member give pleasure—a feeling that may motivate harming rivals (Cikara, Botvinick, & Fiske, 2011).

All these behavioral, developmental, and neuroimaging studies clearly demonstrate that emotional sharing is not automatic and is modulated by top-down processes such those involved in ingroup membership, and this can impact empathic concern and helping behaviors.

Low Empathic Concern Increases Utilitarian Moral Judgments

In examining the complex relationship between components of empathy and moral judgment, an often-used paradigm in psychological and some neuroscience studies of morality is a thought experiment borrowed from philosophy, the trolley dilemma (Foot, 1967; Thomson, 1976).

Participants are told about an out-of-control trolley headed down a track to which a great number of people are tied; there is an alternate track on which one individual is tied. Subjects are then given an option of diverting the trolley: they can pull a lever and the trolley will be diverted to the alternate track, killing the one individual and saving the group. This decision is relatively easy to make, and the majority of participants will chose to divert the trolley. However, another option is presented: rather than pulling a lever, participants have to either let the six die or they can push a large man in front of the trolley, again, sacrificing the one to save the group. This decision, for the majority of participants, is not something that is comfortable, and most refuse to do so. This classic thought problem, comparing impersonal and personal moral decision making, has led to a great deal of inquiry about the nature of individuals who will push the large man in front of the trolley.

Are individuals who make a utilitarian judgment in personal situations more rational and calculating, or are they simply colder and less averse to harming others? Support for a link between empathy and moral reasoning is given by studies demonstrating that low levels of dispositional empathic concern predicts utilitarian moral judgment (e.g., personal, impersonal) (Gleichgerrcht & Young, 2013). A functional neuroimaging study recently

examined the neural basis of such indifference to harming while participants were engaged in moral dilemmas (Wiech et al., 2013). A tendency
toward counterintuitive impersonal utilitarian judgment was associated
both with "psychoticism" (or psychopathy), a trait linked with a lack of
empathic concern and antisocial tendencies, and with "need for cognition," a trait reflecting preference for effortful cognition. Importantly, only
psychoticism was also negatively correlated with activation in the vmPFC,
a brain area implicated in empathic concern and social emotions such as
guilt, during counterintuitive utilitarian judgments. These findings suggest
that when individuals reach highly counterintuitive utilitarian conclusions, it does not need to reflect greater engagement in explicit moral deliberation. It may rather reveal a lack of empathic concern and diminished
aversion to harming others. Lesions of the orbitofrontal cortex (including
the vmPFC) have consistently been associated with increased utilitarian
choices in highly conflicting moral dilemmas more often than control subjects, opting to sacrifice one person's life to save a number of other individuals (Koenigs et al., 2007).

In summary, both fMRI experiments and lesion studies clearly document
the critical role of the vmPFC in moral decision making and empathic concern as well as the importance of this region in processing aversive emotions that arise from perceiving or imagining harmful intentions. Such
information is processed extremely rapidly as demonstrated by high-density EEG/ERPs recordings in individuals viewing intentional interpersonal
harm (Decety & Cacioppo, 2012) and is factored in when they are making
moral judgments.

Beyond the Tribe

In humans as well as in nonhuman animals, empathy and prosocial behaviors are modulated by the degree of affiliation and are extended preferentially toward ingroup members and less often toward unaffiliated others
(Echols & Correll, 2012). Yet humans can and often do act prosocially
toward strangers and extend concern beyond kin or own social group.
Humans have created meta-level symbolic social structures for upholding
moral principles to all humanity, such as Human Rights and the International Criminal Court. Thus, nurture is not confined to the dependent
young of one's own kin system but also extends to the ecosystem and

toward future generations. Such a capacity to help and care for unfamiliar individuals is often viewed as a cognitively complex behavior that depends on high cognitive capacities, social modeling, and cultural transmission (Levine, Prosser, Evans, & Reicher, 2005).

Affective perspective taking can be used as a strategy for reducing group biases. Adopting the perspective of another has been linked to social competence and social reasoning (Underwood & Moore, 1982). A substantial body of behavioral studies has documented that affective perspective taking is a powerful way to elicit empathy and concern for others (Batson, Early, & Salvarini, 1997) and reduce prejudice and intergroup bias. For instance, taking the perspective of an outgroup member leads to a decrease in the use of explicit and implicit stereotypes for that individual and to more positive evaluations of that group as a whole (Galinsky & Moskowitz, 2000). Something of this sort occurred among the rescuers of Jews during the Second World War in Europe. A careful look at data collected by Oliner and Oliner (1988) suggests that involvement in rescue activity frequently began with concern for a specific individual or individuals for whom compassion was felt—often individuals known previously. This initial involvement subsequently led to further contact and rescue activity and to a concern for justice that extended well beyond the bounds of the initial empathic concern. Assuming the perspective of another (like being in a wheelchair) brings about changes in the way we see the other, and these changes generalize to people similar to them, notably members of the same social groups to which they belong (Castano, 2012). Some studies have documented long-lasting effects of such interventions. For instance, Sri Lankan Singhalese participants expressed enhanced empathy toward Tamils even a year after participating in a four-day intergroup workshop (Malhotra & Liyanage, 2005). Affective perspective taking can thus be a powerful way to connect morality and empathic concern.

Empathy and Morality Have Complex Relationships

There is no reason to see empathy and morality as either systematically opposed to one another or complementary. Both have different ultimate and proximal mechanisms. Even the most advanced forms of empathy in humans are built on more basic forms and remain connected to affective communication, social attachment, and parental care that are highly conserved across mammalian species. Empathic concern evolved in the context of parent care and group living and has yielded a variety of group biases that can impact our moral behavior. Interestingly, both empathic concern and moral decision making require involvement of the orbitofrontal/ventromedial prefrontal cortex, a region reciprocally connected with ancient affective systems in brainstem, amygdala, and hypothalamus that bridges conceptual and affective processes. This region, across species, is a critical hub for caregiving behavior, particularly for parenting through reward-based and affective associations (Parson et al., 2013).

Thus, care-based morality "piggybacks" on older evolutionary motivational mechanisms associated with parental care. This explains why empathy is not a direct avenue to morality and can at times be a source of immoral action by favoring self-interest.

It has been argued that moral progress involves expanding our concern from the family and the tribe to humanity as a whole. Yet it is difficult to empathize with seven billion strangers or to feel toward someone one has never met the degree of concern one feels for one's own baby or a friend. Nonetheless, over the course of human history, people have enlarged the range of beings whose interests they value as they value their own (Singer, 1981). One of the recent "inventions" that, according to Pinker (2011), contributed in expanding empathy is the expansion of literacy during the humanitarian revolution in the eighteenth century. In the epistolary novel the story unfolds in a character's own words, exposing the character's thoughts and feelings in real time rather than describing them from the distancing perspective of a disembodied narrator. Research indeed demonstrates that reading literary fiction can improve the capacity to identify and understand others' subjective affective and cognitive mental states (Kidd & Castano, 2013). Self-reported empathic skills significantly change over the course of one week for readers of fictional stories (Bal & Veltkamp, 2013). Another line of research demonstrated that arts intervention—training in acting—leads to growth in empathy and theory of mind (Goldstein & Winner, 2012). Reading, language, the arts, the media all provide rich cultural input, processed by the flexible cognitive resources implemented in our large prefrontal cortex, triggering internal simulation processes (Decety & Grèzes, 2006) that lead to the experience of emotions and potentially elicits both concern and caring for others. After all, there is no reason to see ourselves as marionettes dancing on the strings of evolution. Certain parts of our behavior may be genetically encoded, having been shaped by natural selection in our savanna-dwelling ancestors. Yet, genes are not exactly destiny, and evolution did not stop 30,000 years ago. In just the last few years we have added the ability to function at high altitudes and resistance to malaria to the list of rapidly evolved human characteristics, and the stage is set for many more including expanding our concern to all and upholding justice principles.

To conclude on a provocative note, it may be advantageous for the science of morality, in the future, to refrain from using the catch-all term of empathy, which applies to a myriad of processes and, as a result, yields confusion in both understanding and predictive ability. Rather, emotional sharing, empathic concern, and affective perspective taking are more precise in their scope, evolutionary roots, and neural and developmental mechanisms and will provide for clarity in future studies of moral judgment.

Acknowledgment

The writing of this article was supported by grants from the John Templeton Foundation (The Science of Philanthropy Initiative and Wisdom Research at the University of Chicago).

Note

1. Philosophers and psychologists have long argued about whether there is one "right" answer to such moral questions, be it utilitarian ethics, which advocates saving as many as possible, even if it requires personally harming an individual, or nonutilitarian principles, which mandate strict adherence to rules such as "do not kill" that are rooted in the value of human life and dignity.

References

Bal, P. M., & Veltkamp, M. (2013). How does fiction reading influence empathy? An experimental investigation of the role of emotional transportation. *PLoS ONE, 8*(1), e55341.

Batson, C. D. (2009). These things called empathy: Eight related but distinct phenomena. In J. Decety & W. Ickes (Eds.), *The social neuroscience of empathy* (pp. 3–15). Cambridge, MA: MIT Press.

Batson, C. D. (2012). The empathy-altruism hypothesis: Issues and implications. In J. Decety (Ed.), *Empathy—from bench to bedside* (pp. 41–54). Cambridge, MA: MIT Press.

Batson, C. D., Early, S., & Salvarini, G. (1997). Perspective taking: Imagining how another feels versus imagining how you would feel. *Personality and Social Personality Bulletin, 23,* 751–758.

Batson, C. D., Klein, T. R., Highberger, L., & Shaw, L. L. (1995). Immorality from empathy-induced altruism: When compassion and justice conflict. *Journal of Personality and Social Psychology, 68,* 1042–1054.

Begossi, A., Hanazaki, N., & Ramos, R. (2004). Food chain and the reasons for fish food taboos among Amazonian and Atlantic forest fishers. *Ecological Applications, 14,* 1334–1343.

Bell, D. C. (2001). Evolution of parental caregiving. *Personality and Social Psychology Review, 5,* 216–229.

Ben-Ami Bartal, I., Rodgers, D. A., Bernardez Sarria, M. S., Decety, J., & Mason, P. (2014). Prosocial behavior in rats is modulated by social experience. *eLife, 3,* e1385.

Buckholtz, J. W., & Marois, R. (2012). The roots of modern justice: Cognitive and neural foundations of social norms and their enforcement. *Nature Neuroscience, 15,* 655–661.

Castano, E. (2012). Antisocial behavior in individuals and groups: An empathy-focused approach. In K. Deaux & M. Snyder (Eds.), *The Oxford Handbook of Personality and Social Psychology* (pp. 419–445). New York: Oxford University Press.

Cheng, Y., Chen, C. Y., Lin, C. P., Chou, K. H., & Decety, J. (2010). Love hurts: An fMRI study. *NeuroImage, 51,* 923–929.

Cheng, Y., Hung, A., & Decety, J. (2012). Dissociation between affective sharing and emotion understanding in juvenile psychopaths. *Development and Psychopathology, 24,* 623–636.

Cikara, M., Botvinick, M. M., & Fiske, S. T. (2011). Us versus them: Social identity shapes responses to intergroup competition and harm. *Psychological Science, 22,* 306–313.

Coplan, A. (2011). Understanding empathy. In A. Coplan & P. Goldie (Eds.), *Empathy: Philosophical and psychological perspectives* (pp. 3–18). New York: Oxford University Press.

Cote, S., Piff, P. K., & Willer, R. (2013). For whom do the ends justify the means? Social class and utilitarian moral judgment. *Journal of Personality and Social Psychology, 104,* 490–503.

D'Amato, F. R., & Pavone, F. (1993). Endogenous opioids: A proximate reward mechanism for kin selection? *Behavioral and Neural Biology, 60,* 79–83.

Darwin, C. (1871). *The descent of man and selection in relation to sex.* (1874 ed.). London: John Murray.

Davidov, M., Zahn-Waxler, C., Roth-Hanania, R., & Knafo, A. (2013). Concern for others in the first year of life: Theory, evidence, and avenues for research. *Child Development Perspectives, 7,* 126–131.

Decety, J. (2011a). Dissecting the neural mechanisms mediating empathy. *Emotion Review, 3,* 92–108.

Decety, J. (2011b). The neuroevolution of empathy. *Annals of the New York Academy of Sciences, 1231,* 35–45.

Decety, J., & Cacioppo, S. (2012). The speed of morality: A high-density electrical neuroimaging study. *Journal of Neurophysiology, 108,* 3068–3072.

Decety, J., Chen, C., Harenski, C. L., & Kiehl, K. A. (2013). An fMRI study of affective perspective taking in individuals with psychopathy: Imagining another in pain does not evoke empathy. *Frontiers in Human Neuroscience, 7,* 489.

Decety, J., Echols, S. C., & Correll, J. (2009). The blame game: The effect of responsibility and social stigma on empathy for pain. *Journal of Cognitive Neuroscience, 22,* 985–997.

Decety, J., & Grèzes, J. (2006). The power of simulation: Imagining one's own and other's behavior. *Brain Research, 1079,* 4–14.

Decety, J., & Howard, L. (2013). The role of affect in the neurodevelopment of morality. *Child Development Perspectives, 7,* 49–54.

Decety, J., Michalska, K. J., & Kinzler, K. D. (2012a). The contribution of emotion and cognition to moral sensitivity: A neurodevelopmental study. *Cerebral Cortex, 22,* 209–220.

Decety, J., Norman, G. J., Berntson, G. G., & Cacioppo, J. T. (2012b). A neurobehavioral evolutionary perspective on the mechanisms underlying empathy. *Progress in Neurobiology, 98,* 38–48.

Decety, J., Skelly, L. R., & Kiehl, K. A. (2013). Brain response to empathy-eliciting scenarios in incarcerated individuals with psychopathy. *JAMA Psychiatry, 70*(6), 638–645.

Decety, J., & Sveltova, M. (2012). Putting together phylogenetic and ontogenetic perspectives on empathy. *Developmental Cognitive Neuroscience, 2,* 1–24.

Echols, S., & Correll, J. (2012). It's more than skin deep: Empathy and helping behavior across social groups. In J. Decety (Ed.), *Empathy: From bench to bedside* (pp. 55–71). Cambridge, MA: MIT Press.

Eisenberger, N. (2011). Why rejection hurts: What social neuroscience has revealed about the brain's response to social rejection. In J. Decety & J. T. Cacioppo (Eds.), *The Oxford handbook of social neuroscience* (pp. 586–598). New York: Oxford University Press.

Foot, P. (1967). The problem of abortion and the doctrine of the double effect in virtues and vices. *Oxford Review, 5*, 5–15.

Franklin, R. G., Nelson, A. J., Baker, M., Beeney, J. E., Vescio, T. K., Lenz-Watson, A., et al. (2013). Neural responses to perceiving suffering in humans and animals. *Social Neuroscience, 8*(3), 217–227.

Fumagalli, M., & Priori, A. (2012). Functional and clinical neuroanatomy of morality. *Brain, 135*, 2006–2021.

Galinsky, A. D., & Moskowitz, G. B. (2000). Perspective-taking: Decreasing stereotype expression, stereotype accessibility and in-group favoritism. *Journal of Personality and Social Psychology, 78*, 708–724.

Gleichgerrcht, E., & Young, L. (2013). Low levels of empathic concern predict utilitarian moral judgment. *PLoS ONE, 4*, e60418.

Goldstein, T. R., & Winner, E. (2012). Enhancing empathy and theory of mind. *Journal of Cognition and Development, 13*, 19–37.

Greene, J., & Haidt, J. (2002). How (and where) does moral judgment work? *Trends in Cognitive Sciences, 12*, 517–523.

Haidt, J., & Kesebir, S. (2010). Morality. In S. Fiske, D. Gilbert, & G. Lindzey (Eds.), *Handbook of social psychology* (5th ed., pp. 797–832). Hoboken, NJ: Wiley.

Hamlin, J. K. (2014). The origins of human morality: Complex socio-moral evaluations by pre-verbal infants. In J. Decety & Y. Christen (Eds.), *New Frontiers in Social Neuroscience* (pp. 165–188). New York: Springer.

Harsay, H. A., Spaan, M., Wijnen, J. G., & Ridderinkhof, K. R. (2012). Error awareness and salience processing in the oddball task: Shared neural mechanisms. *Frontiers in Human Neuroscience, 6*, 246.

Insel, T. R., & Young, L. J. (2001). The neurobiology of attachment. *Nature Reviews Neuroscience, 2*, 129–136.

Jackson, P. L., Brunet, E., Meltzoff, A. N., & Decety, J. (2006). Empathy examined through the neural mechanisms involved in imagining how I feel versus how you feel pain: An event-related fMRI study. *Neuropsychologia, 44*, 752–761.

Jeon, D., Kim, S., Chetana, D., Jo, D., Ruley, H. E., Rabah, D., et al. (2010). Observational fear learning involves affective pain system and Ca1.2 CA channels in ACC. *Nature Neuroscience, 13*, 482–488.

Joyce, R. (2006). *The evolution of morality.* Cambridge, MA: MIT Press.

Kidd, D. C., & Castano, E. (2013). Reading literary fiction improves theory of mind. *Science, 342*, 377–380.

Killen, M., & Rutland, A. (2011). *Children and social exclusion: Morality, prejudice, and group identity* (pp. 480–481). New York: Wiley-Blackwell.

Koenigs, M., Young, L., Adolphs, R., Tranel, D., Cushman, F., Hauser, M., et al. (2007). Damage to the prefrontal cortex increases utilitarian moral judgments. *Nature, 446*, 908–911.

Lamm, C., Batson, C. D., & Decety, J. (2007). The neural basis of human empathy: Effects of perspective-taking and cognitive appraisal. *Journal of Cognitive Neuroscience, 19*, 42–58.

Lamm, C., Decety, J., & Singer, T. (2011). Meta-analytic evidence for common and distinct neural networks associated with directly experienced pain and empathy for pain. *NeuroImage, 54*, 2492–2502.

Lamm, C., Meltzoff, A. N., & Decety, J. (2010). How do we empathize with someone who is not like us? *Journal of Cognitive Neuroscience, 2*, 362–376.

Langford, D. J., Tuttleb, A. H., Brown, K., Deschenes, S., Fischer, D. B., Mutso, A., et al. (2010). Social approach to pain in laboratory mice. *Social Neuroscience, 5*, 163–170.

Legrain, V., Iannetti, G. D., Plaghki, L., & Mouraux, A. (2011). The pain matrix reloaded: A salience detection system for the body. *Progress in Neurobiology, 93*, 111–124.

Levine, M., Prosser, A., Evans, D., & Reicher, S. (2005). Identity and emergency intervention: How social group membership and inclusiveness of group boundaries shape helping behavior. *Personality and Social Psychology Bulletin, 31*, 443–453.

Lockwood, P. L., Sebastian, C. L., McCrory, E. J., Hyde, Z. H., Gu, X., De Brito, S. A., et al. (2013). Association of callous traits with reduced neural response to others' pain in children with conduct disorder. *Current Biology, 23*, 1–5.

Malhotra, D., & Liyanage, S. (2005). Long-term effects of peace workshops in protracted conflicts. *Journal of Conflict Resolution, 49*, 908–924.

Marsh, A. A., Finger, E. C., Fowler, K. A., Adalio, C. J., Jurkowitz, I. N., Schechter, J. C., et al. (2013). Empathic responsiveness in amygdala and anterior cingulate cortex in youths with psychopathic traits. *Journal of Child Psychology and Psychiatry, and Allied Disciplines, 54*, 900–910.

Masten, C. L., Gillen-O'Neel, C., & Spears Brown, C. (2010). Children's intergroup empathic processing: The roles of novel ingroup identification, situational distress, and social anxiety. *Journal of Experimental Child Psychology, 106*, 115–128.

Montalan, B., Lelard, T., Godefroy, O., & Mouras, H. (2012). Behavioral investigation of the influence of social categorization on empathy for pain: A minimal group paradigm study. *Frontiers in Psychology, 3*, 389.

Morton, T. A., & Postmes, T. (2011). Moral duty or moral deference? The effects of perceiving shared humanity with the victims of ingroup perpetrated harm. *European Journal of Social Psychology, 41*, 127–134.

Nichols, S. R., Svetlova, M., & Brownell, C. A. (2009). The role of social understanding and empathic disposition in young children's responsiveness to distress in parents and peers. *Cognition Brain Behavior, 13*, 449–478.

Oliner, S. P., & Oliner, P. M. (1988). *The altruistic personality: Rescuers of Jews in Nazi Europe*. New York: Free Press.

Parkinson, C., Sinnott-Armstrong, W., Koralus, P. E., Mendelovici, A., McGeer, V., & Wheatley, T. (2011). Is morality unified? Evidence that distinct neural systems underlie moral judgments of harm, dishonesty, and disgust. *Journal of Cognitive Neuroscience, 23*, 3162–3180.

Parson, C. E., Stark, E. A., Young, K. S., Stein, A., & Kringelbach, M. L. (2013). Understanding the human parental brain: A critical role of the orbitofrontal cortex. *Social Neuroscience, 8*, 525–543.

Pinker, S. (2011). *The better angels of our nature: Why violence has declined*. New York: Penguin Group.

Prinz, J. J. (2008). *The emotional construction of morals*. New York: Oxford University Press.

Rhodes, M., & Chalik, L. (2013). Social categories as markers of intrinsic interpersonal obligations. *Psychological Science, 24*, 999–1006.

Roth-Hanania, R., Davidov, M., & Zahn-Waxler, C. (2011). Empathy development from 8 to 16 months: Early signs of concern for others. *Infant Behavior and Development, 34*, 447–458.

Roy, M., Shohamy, D., & Wager, T. D. (2012). Ventromedial prefrontal-subcortical systems and the generation of affective meaning. *Trends in Cognitive Sciences, 16*, 147–156.

Ruby, P., & Decety, J. (2004). How would you feel versus how do you think she would feel? A neuroimaging study of perspective taking with social emotions. *Journal of Cognitive Neuroscience, 16*, 988–999.

Shamay-Tsoory, S. (2009). Empathic processing: Its cognitive and affective dimensions and neuroanatomical basis. In J. Decety and W. Ickes (Eds.), *The social neuroscience of empathy* (pp. 215–232). Cambridge, MA: MIT Press.

Singer, P. (1981). *The expanding circle: Ethics and sociobiology.* New York: Farrar Straus & Giroux.

Smetana, J. (1981). Preschool children's conceptions of moral and social rules. *Child Development, 52,* 1333–1336.

Sobhani, M., & Bechara, A. (2011). A somatic marker perspective of immoral and corrupt behavior. *Social Neuroscience, 6,* 640–652.

Thomson, J. J. (1976). Killing, letting die, and the trolley problem. *Monist, 59,* 204–217.

Trivers, R. L. (1985). *Social evolution.* Menlo Park, CA: Benjamin Cummings Press.

Turiel, E. (2014). Morality: Epistemology, development and social opposition. In M. Killen & J. G. Smetana (Eds.), *Handbook of moral development* (pp. 3–22). Hove: Psychology Press.

Underwood, B., & Moore, B. (1982). Perspective-taking and altruism. *Psychological Bulletin, 91,* 143–173.

Wiech, K., Kahane, G., Shackel, N., Farias, M., Savulescu, J., & Tracey, I. (2013). Cold or calculating? Reduced activity in the subgenual cingulate cortex reflects decreased emotional aversion to harming in counterintuitive utilitarian judgment. *Cognition, 126,* 364–372.

Wilson, D. S. (2002). *Darwin's cathedral: Evolution, religion and the nature of society.* Chicago: University of Chicago Press.

Xu, X., Zuo, X., Wang, X., & Han, S. (2009). Do you feel my pain? Racial group membership modulates empathic neural responses. *Journal of Neuroscience, 29,* 8525–8529.

Yoder, K. J., & Decety, J. (2014). Spatiotemporal neural dynamics of moral judgments: A high-density EEG/ERP study. *Neuropsychologia, 60,* 39–45.

Young, L., & Dungan, J. (2012). Where in the brain is morality? Everywhere and maybe nowhere. *Social Neuroscience, 7*(1–2), 1–10.

Young, L., & Saxe, R. (2011). Moral universals and individual differences. *Emotion Review, 3,* 323–324.

Zahn, R., de Oliveira-Souza, R., Bramati, I., Garrido, G., & Moll, J. (2009). Subgenual cingulate activity reflects individual differences in empathic concern. *Neuroscience Letters, 457,* 107–110.

Contributors

Scott Atran   CNRS, University of Oxford, CUNY, University of Michigan

Abigail A. Baird   Vassar College

Nicolas Baumard   Ecole Normale Supérieure, Paris

Sarah Brosnan   Georgia State University

Jason M. Cowell   University of Chicago

Molly J. Crockett   University of Oxford

Mark Dadds   University of New South Wales

Jean Decety   University of Chicago

Andrew W. Delton   Stony Brook University

Ricardo de Oliveira-Souza   D'Or Institute for Research and Education

Jeremy Ginges   University of Melbourne

Andrea L. Glenn   University of Alabama

Joshua D. Greene   Harvard University

J. Kiley Hamlin   University of British Columbia

David Hawes   University of Sydney

Jillian Jordan   Yale University

Max M. Krasnow   Harvard University

Ayelet Lahat   McMaster University

Jorge Moll   D'Or Institute for Research and Education

Caroline Moul   University of New South Wales

Alexander Peysakhovich   Harvard University

Laurent Prétôt   Georgia State University

Jesse Prinz   City University of New York

David G. Rand   Yale University

Rheanna J. Remmel   University of Alabama

Regina A. Rini   University of Oxford

Emma V. Roellke   Vassar College

Joshua Rottman   Boston University

Mark Sheskin   Ecole Normale Supérieure, Paris

Thalia Wheatley   Dartmouth College

Liane Young   Boston College

Roland Zahn   Institute of Psychiatry at King's College

Author Index

Abbot, P., 36
Abe, N., 209
Ackerman, P., 92
Addis, D. R., 205
Adolphs, R., 159, 185, 190
Aggleton, J. P., 161
Aharoni, E., 62, 200, 244
Ahlgren, M., 11
Ahmad, N., 92
Ahmadi, S., 228
Aikins, J. W., 166
Akitsuki, Y., 143
Aknin, L. B., 111
Alexander, L., 44
Algoe, B. S., 54
Alicke, M. D., 129
Allen, J. L., 257
Allison, G., 71
Allison, S. T., 94
Allison, T., 161
Almenberg, J., 96
Alpers, B. J., 190
Amit, E., 205
Anderson, E., 205
Anderson, M. L., 269
Anderson, S. W., 62, 199
Andre, J. B., 38, 70
Andreoni, J., 91
Andrews, B. P., 240
Andrews, C., 162
Andrews-Hanna, J. R., 198, 201

Anen, C., 208
Apicella, C. L., 96
Armony, J. L., 160
Arnsten, A. F. T., 222, 224
Aronfreed, J., 144
Aronson, J. A., 96, 208
Arreguín-Toft, I., 72
Atran, S., 69–73, 75–77, 81
Aureli, F., 6, 12
Avnaim-Pesso, L., 221
Axelrod, R., 69, 72, 73, 89

Baars, B. J., 271
Baillargeon, R., 109, 114, 117, 131
Baird, A. A., 157, 168
Baird, J. A., 116
Baker, C., 116
Baker, L., 11
Bakker, T. C. M., 90
Bal, P. M., 296
Baldassari, D., 91
Balslev, D., 161
Banaji, M. R., 162
Bandettini, P., 210
Bandura, A., 164
Barack, D. L., 161
Barclay, P., 90, 91, 116
Barlow, D., 54
Barnes, J. C., 255
Baron, J., 36–37, 39, 41, 42, 44, 70
Barrash, J., 190

Barraza, J. A., 227t, 228
Barrett, H. C., 132
Barter, J. W., 228
Bartlett, M. Y., 9
Bartz, J. A., 225
Basile, B., 61
Baskin-Sommers, A. R., 259
Bastian, B., 76
Batson, C. D., 92, 279, 284, 287, 288, 291, 294
Baumard, N., 38, 42, 70
Baumgartner, T., 202, 209
Baumrind, D., 163
Bear, D. M., 190
Beauchamp, M. H., 148, 150, 151, 153
Beaver, K. M., 255
Bechara, A., 62, 160, 199, 202, 287
Becker, G., 71
Begossi, A., 281
Behlol, M. G., 157
Bekoff, M., 3
Bell, D. C., 286
Bellugi, U., 184
Ben-Ami Bartal, I., 12, 289
Bennett, C. M., 168
Berman, M. E., 226t, 227
Bernardez Sarria, M. S., 290
Berndt, T., 166
Bernhardt, B. C., 168
Berns, G., 71, 79, 80
Berntson, G. G., 289
Berthoz, S., 160
Bertsch, J. D., 259
Bhagwagar, Z., 228
Bick, P. A., 273
Biran, I., 273
Birbaumer, N., 259
Birch, K., 92
Bird, G., 165
Bjork, J. M., 226t, 227
Bjorklund, F., 21
Black, A., 146
Blackmore, S. J., 270

Blair, R. J. R., 62, 151, 152, 159, 162, 200, 201, 239, 243–245, 248, 255, 259, 260
Blakemore, S.-J., 165, 171, 172
Blakeslee, S., 273
Blonski, M., 89
Bloom, P., 21, 109, 112–114, 131
Bode, S., 273
Boehm, C., 9, 29, 108
Boesch, C., 8
Bohns, V. K., 221
Bolger, N., 225
Bond, A. J., 226t, 227, 228
Boone, C., 229
Borelli, J. L., 166
Borg, J. S., 242, 246
Bornstein, G., 229
Bos, P. A., 228
Botvinick, M. M., 292
Boucher, J., 152
Bowles, S. E., 29, 81, 91, 107, 183, 197
Boyd, R. E., 22, 29, 107, 183, 197
Braeges, J. L., 126
Bramati, I. E., 56, 242, 287
Brand, M., 225
Brass, M., 273
Braver, T. S., 59
Brekke, K. A., 94
Brelstaff, G., 270
Briskman, J., 257
Brosnan, S. F., 5, 9–12
Brothers, L., 161, 184
Brown, B., 166
Brown, D. E., 25
Brown, R. E., 108
Brownell, C. A., 109, 285
Bruening, R., 29, 30
Bruhin, A., 208
Brunet, E., 288
Bshary, R., 5
Buchholz, S., 157, 172
Buchsbaum, M., 248
Buckholtz, J. W., 199, 201, 282

Buckley, T., 92

Buckner, R. L., 161, 198, 201, 204, 205

Budhani, S., 259

Buffone, A., 210

Bugnyar, T., 7

Buonocore, M. H., 161

Bureau, J., 54

Burkart, J. M., 10

Burnett, S., 165, 262

Burns, J. M., 274

Bussey, K., 164

Cacioppo, J. T., 289

Cacioppo, S., 135, 282, 293

Calhoun, V. D., 202

Caltran, G., 109

Camerer, C. F., 88, 93, 190, 198, 208

Campbell, A., 171

Campbell, M. W., 12

Campos, J. J., 115

Camprodon, J. A., 135, 202

Caravita, S. C. S., 157, 165, 172

Carey, S., 136

Carlsmith, K., 36, 41

Carlson, S. M., 143, 148

Carney, D. R., 205, 225

Carpenter, M., 109, 110, 115

Carré, J. M., 244

Casebeer, W. D., 171

Cashdan, E., 29

Castano, E., 294, 295

Castellanos, M. A., 12

Cavell, T. A., 200

Chagnon, N., 29

Chaiken, S., 203

Chakroff, A., 136

Chalik, L., 127, 290

Chan, W-T., 105

Chandroo, K. P., 12

Chang, S. W. C., 227t, 228

Chapman, M., 109–110, 124

Chatterjee, A., 273

Cheah, C. S. L., 166

Chen, C. Y., 289, 292

Chen, Q., 12

Cheng, P., 24

Cheng, Y., 286, 291

Cherek, D. R., 227

Chiew, K. S., 59

Choi, J.-K., 81, 91

Chou, K. H., 291

Christensen, J. F., 225

Christner, J., 30

Church, R. M., 12

Churchland, P. S., 171

Ciaramelli, E., 205

Cikara, M., 292

Cima, M., 244

Cimino, A., 25

Claes, M. E., 166

Clark, A., 270

Clark, F., 200

Clark, J. J., 270

Clark, L., 206, 224, 227

Clasen, L. S., 171

Cleare, A. J., 227

Clore, G. L., 129, 221

Clutton-Brock, T., 36

Coccaro, E. F., 227

Cohen, A., 129

Cohen, J. D., 21, 36, 57, 60, 96, 143,
      149, 198, 201–202, 204, 206, 208,
      241, 276

Cohen, L., 269

Collins, D. L., 171

Cooper, J. C., 162

Coplan, A., 284

Cornelissen, G., 94

Correll, J., 291, 293

Cosmides, L., 23–26, 28, 31, 132

Cote, S., 291

Coupe, P., 171

Craig, A. D., 160, 168, 208

Critchley, H. D., 160, 168

Crockett, M. J., 187, 206, 207, 222, 223,
      223t, 224, 226t, 227

Croft, K. E., 205
Cronin, K. A., 9
Cross, D., 136
Crowe, P. A., 166
Cuddy, A. J. C., 24
Cummins, D. D., 132
Curtin, J. J., 259
Cushman, F., 24, 36, 37, 116, 131, 135, 136, 202, 203, 205–207, 223

Dadds, M. R., 257, 261
Dahl, A., 115
Dalai Lama. *See* Tenzen, G.
Dal Bó, P., 89
Damasio, A. R., 62, 81, 162, 184, 198–199
Damasio, H., 62, 81, 162, 199
D'Amato, F. R., 289
Danziger, S., 221, 229
Dapretto, M., 162
Darley, J. M., 21, 36, 41, 57, 60, 143, 149, 201, 202, 241
Darvishzadeh, A., 229
Darwin, C., 3, 19, 70, 81, 107, 197, 280
Davidov, M., 109, 285, 289
Davidson, P., 146, 150, 152
Davis, M., 200
Davis, R., 72, 73
Daw, N. D., 207
Dawes, C. T., 209
Dawkins, R., 20
Dayan, P., 198, 224
De Brigard, F., 205
Decety, J., 12, 109, 129, 135, 143, 147–148, 152–153, 161, 169, 184, 201–202, 282, 284–285, 286, 289, 291, 293, 295
Declerck, C. H., 227t, 229
De Dreu, C. K. W. D., 209, 223t, 225, 227t, 229
Dehaene, S., 269
Dehghani, M., 70, 72, 73
Delgado, M. R., 209

Delton, A. W., 23–25, 31
Denburg, N. L., 190
Dennett, D. C., 270
Dent, J. L., 94
de Oliveira-Souza, R., 56–57, 187, 191, 242, 246, 288
de Quervain, D. J., 208
DeScioli, P., 21, 29–30
DeSteno, D., 9, 129
Detre, G. J., 210
de Waal, F. B. M., 5–9, 11–12, 109
Dewitte, S., 94
Dias, M. G., 127
di Pellegrino, G., 205
Ditto, P. H., 128, 129, 244
Dolan, R. J., 160, 209
Dölen, G., 229
Donald, M. J., 267
Dondi, M., 109
Dooley, J. J., 148, 150, 151, 153
Dougherty, D. M., 226t, 227
Douglas, T., 229
Doya, K., 207
Dreber, A., 89, 90, 96
Dunbar, R. I. M., 185
Duncan, B. D., 92
Duncan, I. J. H., 12
Dunfield, K. A., 110
Dungan, J., 135, 136, 191, 197, 202
Dunn, E. W., 111
Dunne, S., 162
Durkheim, É., 69
Dwyer, S., 131, 221

Early, S., 294
Earp, S. E., 12
Ebitz, R. B., 228
Ebstein, R. P., 229
Echols, S. C., 291, 293
Edwards, C. P., 129, 133
Eisenberg, N., 110, 227t
Eisenberger, N. I., 160, 287
Eisenegger, C., 225, 228

Ekholm, S., 248
Eliade, M., 69
Ellingsen, T., 94, 95
El Mouden, C., 35
Elson, S. B., 80
Emonds, G., 229
Eng, S. J., 127
Engel, C., 94
Engell, A. D., 60, 143, 149, 202
Eslinger, P. J., 56, 143, 147, 152, 161, 198, 199
Espinosa, P., 54
Esposito, J., 75
Etcoff, N. L., 23
Evans, D., 294

Fabes, C., 166
Fair, D. A., 161, 167
Fanning, J. R., 227
Fehr, E. E., 10, 94, 96, 107, 183, 197, 202, 208, 209, 222, 225, 228
Feierabend, A., 202
Feinberg, M., 90
Fessler, D. M. T., 91, 127
Finch, D. M., 160
Fischbacher, U., 94, 96, 202, 209, 228
Fiske, A. P., 22, 70
Fiske, S. T., 23, 24, 292
Fitzpatrick, C., 240
Flack, J. C., 5, 9
Fletcher, P. C., 161
Flicker, L., 54
Foot, P., 203, 292
Ford, J. H., 205
Forsman, A., 248
Frank, R. H., 11, 107, 162, 197, 209
Franklin, R. G., 287
Franklin, S., 271
Fraser, O. N., 7
Frechette, G. R., 89
Frederickson, N., 151
Freeman, H., 10
Freud, S., 144

Frey, B. S., 96
Freyd, J. J., 271
Frick, P. J., 255
Friston, K., 185, 187
Frith, C. D., 198, 202, 209
Frith, U., 161, 198, 202
Fudenberg, D., 89, 90
Fumagalli, M., 282
Furey, T., 162

Gächter, S., 94, 95
Galaburda, A. M., 162
Galef, B. C., 10
Galinsky, A. D., 294
Gallese, V., 269
Gardner, A., 35
Garrett, A. S., 161
Garrido, G., 287
Garrido, G. J., 242
Gazzaniga, M. S., 184, 211n1
Gee, L., 91
Geraci, A., 117, 131
Giedd, J. N., 170–171
Gilbert, D. T., 269
Gilbert, F., 262
Gillen-O'Neel, C., 290
Gilligan, C., 163, 164, 172
Gimbel, S., 80
Giner-Sorolla, R., 54, 136
Ginges, J., 69–70, 72, 73, 81
Gini, G., 157, 172
Gino, F., 221
Gintis, H. E., 29, 107, 183, 197
Giovanello, K. S., 205
Girvan, M., 9
Gleichgerrcht, E., 292
Glenn, A. L., 62, 201, 205, 206, 240, 244, 245
Glick, P., 24
Glover, G., 198
Gneezy, U., 96
Godefroy, O., 290
Goebel, R., 210

Gogtay, N., 168, 170
Goldstein, T. R., 295
Gomes, C. M., 8
Gomez, A., 76
Gomila, A., 225
Goodman, N., 116
Goodnow, J. J., 133
Gordon-McKeon, S., 205
Grabowski, T., 162
Grafenhain, M., 111
Grafman, J., 56–57
Grafton, S., 201, 242, 246
Graham, J., 129, 244
Grant, C. M., 152
Grattan, L. M., 198, 199
Gray, J., 69
Grayson, A., 152
Green, C. D., 188
Green, M. C., 80
Greenberg, F., 184
Greene, D., 96, 111
Greene, J. D., 21, 36, 37, 43, 57–61, 70,
    93, 107, 129–131, 143, 149, 150,
    152, 161, 197, 201–207, 209, 210,
    211n2, 221, 223, 241, 243, 276, 281
Greer, L. L., 209, 225
Greif, A., 91
Grèzes, J., 160, 296
Grimm, V., 96
Groff, P. R., 188
Grossman, G., 91
Grusec, J. E., 133
Guemo, M., 23, 24
Gummerum, M., 151
Gurven, M., 22
Guyer, A. E., 165, 170

Haidt, J., 4–5, 21, 44, 52, 54, 55, 107,
    123, 127–130, 132–134, 161, 184,
    187, 189, 197, 203, 221, 280, 281
Haley, K. J., 91, 127
Hall, L., 129
Hamann, S., 56, 201, 242

Hamlin, J. K., 21, 24, 109, 111, 113–117,
    131, 136, 282
Hammerstein, P., 38
Han, S., 292
Hanazaki, N., 281
Handgraaf, M. J. J., 209, 225
Hanoch, Y., 151
Hansen, L. K., 161
Happe, F. G. E., 262
Hare, B., 8
Hare, R. D., 109, 200, 201, 239, 240, 242
Hare, T. A., 207, 208
Harenski, C. L., 56, 62, 170, 201, 242,
    247, 289
Harenski, K. A., 201, 247
Hariri, A. R., 244
Harlan, E. T., 148
Harris, J., 230
Harris, P. L., 127, 132–134
Harsay, H. A., 283
Hastings, P. D., 109
Hauge, K. E., 94
Hauser, M. D., 62, 108, 124, 130–131,
    135, 198, 201–202, 206, 221, 224,
    244
Haushofer, J., 225
Haxby, J. V., 210
Hay, D. F., 110
Hayden, B. Y., 161
Haynes, J. D., 273
Heekeren, H. R., 201, 242
Heider, F., 112
Heilbronner, S. R., 161
Heinrichs, M., 209, 228
Heinze, H. J., 273
Helkama, K., 163, 164
Helwig, C. C., 124, 126, 136, 143–145,
    152
Henrich, J., 70, 81, 94, 95, 183
Hepach, R., 111
Herrington, J. D., 60
Herrmann, B., 94–95
Heyd, D., 43

Heyes, C. M., 10
Highberger, L., 279
Hildebrandt, C., 124, 136
Hinson, J. M., 173
Hobbes, T., 81
Hoebel, A. E., 42
Hoffman, M., 91
Holbrook, C., 127
Holloway, R. L., 185
Holman, E. A., 210
Holmes, A., 10
Holmes, O. W., Jr., 273
Holyoak, K., 24
Hopfensitz, A., 91
Hopper, L., 10
Hornak, J., 162
House, B. R., 133
Howard, L. H., 148, 282
Howland, E. W., 258
Hrdy, S. B., 10
Hsu, M., 208
Hu, J., 160
Hu, L., 160
Hu, S., 171
Huang, K. W., 229
Huebner, B., 57, 65, 221
Huebner, E. S., 166
Hughes, J. N., 200
Humayun, S., 257
Hume, D., 51–55, 107
Hung, A., 286
Hussar, K. M., 127
Huys, Q. J. M., 224
Hyde, L. W., 244
Hynes, C., 201, 242, 246
Hysek, C. M., 226t, 228

Iannetti, G. D., 285
Ignacio, F. A., 242
Iliev, R., 70
Imada, S., 54
Immordino-Yang, M., 81
Ingvar, M., 161

Insel, T. R., 209, 224, 287
Isaac, G., 189
Israel, S., 227t, 229
Iyer, R., 244

Jackson, P. L., 184, 288
Jaeggi, A. V., 9
James, W., 273
Jameson, T. L., 173
Jampol, N., 136
Jensen, L. A., 134
Jeon, D., 289
Jetten, J., 76
Jin, R. K., 131
Johansson, L.-O., 94
Johansson, P., 129
Johansson-Stenman, O., 94
Jones, A. P., 255, 262
Jones, L., 200
Jordan, A. H., 129, 221
Jorgensen, M., 203
Joseph, C., 134, 184, 187
Joyce, R., 281
Jurney, J., 39

Kagan, J., 106, 168
Kahle, S., 109
Kahn, P. H., Jr., 43
Kahn, R. E., 255
Kahn, V., 168
Kahneman, D., 36, 93, 187, 203
Kaini, A., 157
Kalsoom, F., 157, 163, 167
Kandel, E. R., 268
Kant, I., 71, 107, 203
Kanwisher, N., 135, 161, 169
Karlan, D., 91
Katz, L. D., 108
Kaufman, M., 198
Kayani, M. M., 157
Kelemen, D., 129
Kelley, E., 110
Kelly, D., 127

Keltner, D., 90
Kensinger, E. A., 162
Kerr, D. C., 169
Kesebir, S., 280
Kesek, A., 143
Kidd, D. C., 295
Kiebel, S., 209
Kiehl, K. A., 62, 200–202, 240, 244, 247,
    248, 289
Killcross, S., 257
Killen, M., 106, 114, 117, 123–124, 136,
    149, 280
Kim, S. H., 201
King, R. A., 134
Kinsbourne, M., 273
Kinzler, K. D., 129, 143, 169, 202, 282
Kirschbaum, C., 228
Kjaer, T. W., 161
Klein, T. R., 279
Knafo, A., 289
Knobe, J., 129
Knoepfle, D. T., 208
Knudsen, B., 110
Knutson, B., 198, 201, 207
Kochanska, G., 148
Koenigs, M. L., 61, 62, 201, 205, 206,
    209, 242, 244, 247, 293
Kogut, T., 44
Kohlberg, L., 107, 123, 128, 144, 163,
    203
Kohlers, P. A., 271
Koleva, S., 244
Koller, S. H., 127
Konner, M., 12
Konow, J., 40
Kosfeld, M., 209
Koski, S. E., 7, 12
Koster-Hale, J., 135, 202
Koven, N. S., 205
Kozinski, A., 221
Krach, S., 160
Kraft-Todd, G. T., 94
Krajbich, I., 190

Krakauer, D. C., 9
Krambeck, H., Jr., 90
Krasnow, M. M., 26, 31
Kriegeskorte, N., 210
Kringelbach, M. L., 162
Kristel, O. V., 80
Krueger, F., 56–57
Kruepke, M., 201, 209
Krumme, C., 90
Kuhlmeier, V. A., 110, 112–113, 131
Kuhse, H., 36
Kupanoff, L., 166
Kurzban, R., 21, 29–30
Kymlicka, W., 44

Laakso, M. P., 246
LaBounty, J., 169
LaCasse, L., 248
Lacetera, N., 91
Ladavas, E., 205
Lafreniere, M., 54
Lahat, A., 143, 145, 148–153
Lahvis, G. P., 12
Lakoff, G., 269
Lalonde, F., 171
Lambeth, S. P., 11
Lamm, C., 285, 288
Lane, J. A. S., 94
Lane, J. D., 169
Lane, S. D., 227
Langford, D. J., 12, 289
Larson, R., 166
Lau, A., 126
Le, D., 116
Lee, J. J., 30
Legrain, V., 285
Lelard, T., 290
Lenhoff, H. M., 184
Lepper, M. R., 96, 111
Lerner, J. S., 36, 80
Leshner, S., 70
Leslie, A. M., 129
Levav, J., 221

Levenson, M., 240
Levine, M., 294
Lewis, C. S., 183–184
Lewis, R. L., 54
Lieberman, D., 202
Lieberman, M. D., 162, 209, 227
Lilienfeld, S. O., 240
Lin, C. P., 291
Linden, D. J., 269
Lishner, D. A., 92
Liszkowski, U., 110
Liu, B. S., 129
Liu, S., 269, 270
Liyanage, S., 294
Lockwood, P. L., 287
Loewenstein, G., 44
Lou, H. C., 161
Lowenberg, K., 206
Lowry, L., 54
Ludwig, A.-C., 225
Lundstrom, B. N., 161
Luo, Q., 143, 152
Lutz, K., 202

Ma, C. Q., 166
Macis, M., 91
Mack, A., 270
Mackay, L., 151
Macrae, C. N., 162
Maddock, R. J., 161, 204
Madl, T., 271
Maguire, E. A., 161
Mahajan, N., 114, 116
Mahapatra, M., 127
Malenka, R. C., 229
Malhotra, D., 295
Malle, B. F., 116
Mancini, F., 61
Maney, D. L., 12
Marazziti, D., 183
Marois, R., 282
Marsh, A. A., 200, 226t, 227, 260, 262, 287

Marsh, D. M., 227
Maskin, E. S., 89
Mason, M. F., 205, 225
Mason, P., 12, 290
Masten, C. L., 290
Mattioli, F., 205
Maudsley, H., 184
Maurer, M., 116
May, J. S., 255
Mayr, E., 19
McCarthy, G., 161
McCloskey, M. S., 227
McColl, A., 81
McComb, K., 11
McConnell, M. A., 91
McCullough, M. E., 228
McGraw, A. P., 70
Medin, D., 69, 70, 72
Melis, A. P., 8
Meltzoff, A. N., 289
Mendez, M. F., 205
Mengel, F., 96
Meyer, D. E., 54
Meyer-Lindenberg, A., 199
Michalska, K. J., 129, 143, 169, 202, 282
Mikhail, J., 124, 130–131, 223
Milgram, S., 188
Milgrom, P. R., 91
Milinski, M., 90
Mill, J. S., 71, 203
Miller, B. L., 190
Miller, E. K., 198, 202, 204
Miller, J., 36
Miller, J. G., 109, 127
Miller, M. B., 202
Miller, P. A., 110
Miller, R. S., 54
Mills, K. L., 171
Mitchell, G., 36
Mitchell, J. P., 162, 198, 202, 208
Moccia, R. D., 12
Moeller, F. G., 226t, 227
Moffitt, T. E., 191, 200, 255

Mogahed, D., 75
Moll, J., 56–57, 143, 152, 161, 184, 187,
    188, 190–191, 198, 201, 202, 208,
    242, 243, 246, 288
Monson, J., 151
Montague, P. R., 198
Montalan, B., 291
Montoya, E. R., 223t, 225, 228
Moore, B., 294
Moore, M., 44
Moran, J. M., 202
Morelli, S., 206
Moretto, G., 205
Morgenstern, O., 71
Morishima, Y., 208
Morton, T. A., 290
Moses, L. J., 116
Moskowitz, G. B., 294
Moss, C. J., 11
Motzkin, J. C., 244
Moul, C., 257
Mouras, H., 290
Mouraux, A., 285
Muccioli, M., 205
Mulvey, K. L., 136, 149
Murray, B. D., 162
Murray, D., 205
Murray, K. T., 148
Muuss, R. E., 163, 164

Naef, M., 228
Nelson, A. K., 157, 172
Nelson, E. E., 165, 170
Nelson, K., 270
Nemirow, J., 25
Neumann, C. S., 244
Newman, D. G., 36
Newman, J. P., 201, 209, 244, 258–260
Nichols, S., 129, 132–134
Nichols, S. L., 258
Nichols, S. R., 109, 285
Nielsen, F. A., 161
Nieuwenhuis, S., 150

Nisan, M., 127
Nisbett, R., 273
Nisbett, R. E., 96, 111
Noe, R., 38
Nord, E., 36
Norman, G. J., 289
Norman, K. A., 210
North, D. C., 91
Nosek, B. A., 129
Nowak, M. A., 30, 88–91, 93, 94, 96,
    108, 161, 183, 209
Ntoumanis, N., 54
Nucci, L. P., 107, 125, 126, 128, 133,
    143–146, 149, 150, 152
Nuñez, M., 132
Nystrom, L. E., 21, 36, 57, 60, 96, 143,
    149, 201, 202, 206, 208, 241

Oberholzer-Gee, F., 96
Ochsner, K. N., 162, 225
Ockenfels, P., 89
O'Connell, L., 110
O'Doherty, J. P., 162, 208
Ohman, A., 160
Ohtsuki, H., 89
Oliner, P. M., 294
Oliner, S. P., 294
Oliveira-Souza, de, R., 56, 161, 198
Olson, C. R., 160
Olson, S. L., 169
O'Neill, P., 203
O'Regan, J. K., 270
Ostrom, E., 91
Ostrowsky, K., 160
Ouss, A., 91
Oxford, M., 200

Panksepp, J. B., 12
Park, H.-J., 185, 187
Park, J. H., 21
Parkinson, C., 197, 269, 270, 282
Parson, C. E., 295
Pascual-Leone, A., 135, 202

Patterson, C. M., 258
Paul, J. F., 255
Pavone, F., 289
Paxton, J. M., 206, 209
Pearson, J. M., 161
Pedersen, E. J., 26
Pedroni, A., 227t
Pellizzoni, S., 131
Pelphrey, K. A., 170
Penton-Voak, I. S., 21
Perkins, A. M., 206
Persson, I., 229
Pessoa, L., 198, 200
Petersen, M. B., 28
Petersen, S. E., 167
Peterson, R., 198
Petersson, K. M., 161
Petrinovich, L., 203
Peysakhovich, A., 91, 95
Pfaff, D. W., 183
Pfeifer, J. H., 162
Pfeiffer, T., 90
Phelps, E. A., 209
Phillips, M. L., 160, 162
Piaget, J., 4, 107, 115, 123, 136, 144, 202
Piazza, J., 127
Piff, P. K., 291
Pihl, R. O., 227
Pilleri, G., 190
Pinker, S., 30, 210, 295
Pizarro, D. A., 128
Plaghki, L., 285
Platt, M. L., 161, 228
Plomin, R., 200, 255
Plum, F., 190
Poeck, K., 190
Polyn, S. M., 210
Porges, E. C., 201, 202
Postmes, T., 290
Poulin, M. J., 210
Power, J. D., 167, 171
Pozzoli, T., 157, 172

Prehn, K., 247
Premack, A. J., 112, 131
Premack, D., 112, 117, 131
Preston, S. D., 11
Price, D. D., 160
Price, S. A., 10
Prinstein, M. J., 166
Prinz, J. J., 54, 57, 60, 65, 281
Priori, A., 282
Proctor, D., 9
Prosser, A., 294
Pruessner, J. C., 171
Puce, A., 161
Pujol, J., 143, 146–147, 152, 191, 201, 246, 247

Quartz, S. R., 208

Rabin, J., 190
Radke-Yarrow, M., 109–110, 124, 134
Raichle, M. E., 198, 201, 204
Raine, A., 62, 200–201, 205, 206, 240, 243, 245, 248
Ramachandran, V. S., 273
Ramos, R., 281
Rand, D. G., 88–91, 93–96, 209
Rangel, A., 198, 201, 207, 208
Rankin, K. P., 190
Rappaport, R., 69
Rawls, J., 42, 81, 107
Ray, J. V., 255
Reeves, A. G., 190
Reicher, S., 294
Reiman, E. M., 162
Reniers, R. L. E. P., 246
Rensink, R. A., 270
Rhee, S. H., 110
Rheingold, H. L., 110
Rhodes, M., 127, 290
Richards, M. H., 166
Richardson, C. B., 136, 149
Richardson, J., 36
Richell, R. A., 259

Richerson, P. J., 22
Ridderinkhof, K. R., 150, 283
Ridley, M., 21, 26
Riggs, K. J., 152
Rilling, J. K., 96, 208, 227t, 228, 246
Ring, B., 161
Ritov, I., 41, 42, 44
Rizzo, M. T., 114, 117, 124
Robbins, T. W., 206, 222, 224, 227
Robertson, T. E., 23–25
Robinson, P., 36, 41
Roch, S. G., 94
Rock, I., 270
Rodgers, D. A., 289
Rogers, R. D., 61, 228
Rolls, E. T., 162, 247
Romero, T., 12
Roskies, A., 61
Roth-Hanania, R., 109, 285, 289
Rotshtein, P., 160
Rottman, J., 129
Roy, M., 287
Rozin, P., 29, 54, 129
Ruby, P., 288
Ruderman, A. J., 23
Rudolph, K. D., 166
Ruff, C. C., 94, 208
Russell, P. S., 136
Rustichini, A., 96
Rutland, A., 280
Rutte, C., 9

Sachdeva, S., 70, 72
Sageman, M., 75
Salvarini, G., 294
Salwiczek, L., 5
Samuelson, C. D., 94
Sanfey, A. G., 96, 208, 209, 228
Sarlo, M., 206
Sasaki, T., 22
Satpute, A. B., 209
Saver, J. L., 190, 199
Savulescu, J., 229

Saxe, R. R., 135, 136, 161, 169, 170, 202, 281
Schacter, D. L., 198, 201, 205
Schaich Borg, J., 201–202, 205, 206
Schapiro, S. J., 11
Schelling, T. C., 38
Schkade, D., 36
Schlaggar, B. L., 167
Schmidt, H., 242
Schmidt, K. M., 10
Schmidt, M. F. H., 117, 131
Schnall, S., 129, 221
Scholz, J., 170
Schooler, J. W., 272
Schuck, R. K., 115
Schug, R. A., 62, 201, 205, 206, 245
Schulkin, J., 191
Schultz, W., 198
Schultz-Darken, N., 10
Schulz, J. F., 94
Schumacher, J. A., 227
Schunk, D., 208
Schwartz, J. A., 255
Schwartz, S. H., 189
Schwintowski, H., 242
Scott, S., 257
Sebastian, C., 165
Seidel, A., 54, 60
Sell, A., 28
Semmann, D., 90
Shalvi, S., 209, 225
Shamay-Tsoory, S. G., 190, 287
Shane, M. S., 201, 247
Shapira, J. S., 205
Shaw, L. L., 279
Shaw, P., 168
Sheikh, H., 70, 71
Sheketoff, R., 136
Shenhav, A., 60, 202, 206, 207
Shiffrar, M., 271
Shikaki, K., 69
Shin, L. M., 160
Shohamy, D., 287

Shweder, R. A., 127, 129, 133
Siegal, M., 131
Siegel, J. Z., 224
Sigmund, K., 89, 90, 183
Silfver, M., 163, 164
Simion, F., 109
Simmel, M., 112
Simmonds, M. P., 11
Simner, M. L., 109
Simon, V. A., 166
Singer, P., 36, 43, 209, 295
Singer, T., 168, 208, 209, 285
Singh, L., 29
Sinnott-Armstrong, W., 62, 200–202, 242, 244, 246
Skelly, L. R., 289
Skinner, B. F., 144
Sloane, S., 117, 131
Sloman, S. A., 93
Small, D. A., 44
Small, D. M., 161
Smallwood, J., 272
Smetana, J. G., 106, 123–126, 128, 133, 134, 143–145, 150–152, 290
Smith, A., 3, 107, 189
Smith, C., 69
Smith, M., 200
Sneddon, L. U., 13
Snidman, N., 168
Snozzi, R., 228
Sobhani, M., 288
Soderstrom, H., 248
Sommerville, J. A., 117, 131, 161
Sommerville, R. B., 21, 36, 57, 201, 241
Soon, C. S., 273
Sousa, P., 127
Spaan, M., 283
Spagnolo, G., 89
Sparks, A., 91
Spears Brown, C., 290
Sperber, D., 70
Spiga, R., 227
Spranca, M., 70

Sripada, C. S., 133
Starcke, K., 223t, 224, 225
Stellar, J., 90
Sterck, E. H. M., 7, 12
Stich, S., 127, 133
Strandberg, T., 129
Street, A., 36
Strohminger, N., 54
Sugiyama, L. S., 22
Sullivan, J., 10
Sunstein, C., 36, 37, 44
Surian, L., 117, 131
Svedsäter, H., 94
Svetlova, M., 109, 284–285
Swann, A. C., 227
Swann, W., 69, 76, 77
Swarr, A., 166
Swerdlow, R. H., 274

Taber-Thomas, B. C., 62
Tabibnia, G., 209, 227
Taborsky, M., 9
Talbot, C., 10, 11
Tangney, J., 54
Tannenbaum, D., 128
Tarnita, C. E., 94
Taylor, J., 198
Taylor, S. E., 23, 165–166, 171, 225
Tenenbaum, J., 116
Tenzen, G. (Dalai Lama), 276
Terbeck, S., 223t, 224
Terburg, D., 228
Tetlock, P. E., 36, 69–70, 80
Thomas, B. C., 205
Thompson, S., 165
Thomson, J. J., 203, 293
Thöni, C., 94
Thornton, L. C., 255
Tinghög, G., 94
Tisak, M. S., 128
Titmuss, R., 96
Tomasello, M., 8, 10, 107, 109–111, 115
Tonnaer, F., 244

Tooby, J., 23–26, 28, 31, 132
Tovar-Moll, F., 242
Towsley, S., 168
Tran, L., 90
Tranel, D., 62, 190, 199, 205, 209
Trivers, R. L., 7, 38, 280–281
Trope, Y., 203
Troscianko, T., 270
Tse, W., 226t, 228
Tsoi, L., 135
Tullberg, M., 248
Turiel, E., 107, 123–128, 133, 136, 143–146, 149, 152, 203, 281

Uchida, S., 22
Ugazio, G., 94
Uhlmann, E. L., 128
Ullman, T., 116
Underwood, B., 294
Ungar, L., 206
Unger, P. K., 43
Utikal, V., 94

Vaish, A., 109, 111, 115
Valdesolo, P., 129
Vallerand, R., 54
van den Wildenberg, W., 150
van Honk, J., 227t, 228
Van Horn, J., 201, 242, 246
Van Kleef, G. A. V., 209, 225
van Leeuwen, F., 21, 24
Vann, S. D., 161
van Roosmalen, A., 6–7
van Schaik, C. P., 9, 10
Van Slyke, J., 70
Van Vugt, M., 228
Varese, F., 44
Veit, R., 246
Veltkamp, M., 296
Viding, E., 165, 200, 244, 255, 262
Villringer, A., 242
Vinkers, C. H., 227t, 228
Vogt, B. A., 160

von Dawans, B., 227t, 228
von Grunau, M., 271
von Neumann, J., 71

Wager, T. D., 287
Wagner, E., 109–110, 124
Wainryb, C., 124
Walker, L. J., 157, 163
Wang, P. P., 184
Wang, X., 291
Warlop, L., 94
Warneken, F., 10, 107, 109–111
Wartenburger, I., 242
Watson, J., 136
Watson, K. K., 228
Waytz, A., 208
Weber, M., 71
Wedekind, C., 90
Wegner, D. M., 273
Weisel, O., 229
Wellman, H. M., 136, 169
West, M. J., 110
West, S. A., 35, 36
Whalen, P. J., 200
Wharton, S., 136, 205
Wheatley, T., 129, 269, 270, 273
Whitehouse, H., 76
Whitfield-Gabrieli, S., 170
Whitney, P., 173
Wickler, W., 5
Wiech, K., 293
Wiens, S., 160
Wijnen, J. G., 283
Wikkelsö, C., 248
Willer, R., 90, 291
Williams, G. C., 20
Williams, K. D., 165
Williamson, S., 240
Wilson, D. S., 280
Wilson, T. D., 273
Winner, E., 296
Winslow, J. T., 224
Winston, J., 209

Wong, S., 240
Wood, R. M., 226t, 228
Woodward, A., 136
Wynn, K., 21, 109, 112–114, 131

Xu, X., 291

Yaish, M., 44
Yang, Q., 63, 64
Yang, Y., 200, 201, 243
Yaralian, P. S., 201
Yeung, N., 150
Yoder, K. J., 282
Yoeli, E., 91
Young, L., 36, 37, 131, 135–136, 191, 197, 201–202, 281
Young, L. J., 62, 209, 287

Youssef, F. F., 223t, 224
Yu, H., 160

Zahm, D. S., 190
Zahn, R., 56–57, 287
Zahn-Waxler, C., 109–110, 124, 134, 285, 289
Zak, P. J., 209, 228
Zaki, J., 208, 225
Zeier, J., 201
Zelazo, P. D., 126, 134, 136, 143, 145, 149, 152
Zelikow, P., 71
Zhong, C.-B., 221
Zhou, X., 160
Zuo, X., 292
Zwolinski, J., 166

Subject Index

Abortion, 77–78, 78t
Adaptation, 19–21
  moral behavior and, 31
Adolescent females, 157–158, 172–173
  female brain development and moral
    reasoning, 167–171
  female puberty and peer relationships,
    165–167
Affective perspective taking, 125,
    283–285, 288–289, 294–296. *See also*
    Empathy
Affective resonance, 284. *See also* Emo-
    tional contagion
Affective sharing. *See* Emotional sharing
Affiliation, 209, 225. *See also* Kinship;
    Tribe
Agency, moral, 188–189
Altruism, 189, 208
  parochial, 81
  reciprocal, 7, 38, 280–281 (*see also*
    Reciprocity; Tit-for-tat (TFT) strategy)
Amygdala, 148, 159, 160, 168, 204–206,
    209
  affective perspective taking and, 289
  emotional experience and, 159, 168
  psychopathy and, 200–201, 205, 244–
    248, 255, 262
  sacred values (SVs) and, 80
  ventromedial prefrontal cortex
    (VMPFC) and, 80, 169, 200, 201,
    206–207, 245, 288

Animals and morality, 3–4
  evidence for moral behavior in ani-
    mals, 6–11
  how to compare moral behaviors in
    humans and animals, 5
  how to study morality in other spe-
    cies, 4–6
  implications for humans, 13–14
Anterior cingulate cortex (ACC), 162,
    287, 292
Anterior insula, 159–160, 168–170
Antisocial behavior, 254–255. *See also*
    Psychopathy
Antisocial personality disorder (APD/
    ASPD), 200, 201, 239–240, 248. *See
    also* Psychopathy
  morality and moral deficits in,
    243–248
Associative learning, 257–259
Attention, deficits in the allocation of,
    259–261
Autism, 152, 202

Backfire effect, 70, 72, 73
Behavioral regulation group (neural
    groups), 162, 170–171
Biological market, 38
Brains
  bad, 199–201
  cooperative, 208–210
  good, 201–202

puzzled, 202–208
Brain structures and pathways engaged in organization of social life and behavior, 186f, 282–283, 283f
By-products, 20

Callous-unemotional (CU) traits, 255–256
development of, 256–257
Citalopram, 206, 224, 228
Cognitive construction of moral rules, 106. *See also* Moral cognition is not always conscious
Cognitive view of morality, 4
Communal sharing, 22. *See also* Resource pooling; Sharing
Condemnation and coordination, 28–30
Conditioned-fear response, 259–260
Conduct disorder (CD), 254
Consequentialism, 222–224
Consolation, 12
Constructivism posited by social domain theory, 123, 125–126, 128
Constructivist processes as the motor of moral development, 123, 125–126
Contact principle, 131
Contagion, emotional, 284, 285
Contractualism and partner choice, 37–39, 39f
Conventional domain. *See* Moral and conventional domains
Convergent vs. homologous traits, 5
Cooperation, 87–88, 95–97
mechanisms that promote, 88–91
proximate psychology of, 91–95
Cooperative behaviors
defined, 87
ultimate vs. proximate explanations for, 92
Cooperative dispositions, 26–27
Coordination, 29–30
Cost, defined, 87

Crowding out/crowding in intrinsic motivation by extrinsic incentives, 96

Decision making, moral
morality and, 241–243
neuromodulation of, 226–227t, 226–229
pathways to, 187
Default mode network (DMN), 201–202, 204–205, 208
Deontology, 58, 223–224
Devoted action, theoretical model of, 79, 79f
Devoted actor hypothesis, 69
Devoted actors, 69–71
as deontic actors, 71–74
as fused actors, 74–79
neural aspects of, 79–81
vs. rational actors, 70–71
Dorsolateral prefrontal cortex (DLPFC), 58–61, 204, 246
Double effect, doctrine of, 131
Dual process perspective, 93
Dual-process theory, 57–61, 203–205, 209

Emotional blunting, 205
Emotional contagion, 284, 285
Emotional experience group (neural regions), 159, 167–169
Emotional sharing, 262, 285, 286f, 287, 292
Emotion deficits, 61–63
Emotion dysregulation, 254
Emotion recognition, 260–261
Emotions
Hume on, 53–54 (*see also* Sentimentalism)
judgments before, 63–64
moral, 4–5, 189 (*see also* Empathy)
and the moral brain, 55–57
nature of, 53

Empathic concern, 92, 109–110, 287–288. *See also* Psychopathy
and utilitarian moral judgments, 292–294
Empathy, 125, 184, 279–280, 296. *See also* Affective perspective taking; Callous-unemotional (CU) traits; Mentalizing; Psychopathy
affective (*see* Emotional sharing)
in animals, 11–13
complex relationships between morality and, 294–296
the experience of, 284
as limited resource, 289–292
three pillars of, 284–289
Empathy-altruism hypothesis, 92
"The enemy of my enemy is memory friend," 116–117
Evolution, 19–21. *See also* Morality evolved
Executive function (EF), 143, 151–153
social domain theory and, 149–151, 153
underlying role in moral development, 148–151
Extrinsic undermining, 96
Extrinsic vs. intrinsic motivations, 96, 110–111

Fairness. *See also* Justice
examples of morality as, 39–44
Fear-potentiated startle, 259–260
Female adolescents. *See* Adolescent females
Feminist perspective on moral development, 163–164
Food sharing, 8–10, 12
Footbridge dilemma. *See* Trolley problem/footbridge dilemma
Free rider problem, 22–25
Free will, 268, 273, 276
we are not as free as we think, 272–276

Frontoparietal control network, 201, 202, 204, 206, 207, 209
Frontopolar cortex (FPC), 191, 246
Frontostriatal pathway, 201, 202, 207–209
Frontotemporal dementia and, 191, 205
Functional outcome of behavior, 7
Fusion. *See* Identity fusion

Gender differences. *See* Adolescent females
Golden Rule, 70
Go/No-Go task, 63–64
Grand illusion, 270–271
Group cooperation, moral concepts for, 21–25
Group selection, utilitarianism and, 35–37, 39, 39f

Habituation, 112–113
Helping, 39–40
and hindering, 112–115, 117
limited requirement to help others, 43–44
Helping norm, 59
Helping rule, 59, 60
Hippocampus, 171
Homologous vs. convergent traits, 5
Human, what it feels lack to be, 270–273
Human exceptionalism, the pull of, 269–270
Human nature, laws of, 183–184
Humility and humanity, toward, 274–276

Identity fusion, 76
Illusions, 276. *See also* Grand illusion
we do not see what we think we see, 270–272
Illusory motion, 271
Impersonal moral dilemmas. *See* Personal vs. impersonal moral dilemmas

Impulsivity, 254
Incentives, nature of, 96
Inequity, 10–11
Infants
  predisposition toward moral behavior,
    109–110
  as prosocial, 110–112
Inferior parietal cortex, 161
Innate primitives, 130–132
Institutions, role in cooperation, 91
Intentional vs. accidental/ignorant
    harmful acts, 116, 135, 147, 202, 282
Intentions (as developmental mecha-
    nisms), role of, 135–136
Interoception, 159–160, 168
Intractible conflicts, 72–74
Intuitions, spontaneous emotional, 4–5
Islam. See Muslims
Israeli-Palestinian conflict, 72–74

Judgment. See also Moral judgment
  types of, 187
Justice. See also Retribution
  the need to restore, 41–42
  phenomenology vs. neuroscience and,
    273–274

Kinship, 21, 71, 78, 289, 293

Marzook, Mousa Abu, 73
Medial prefrontal cortex, 162
Mencius, 105
Mentalizing, 169
  defined, 169
Mentalizing group (neural regions),
    160–161, 169–170
Middle East, conflicts in, 72–74
Moral and conventional domains, dis-
    tinction between, 127–128
Moral behavior(s). See also specific topics
  building blocks/precursors to, 5–6
  categories of, 6–11
Moral brain. See also specific topics

circa 2015, 159–162
  infantile origins (see Moral origins)
  paradox of the, 197–199
"Moral center," 184, 282
Moral cognition is not always con-
    scious, effortful, and reflective,
    128–129
Moral compass, 183, 184
Moral development
  mechanisms of, 123–124, 136
  theories of, 162–165
Morality, 280–283. See also specific topics
  definition and scope of the term, 280
  multidimensional structure of, 184,
    185f
  as a phenomenologically distinctive
    experience, 183–184
  primary virtue of, 70
  views of the origins of, 35–39
Morality evolved, 107–109
Moral judgment, 187. See also specific
    topics
  in animals, 6–7 (see also Animals and
    morality)
  neuromodulation of, 222–225, 223t
Moral neural network, 243
Moral origins
  empirical evidence, 109–117
  historical overview, 105–107
  theoretical perspective, 107–109
Moral rationalism, Hume's critique of,
    51–53
Moral reasoning, 4
Moral sense, 183. See also specific topics
Moral "sprouts"/beginnings, 105–107
Moral system(s), 37, 184, 280
  defined, 5
Motion illusions. See Illusory motion
Motivations, moral
  as the steam of moral conduct,
    189–191
Muslims, 74–75. See also Israeli-Palestin-
    ian conflict

Mutualism and mutualistic theory, 38, 41–44

Natural selection, 19–21
Nature, laws of, 183–184
Netanyahu, Benjamin, 73–74
Neural components of morality, 187–192. *See also specific topics*
Neural correlates of moral judgments in children and adolescents, 146–148
Neurocognitive development of moral judgments, 143, 152–153. *See also* Executive function
future directions, 151–152
Neurocognitive structure of morality, fundamental, 188–189
Neuromodulators, 222
shaping moral cognition, 222–229
normative implications, 229–231
Neurons, 268
Norepinephrine, 224

Omission-commission effect, 30
Oppositional defiant disorder (ODD), 254
Orbitofrontal cortex (OFC), 242–243, 245, 246
Overriding, rational, 60–61
Oxytocin, 209–210, 225, 227t, 228–229

Parietal operculum, 161
Parochial altruism, 81
Partner choice, predictions of, 39, 39f
Passions, 54. *See also* Sentimentalism
Personal vs. impersonal moral dilemmas, 58–60, 204, 211n2, 223t, 225, 241, 243, 292–293
trolley problem/footbridge dilemma and, 58, 203–204, 223–224, 241, 245
Perspective taking. *See* Affective perspective taking
Phenomenology vs. neuroscience, and implications for justice, 273–274

Phi phenomenon, 271
Plato, 105
Political violence, 72–76
Posterior cingulate cortex (PCC), 161
Posterior medial cortices (PMC), 80–81
Precuneus, 160, 161
Predispositions toward prosocial behavior. *See* Moral "sprouts"/beginnings
Prisoner's dilemma (PD) games, 27, 28, 88, 228, 241
Prosocial behavior. *See also* Cooperative behaviors
in animals, 9–10
in infants, 110–112
Psychopathologies associated with immoral behavior, 254–256
Psychopathy, 190, 199–201, 255–260
amygdala and, 200–201, 205, 244–248, 255, 262
and capacity for making moral judgments, 62–63
factors in, 240, 246, 255
measures of, 240, 246
from minor deficits to, 261–262
morality and moral deficits in, 243–248
Psychopathy Checklist-Revised (PCL-R), 240, 246
Puberty, female, 165–167. *See also* Adolescent females
Public goods game (PGG), 88, 93–95
Punishment, 28, 116, 258, 274. *See also* Retribution
and the need to restore justice, 41–42

Rational actors vs. devoted actors, 70–71. *See also* Devoted actors
Rationalism, moral, 51–53
Rational overriding, 60–61
Reason/reasoning. *See also* Moral cognition
role in moral judgments, 51–53
Reciprocity. *See also* Altruism: reciprocal

in animals, 7–9
  direct, 89–90
  indirect, 90–91
Reciprocity theory, 38
"Relational people," 165, 167
Resource pooling, 22–25
Resources, distributing scarce, 40–41
Response-reversal paradigm, 258–259
Retribution, 36, 41, 42, 274, 276. See
  also Punishment
Right temporoparietal junction (RTPJ),
  135

Sacred values (SVs)
  backfire effect and, 70
  defined, 69
  devoted actors and, 70–71, 78–80, 82
  interpretations of, 80
  intractable political conflicts and,
    72–74
  measures of, 76–77
  nature of, 69
  revolutionary movements and, 72
Selective serotonin reuptake inhibitors
  (SSRIs), 206
Self, morality as a core dimension of
  inner, 191–192
Self-other distinction, 162
Sentimentalism, 51–55
  neuroscientific objections to, 57–64
Serotonin, 206, 224, 226t, 227–230
Sharing food, 8–10, 12
Sharing system, 22, 23
Social affiliation. See Affiliation
Social brain, morality as a product of
  the, 184–187
Social domain theory (of moral devel-
  opment), 129, 143–146, 281
  challenges to, 123, 126–130, 133, 136
  constructivism posited by, 123, 125–
    126, 128
  executive function (EF) and, 149–151,
    153

starting state and, 124
Social exchange, moral concepts for,
  25–28
Social heuristics hypothesis (SHH),
  93–95
Socialization and moral development,
  106
Sociocultural learning as the motor of
  moral change and differentiation,
  129, 132–136
Sociopathy. See also Psychopathy
  "acquired," 190
Spontaneous emotional intuitions, deci-
  sions based on, 4–5
Starting state, 124
Switch case, 36, 131, 203–204, 206.
  See also Trolley problem/footbridge
  dilemma

"Taste bud theory" (of moral develop-
  ment), 123–124, 130–136
Temporoparietal junction (TPJ), 161,
  202
"Tend and befriend" model, 165–166
"Tend and befriend" pattern, 166, 225
Terrorism, attitudes toward, 72–76
Testosterone, 225
Tit-for-tat (TFT) strategy, 89, 280–281
Traits, homologous vs. convergent, 5
Tribe, beyond the, 293–294
Trolley problem/footbridge dilemma,
  222
  amygdala dysfunction and, 245
  contractualist analysis and, 45
  dorsolateral prefrontal cortex (DLPFC)
    and, 204
  frontotemporal dementia and, 205
  innate primitives and, 131
  personal vs. impersonal moral dilem-
    mas and, 58, 203–204, 223–224, 241,
    245
  utilitarianism and, 35, 36, 45, 203–
    205, 293

ventromedial prefrontal cortex
(VMPFC) and, 205, 206
vmPFC and, 62
Trust games, 27

Ultimatum game, 208–209
Unconscious processes, 271
we are not aware that we are not
aware, 272
Utilitarianism, 58
group selection and, 35–37
Utilitarian moral judgments, low empathic concern, 292–293

Ventrolateral prefrontal cortex (VLPFC),
80
Ventromedial prefrontal cortex
(VMPFC), 242, 247, 289
affective responses and, 287, 288
amygdala and, 80, 169, 200, 201, 206–
207, 245, 288
dysfunction in, 61, 62, 199, 202, 205,
207, 209, 245, 288, 293–294
empathic concern and, 287–288
integrative role for, 207, 209
psychopathy and, 199–202, 205, 245,
247, 255, 289
psychoticism and, 293
utilitarian judgments and, 61, 80,
205–207, 242, 293–294
Victimless actions, personal experience
cannot account for moral judgments
of, 129
Violence, political, 72–76

www.ingramcontent.com/pod-product-compliance
Lightning Source LLC
Chambersburg PA
CBHW061001280326
41935CB00009B/788